後疫情時代印太戰略情勢下的臺灣安全戰略選擇

翁明賢 主編

淡江大學出版中心

主編序

　　自從2011年美國前國務卿希拉蕊克林頓（Hillary Clinton）在夏威夷發表美國的太平洋世紀一文以來，華盛頓採取亞太再平衡戰略，重新思考亞洲太平洋未來的發展，尤其是中國的崛起，勢必成為美國國家安全主要威脅來源與因應之重點。2017年川普（Donald Trump）繼任歐巴馬（Barack Obama）為第40任美國總統，同年12月公布其任內第一份國家安全戰略報告，直接點出美國的全球戰略重心開始在於印太區域:從美國西海岸到印度西海岸的區域，並藉此大旗:印太戰略，建立許多跨國合作機制，例如在經貿領域的多國合作構想，或是在資訊科技方面，提出清潔網路與藍點計畫的倡議，以及印太軍事戰略構想。

　　換言之，美國川普總統開始落實印太戰略的布局，一方面，透過以美國為優先的戰略思考，要求印太各國跟隨華盛頓從地緣戰略角度，回應北京自2013年對於歐亞兩洲的一帶一路國際合作倡議的政經挑戰，另一方面，美國則是挑動與中國之間的高科技、經貿與智慧財產權之爭，主要目的還在於重振美國經濟與全球影響力。只是，川普所強調的美國優先理念，卻帶來歐亞國家的憂慮。是以，2017年慕尼黑國際安全研討會提出整體國際戰略格局進入〔後真相、後秩序、後西方〕:三後世界，而2020與2021年之間國際政經互動發展結果，具體呈現此種三後詭譎多變的全球戰略情勢。

　　基本上，本書係集合2020與2021年本所兩屆淡江戰略學派年會，參與學者發表有關印太戰略與新冠肺炎對全球戰略情勢影響的論文佳作。例如有關美中或是相關國家例如澳洲等國在南海地區的航行自由權之爭，東協國家如何因應此種大國權力競逐問題，或是中國基於中國夢所啟動一帶一路倡議在全球引發的債務陷阱，或是有關區域戰略問題，例如歐洲難民問題，以及新冠肺炎下美中兩國各自內部發展，及其相互之間的戰略對峙問題。透過上述佳作提供給讀者一個重要啟示在於:面對美中兩強在印太戰略競逐下，台灣的戰略選擇不外乎是:抗衡、扈從、避險或是中立，如何內部一致。

　　最後，感謝本校出版中心主任林雯瑤與編輯黃佩如的大力鼎助，由於出版中心本著學術無價、知識推廣的重要性，持續協助本所歷年淡江戰略學派年會之後的論文出版工作，使得本校與本所國際事務與戰略研究的推廣得以綿延久長、影響社會各界深遠。當然本所助理陳秀真小姐指導聯繫與碩士生莊茗雲Miko不厭其煩、細心協助編排、校稿與聯絡出版事宜，才得以順利出版，在此特以高度致謝。

翁明賢

淡江大學國際事務與戰略研究所教授兼所長

謹誌於2022.03.30淡水校園

驚聲大樓1202所長室

目次

美中印太戰略競逐：整合規則建構主義的研究途徑

翁明賢 *

壹、前言

一、美中阿拉斯加高層對話：兩種規則的對立？

從2010年中國的「國民生產總額」（GDP）首次超過日本，次於美國居於世界第二之後，「中國崛起」（China's rise）就不是一個「名詞」，而是一個現在進行式。2017年慕尼黑國際安全會議中，主辦單位提出三個概念：「後真相、後西方、後秩序」（Post-truth, Post-west, Post-order），[1] 其中的「後西方」其實就在暗指中國的興起，即將取代美國成為「全球秩序」的影響者。

2017年之後，經由非建制派川普（Donald Trump）總統，基於「美國優先」（America First）執政四年之後，新任拜登（Joe Biden）總統透過「暫行國家安全戰略指南」（Interim National Security Strategic Guidance），強調三個回來「美國回來了、外交回來了、聯盟回來了」（America is back, Diplomacy is back, Alliances are back），[2] 將美中戰略競逐場域聚焦於「自由與開放印太」（Free and Open Indo-Pacific）。其後，美國總統在其第一場公開記者會，以及3月3日，國務卿布林肯在國務院外交政策演講，提出八大外交戰略面向，其中「因應中國」成為主要戰略重點。

2021年3月18日至19日，美中兩國外交與國安高層在阿拉斯加一場對話，[3]

---

* 淡江大學國際事務與戰略研究所教授兼所長

[1] "Munich Security Report 2017: "Post-Truth, Post-West, Post-Order?" Error Interact", ORBIS , https://espas. secure.europarl.europa.eu/orbis/document/munich-security-report-2017-post-truth-post-west-post-order (2021/05/05)

[2] "Interim National Security Strategic Guidance", MARCH 03, 2021 • STATEMENTS AND RELEASES, The White House, https://www.whitehouse.gov/briefing-room/statements-releases/2021/03/03/interim-national-security-strategic-guidance/(2021/05/05)

[3] "Secretary Antony J. Blinken, National Security Advisor Jake Sullivan, Director Yang And State Councilor

不僅在開場白時「針鋒相對」、「戰略亮劍」，會後也沒有共同發表任何聲明。美國國務卿布林肯（Antony Blinken）開宗明義點出：「美國承諾以外交為途徑增強美國國家利益，以及強化以規則為基礎的國際秩序」，而且「此種制度並非抽象，可以協助國家和平解決爭端、有效多邊協調，參與全球貿易事務，確保每一個國家都遵守同樣規則。一個缺乏國際規則的世界呈現贏者全拿，但是，會讓我們會面對一個更加暴力與不穩定的世界。」[4]

針對中國方面，布林肯直接指出：「今天我們雙方有機會討論主要優先順序，不僅內部事務，還有全球議題，也進而讓中國瞭解美國的企圖與途徑。」同時，特別強調：「美國與中國的關係應該是競爭性的，如果可以，也可以合作性的，如果必要，也必須是敵對性的」[5]，布林肯特別強調：「我們政府承認以外交方式來強化美國的國家利益，以及增強以規則為基礎的國際秩序。」[6]

反之，中國方面，中央外事辦公室主任楊潔篪則表達對於美國對中國的種種批評，表達「美國利用其軍事力量與金融霸權實施長臂管轄、壓制其他國家」、「濫用國家安全概念，阻礙正常的貿易往來，甚至煽動一些國家攻擊中國」，[7]並且直言：「從中國角度言，美國沒有資格從實力原則來跟中國對

Wang At the Top of Their Meeting, REMARKS, ANTONY J. BLINKEN, SECRETARY OF STATE, ANCHORAGE, ALASKA", U.S. Department of State, https://www.state.gov/secretary-antony-j-blinken-national-security-advisor-jake-sullivan-chinese-director-of-the-office-of-the-central-commission-for-foreign-affairs-yang-jiechi-and-chinese-state-councilor-wang-yi-at-th/(2021/05/05)

4　"Secretary Antony J. Blinken, National Security Advisor Jake Sullivan, Director Yang And State Councilor Wang At the Top of Their Meeting, REMARKS, ANTONY J. BLINKEN, SECRETARY OF STATE, ANCHORAGE, ALASKA", U.S. Department of State, https://www.state.gov/secretary-antony-j-blinken-national-security-advisor-jake-sullivan-chinese-director-of-the-office-of-the-central-commission-for-foreign-affairs-yang-jiechi-and-chinese-state-councilor-wang-yi-at-th/(2021/05/05)

5　原文為 "I said that the United States relationship with China will be competitive where it should be, collaborative where it can be, adversarial where it must be", "US and China publicly rebuke each other in first major talks of Biden era", The Guardian, https://www.theguardian.com/world/2021/mar/19/us-china-talks-alaska-biden-blinken-sullivan-wang(2021/03/31)

6　"Secretary Antony J. Blinken, National Security Advisor Jake Sullivan, Director Yang And State Councilor Wang At the Top of Their Meeting, REMARKS ANTONY J. BLINKEN, SECRETARY OF STATE, ANCHORAGE, ALASKA, MARCH 18, 2021", U.S. Department of State, https://www.state.gov/secretary-antony-j-blinken-national-security-advisor-jake-sullivan-chinese-director-of-the-office-of-the-central-commission-for-foreign-affairs-yang-jiechi-and-chinese-state-councilor-wang-yi-at-th/(2021/04/22)

7　「美國沒資格居高臨下同中國說話！」後川普時代的中美首次交鋒：布林肯與楊潔篪相互指責、毫

話」。[8] 另外，外交部長王毅則說：「中方過去、現在、將來都絕不會接受美國的無端指責，同時我們要求美方徹底放棄干涉中國內政的霸道行徑。美國的這個老毛病要改一改了。」[9] 同時，於2021年4月23日，中國外交部長王毅與美國外交關係協會Richard Haass透過視訊對話，表達中美關係發展的戰略視野的期待：強化溝通、加深合作、縮短差異與避免對抗。[10]

在阿拉斯加對話之後，雙方馬不停蹄進行全球與國的「拉幫結派」，美國國務卿布林肯前往布魯塞爾參加北約組織外交部長會議，闡述美國與歐盟北約國家戰略合作關係，中國外長王毅也「不遑多讓」與俄羅斯外長會面，發表中俄兩國持續戰略協作關係，並且前往中東國家與伊朗簽署合作關係。4月30日，美國國務院發出新聞稿指出從5月3日至5日，布林肯參加在倫敦舉行的G7國家外交與發展部長會議，布林肯強調只要攸關國際安全與全球經濟秩序所在，美國將會以美國人民利益為中心考量。[11]

2021年4月29日，美國總統拜登首赴國會參眾兩院聯席會議發表演說時，[12] 強調「我們現在處於與中國與其他國家競爭，贏得21世紀。」（We're in a competition with China and other countries to win the 21st Century.）[13] 提到他在

無交集」，風傳媒，https://www.storm.mg/article/3548878?page=1。（檢索日期：2021/05/06）

[8] 英文翻譯為："Let me say here that in front of the Chinese side, the United States does not have the qualification to say that it wants to speak to China from a position of strength", "US and China publicly rebuke each other in first major talks of Biden era", The Guardian, https://www.theguardian.com/world/2021/mar/19/us-china-talks-alaska-biden-blinken-sullivan-wang(2021/03/31)

[9] 「美國沒資格居高臨下同中國說話！」後川普時代的中美首次交鋒：布林肯與楊潔篪相互指責、毫無交集」，風傳媒，https://www.storm.mg/article/3548878?page=2。（檢索日期：2021/05/06）

[10] "Chinese FM: U.S. yet to find the right way to deal with China", CGTN, https://newsaf.cgtn.com/news/2021-04-24/Chinese-FM-held-talks-with-U-S-Council-on-Foreign-Relations-ZHOuyTCCuA/index.html (202104/29)

[11] "G7 Foreign and Development Ministers' Meeting and the Secretary's Travel to Ukraine", U.S. Department of State, https://www.state.gov/g7-foreign-and-development-ministers-meeting-and-the-secretarys-travel-to-ukraine/ (2021/05/05)

[12] "Remarks as Prepared for Delivery by President Biden — Address to a Joint Session of Congress, APRIL 28, 2021 • SPEECHES AND REMARKS", The White House, https://www.whitehouse.gov/briefing-room/speeches-remarks/2021/04/28/remarks-as-prepared-for-delivery-by-president-biden-address-to-a-joint-session-of-congress/(2021/04/29)

[13] "Remarks as Prepared for Delivery by President Biden — Address to a Joint Session of Congress, APRIL 28, 2021 • SPEECHES AND REMARKS", The White House, https://www.whitehouse.gov/briefing-room/speeches-remarks/2021/04/28/remarks-as-prepared-for-delivery-by-president-biden-address-to-a-joint-session-of-congress/(2021/04/29)

入主白宮之後，跟中國大陸國家主席習近平進行的2小時對話，「我告訴習主席，我們將在印太地區維持強大的軍事存在，就像我們跟北約組織在歐洲做的一樣。不為挑起衝突，而是防範衝突」。拜登在演講表示，美國正跟中國與其他國家競爭，以贏得21世紀，還指美國正花GDP的不到1%在研發，中國與其他國家正快速趕上。他說，習近平「極其認真」（Deadly earnest）要成為世上最顯著與重要的國家，「他跟其他人，獨裁者們，認為民主無法在21世紀跟獨裁政體競爭」，還指兩人在這一點無法達成共識。拜登稱他告訴習近平，「我們歡迎競爭，我們不會尋求衝突。但我非常清楚地表示，我將全面捍衛美國利益。[14]

中國方面，在一場在回答有關拜登國會聯席會議講話的記者會上，中國外交部發言人汪文斌表示：「中國致力於讓老百姓過上幸福日子，而美方一些人言必稱中國，說到底是冷戰零和思維和意識形態偏見在作祟，是缺乏自信的表現。希望美方對中國不要有酸葡萄心理，以更平和理性的心態看待中國的發展。」[15] 2021年4月30日，一篇CNN文章標題：「美中兩強爭霸開展，中國有信心勝出」，提出「當美國總統拜登強調美國會在21世紀勝出，已經勾勒出與中國領導人習近平的一場新強權競爭，這是一場與專制對抗的世紀，民主將會戰勝專制，美國將會勝出。」[16]

---

[14] "Remarks as Prepared for Delivery by President Biden — Address to a Joint Session of Congress, APRIL 28, 2021 • SPEECHES AND REMARKS", The White House, https://www.whitehouse.gov/briefing-room/speeches-remarks/2021/04/28/remarks-as-prepared-for-delivery-by-president-biden-address-to-a-joint-session-of-congress/(2021/04/29)

[15] 「美媒：中國在『大國競爭』中愈發自信」，中國新聞評論網，http://hk.crntt.com/doc/1060/7/5/2/106075263.html?coluid=7&kindid=0&docid=106075263&mdate=0502155310（檢索日期：2021/05/02）

[16] "The great power race between the US and China is on. And Beijing is confident of winning", CNN, https://edition.cnn.com/2021/04/30/china/biden-xi-china-us-mic-intl-hnk/index.html (2021/05/02)

二、美國拜登暫行國安指導：全面因應崛起中國？

　　針對雙方在阿拉斯加的亮劍過程引發一個值得深入思考的議題，為何美方從「規則」角度切入，劍指北京當局的違反國際規則的意涵十足，中方則毫不保留直言相向，從「實力」原則，闡述雙方不同問題的立場？美中此一阿拉斯加對話之後開展大國戰略競逐新時代，凸顯出基於維護以「規則」為基礎的國際秩序，因應一個崛起的中國，並透過民主制度與專制體制的理念價值對抗，成為「美國回來了」（America is Back）的戰略主軸，也帶來其他中等國家選邊站的疑慮。因此，國際關係與戰略學界該如何進一步分析與預測未來美中全球、印太與兩國之間的戰略競爭，應該是一個刻不容緩的議題。

　　1991年「冷戰」（Cold War）終結，全球戰略情勢邁向「後冷戰」（Post-Cold War），國際關係理論也從傳統「現實主義」（realism）、「新現實主義」（neo-realism）與「新自由制度主義」（Neo-liberalism）獨擅其場，轉向從「建構主義」角度，強調「文化」、「規則」與「身份」對於行為體之間互動，所產生的「利益」與「政策作為」的「關鍵變項」邏輯思考。學者奧魯夫（Nicholas Onuf），他是第一個將建構主義引進分析國際關係的美國學者，他主張社會是由人互動的結果，在互動過程中，透過「規則」（rule）形成「慣例」，再累積上述「規範」形成「制度」（institutions），成為「規則建構主義」的論述主軸。

三、本文研究主軸：命題、假設、步驟與預期成果

　　亦即，奧魯夫提出的「規則建構主義」：「行為體」之間透過互動，建立「規則」（rule）：即是「言語行為」（speech act）的體現，「規則」再進而影響行為體的互動過程，從而形成「制度」：一種「統治形態」（a condition of rule），累積不同「制度」運作，從而形成一個「國際社會」。[17] 換言之，從何種「規則」的角度切入，整理美國拜登總統上台以來，透過正式公布的

---

[17] Nicholas Greenwood Onuf, Making Sense, Makig Worlds: Constructivism in social theory and international relations(London and New York: Rouledge Taylor & Francis Group, 2013), pp.1-50.

「暫行國家安全戰略指南」，以及其外交與國家安全機制實際上對中國的政策與作為，藉以分析有別於以往川普總統的「美國優先」主軸下的「單邊主義」作為，呈現拜登以「多邊主義」為號召，配合民主、人權、自由等價值鎖型塑的國家安全戰略內涵。

　　本文研究分成一、前言、二、規則與言語行為的整合、三、美中印太相互規則建構，四、美中印太相互制度檢證，以及五、結語，本文「解釋」美中在印太戰略競逐下的實際「統治狀態」，及其未來可能發展趨勢，從而思考在不同「國際秩序」思考下，基於不同「原則」的建構過程，台灣的國家利益之下，可行戰略優先順序與政策作為。

貳、規則與身份概念的整合

一、規則的定義與內涵

　　建構主義是1990年代興起的一個與傳統主流國際關係理論辯論的研究範式，主要討論人的意識及其在國際生活中的作用，建構主義認為認同與利益不僅僅是由社會建構而成的，也是如同韋伯強調的人的能力與意願所產生的其他觀念因素同屬一個範疇。因此，在社會聚合的所有層面上，人的行為都受到一定程度的制約。[18] 建構主義的發展過程中，出現許多流派之分，其中，美國學者溫特（Alexander Wendt）的社會建構主義，強調「身份」（identity）決定「利益」的制約過程，成為主流建構主義學者的代表。溫特提出兩個建構主義的基本原則，第一、人類關係的結構主要是由「共有觀念」（shared ideas）而非由「物質力量」（material forces）所決定的。第二、有目的的行為體的「身份」（identities）與「利益」（interests），係由這些共有觀念所建構的，而非然固有的（given by nature）。[19]

---

[18] 約翰杰拉爾德魯杰，「什麼因素將世界維繫在一起？新功利主義與社會建構主義的挑戰」，彼得卡贊斯坦（Peter Katzenstein）、羅伯特基歐漢（Robert Keohane）、斯蒂芬克拉斯納（Stephen Krasner）編，秦亞青、蘇長和、門洪華、魏玲譯，世界政治理論的探索與爭鳴（Exploration and Contestation in the Study of World Politics）（上海：上海人民出版社，2018），頁256。

[19] Alexander Wendt, Social Theory of International Politics (Cambdirge: Cambridge University Press, 1999), p.1.

　　至於另一位代表學者魯杰總結建構主義議程的目的在於開拓國際關係領域中相對狹窄的理論空間，透過三個研究方法，第一、首先思考行為體的身份與利益問題。其次，進一步深入分析，聚焦於研究「社會行動」與「社會秩序」的主體間基礎，第三、思考行為體的時空維度問題，建立英國社會學者紀登斯（Giddens）所倡導的「二元性結構」（duality of structure），此種結構一方面限制了社會行動，同時又被社會行動所構建與再建構的過程。[20]

　　另外一位少被提及的美國學者Onuf強調規則、主體間理解，借鑑語言學理論，雖然出現比較早，相對比較少人關切。事實上，Onuf是第一個將「建構主義」引進國際關係研究的美國學者，他在1998年整理他以往於1989年的著作，[21] 創作了一篇「建構主義宣言」（A Constructivist Manifesto），[22] 之後加以改寫有收錄在一本專著。[23] 總結Onuf的建構主義邏輯命題在於「人是社會存在的現象」，基於人所具有的「社會關係」，從而「建構」（construct）或是「構建」（make）了人，進而使我們成為目前這樣的存在實體，同時，我們也建構現在世界的樣貌。在此一相互人與社會相互建構過程，透過「彼此互動」與「相互交談」，是以「言語即行動」（Saying is doing）。[24]

　　是以，建構主義認為將「人」與「社會」結合的第三要素：「規則」（rule），經由此種社會規則建構出某種過程，使得人與社會持續不斷互補構建對方。依此「規則」被定義為一種人們應該做什麼的陳述，「什麼」：「係指我們可以確定是類似的、並可能預期情況下人的行為準則」，而「應該」：「則是告訴我們要使自己的行為符合那些標準」。面對「規則」出現以下反

---

[20] 約翰杰拉爾德魯杰，「什麼因素將世界維繫在一起？新功利主義與社會建構主義的挑戰」，彼得卡贊斯坦（Peter Katzenstein）、羅伯特基歐漢（Robert Keohane）、斯蒂芬克拉斯納（Stephen Krasner）編，秦亞青、蘇長和、門洪華、魏玲譯，世界政治理論的探索與爭鳴（Exploration and Contestation in the Study of World Politics），頁263。

[21] Nicholas Greenwood Onuf, World of Our Making: Rules and Rule in Social Theory and International Relations. Columbia: University of South Carolina Press, 1989.

[22] Nicholas Greenwood Onuf, "A Constructivist Manifesto," in Kurt Burch and Robert A. Denemark ed., Constituting International Political Economy (Boulder, CO: Lynne Rienner, 1997), pp. 7-17.

[23] Nicholas Greenwood Onuf, Making Sense, Making Worlds-Constructivism in social theory and international relations (London and New York: Rouledge Taylor & Francis Group, 2013), pp.1-50.

[24] Nicholas Greenwood Onuf, Making Sense, Making Worlds-Constructivism in social theory and international relations, p.4.

應：遵守、破壞或置之不理等等，就是一種實踐過程。[25]

另外，「規則」的出現，也確立了社會的積極參與者：「行為體」（agent），人之所以為「行為體」，係基於社會通過多種規則，讓行為體可以參與許多規則，在此規則情境下發揮作用。因此，「規則」可以協助「行為體」在不同的社會情境下做出「選擇」，從「行為體」的角度言，「規則」界定了所有相關的情境。[26]

Onuf認為，不管規則是透過「刻意」或是「偶然」情況下，及其相關的實踐活動，都會促成與行為體目標相符合一個穩定發展模式：「制度」（institution），讓「人」成為「行為體」，並且建構了某種行為體在其間理性行事的環境。是以，Onuf認為人如果為社會中的「行為體」或是行為的參與者，可以稱之為「行為體」（agent），透過「行為體」之相互建構「規則」（rule），經由「規則」形成「制度」（system），從而構建「社會」（system）。行為體經由互動產生許多規則，「接受」或是「否認」規則，都屬於社會實踐過程，引導後續「社會」的成型。[27]

Onuf主張「無政府狀態」（anarchy）應該是一種「統治狀態」（a condition of rule）：在其中沒有一個國家或是國家集團能夠統治他國家或國家集團，亦即：不存在某種凌駕於國家之上，並統治國家的機構。相似於我們論述國家具有主權的觀點。是以，從「規則」切入分析國際關係，只能描述為統治狀態的某種關係形式。在此種高度制度化的產物，可以分辨出作為「統治者」（rulers）的特定行為體。易言之，存在「規則」，就存在「統治狀態」：運用規則來控制其他行為體，並取得相對於被控制者的優勢利基。

同樣，中國學者秦亞青點出建構主義質疑國際社會的基本特質：「無政

---

[25] Nicholas Greenwood Onuf, Making Sense, Making Worlds-Constructivism in social theory and international relations, p.4.

[26] Nicholas Greenwood Onuf, Making Sense, Making Worlds-Constructivism in social theory and international relations, p.5.

[27] Nicholas Greenwood Onuf, Making Sense, Making Worlds-Constructivism in social theory and international relations, p.5.

府狀態」（Anarchy），而認為此一「無政府狀態」是社會行為體互動行為的結果，並非國際體系固有的客觀事實，因此，建構主義學者主張國際社會存在的「規範」與「規則」可以抑止無政府性質。[28] 最具代表性美國學者溫特（Alexander Wendt）提出「行為體」或是「能動者」（agent）之間透過「有意義互動」（meaningful interaction），產生「共有理解」（shared ideas），形成三種國際無政府文化：霍布斯、洛克、康德文化，從主導型為體之間的「利益」與「政策」。溫特強調透過「武力」、「利益」與「合法性」驅動無政府文化內化的三個客觀決定因素。

「無政府狀態」是一種「統治狀態」，「規則」無法主導「行為體」處理其關係的方式產生直接決定作用，因為，整體社會還存在許多隱藏的運作規則（潛規則）？一定程度也影響行為體的選擇。Onuf主張建構主義學者採取「社會安排」（social arrangement），取代「結構」（structure）一詞，因此，國際無政府狀態就是一種廣泛意義上的「社會安排」：一種制度，在此範圍內，許多制度相互聯繫，規則創造了多種統治狀態，提供行為體許多「選擇」與「社會安排」。[29]

從Onuf提出國際關係的建構主義研究方法論的角度言，人建構了社會，並且透過「規則」，建構一個人與社會互動的實踐過程，構成以下的人與社會關係示意圖。

圖一：規則建構主義人與社會關係示意圖

---

[28] 秦亞青，「譯者前言 國際關係理論的爭鳴、融合與創新」，彼得卡贊斯坦（Peter Katzenstein）、羅伯特基歐漢（Robert Keohane）、斯蒂芬克拉斯納（Stephen Krasner）編，秦亞青、蘇長和、門洪華、魏玲譯，世界政治理論的探索與爭鳴（Exploration and Contestation in the Study of World Politics）（上海：上海人民出版社，2018），頁6。

[29] Nicholas Greenwood Onuf, Making Sense, Making Worlds-Constructivism in social theory and international relations, p.7.

二、言語行為及其類型

奧魯夫（Nicholas Onuf）提出透過「規則」（rule）形成「制度」（regime），「制度」構建「社會」（society）的邏輯思考。其中，透過行為體之間的「言語行為」（speech act）：「通過語言的方式，促使其他人採取相應行為的做法」，[30] 從語言學的專業角度亦可解釋為：「言語行為就是一種話語形式，具有溝通的功能」（A speech act is an utterance that serves a function in communication.）。[31]

上述「言語行為」出現之後，形成三種：斷言性（assertive speech acts）、指導性（directive acts）、承諾性（commissive speech acts）的「言語行為」模式，「規則」也是通過「言語行為」的一般形式表現出來。首先，通過「斷言性言語行為」出現的「規則」，將世界的本來面貌（世界的特徵與運作方式）呈現在行為體面前，告訴行為體如果忽視相關的信息會導致何種可能的後果，「主權原則」屬於上述「斷言性言語行為」得明顯例子。「指導性規則」（instruction-rules），類似運用設備的說明書、增補委員會席位的說明書，或是呈遞外交文書的指南，換言之，提供信息本身並不具備規範效應，但是，告訴行為體應該如何應對這些信息，則具有一定程度的規範性。[32]

第二、「指令性言語行為」，具有強制性，出現一種「指令性規則」（directive-rules），通過告訴行為體必須做的事情，也會提供行為體瞭解如果忽視這一規則，可能導致何種後果的信息。[33]

---

[30] Nicholas Greenwood Onuf, Making Sense, Making Worlds-Constructivism in social theory and international relations, p.10.

[31] "What is a Speech Act?", "A speech act is an utterance that serves a function in communication. We perform speech acts when we offer an apology, greeting, request, complaint, invitation, compliment, or refusal. A speech act might contain just one word, as in "Sorry!" to perform an apology, or several words or sentences: "I'm sorry I forgot your birthday. I just let it slip my mind." Speech acts include real-life interactions and require not only knowledge of the language but also appropriate use of that language within a given culture." Center for Advanced Research on Language Acquisition, CARLA, https://carla.umn.edu/speechacts/definition.html(2021/04/09)

[32] Nicholas Greenwood Onuf, Making Sense, Making Worlds-Constructivism in social theory and international relations, pp.10-11.

[33] Nicholas Greenwood Onuf, Making Sense, Making Worlds-Constructivism in social theory and international relations, p.11.

　　第三、「承諾性言語行為」牽涉行為體所作的諾言，說話人做出了為聽眾所接受的承諾，形成一種「承諾性規則」（commitment-rules），行為體可以從「效應上」來區分此種規則，亦即行為體瞭解他們自己相對於其他行為體所具有的「權利」與「義務」。是以，通過上述三種方式，使得說話者與聽眾之間具有「互動關係」，也同樣是合於「規則」，主要在於「規則」的運行與「言語行為」都是借助三種途徑：通過給行為體以指導、指令和承諾的方式完成某種行動。行為體構建出規則，並運用這些「規則」以達成指導、指令與承諾的效力。[34]

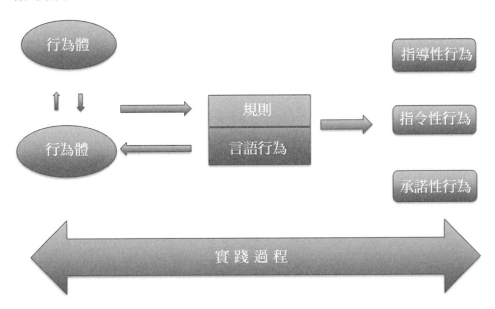

　　基本上，「行為體」面對「規則」作出回應時，其心中存在具體目標的，而「制度」正好有助於其利益的實踐。Onuf提出一個檢視「規則」的途徑，亦即「規則」可以透過三種「言語行為」（speech act）加以觀察：指導性、指令性、承諾性等言語行為，從而讓行為體之間的互動有了依據準則，從而形成規範社會的「規則」。基本上，行為體對於「規則」做出回應時，係透過具體的

---

[34] Nicholas Greenwood Onuf, Making Sense, Making Worlds-Constructivism in social theory and international relations, p.12.

目標：「制度」，有助於其利益的實現。[35]

三、規則產生統治途徑

Onuf指出當「指導性規則」處於明顯優勢時，行為體之間處於一種被規則與相關實踐活動所編織成的網絡，「權力平衡」屬於其中的例證之一。與其相應的規則賦予一些大國一種高高在上的「地位」，此一規則運作的前提在於這些大國擁有的資源條件近乎相同。國家行為體依照有關指示行事，無論這些行為體的近期目標如何，其結果都應該是大國之間建立一種動態並且相互平穩的「聯盟均衡」（balance of alliances）。基於行為體將「權力平衡」視為一種符合其自身利益的制度，也會透過「平衡」名義，推動相同後果的產生。[36]

在「指導性規則」下的國際「統治形式」：觀念與信念來主導支配作用，行為體實際上是通過促使其他行為體接受其觀念與信念以實現統治行為，他們為此採用樹立榜樣與灌輸的方法，這是一種「霸權」（hegemony）的統治形式。每一個在霸權統治下的等級在構成社會的等級網路都有清晰的界線與確定的位置。[37]

當「指令性規則」處於明顯優勢時，行為體之間所處的狀態可以被稱之為一種命令性的鏈條，類似一個公司或是正式組織，而「勢力範圍」就是一種初始化的制度。與其相對應的非成型化規則，分配給每一個行為體某種「辦公場所」（office），一如在規則化組織內部的處所。在「勢力範圍」制度下，可以發揮實現作為最高決者者的主導性大國，控制該勢力範圍內較弱小國家的願望。

在此指令性規則為社會明顯標誌下，有關的事物場所係通過某個指令性

---

[35] 溫都爾卡・庫芭科娃（Vendulka Kubalkova）、尼古拉斯 奧魯夫（Nicholas Onuf）、保羅 科維特（Paul Kowert）主編，肖鋒譯、張志洲校，建構世界中的國際關係（International Relations in a Constructed World）（北京：北京大學出版社，2006），頁85。

[36] Nicholas Greenwood Onuf, Making Sense, Making Worlds-Constructivism in social theory and international relations, p.15.

[37] Nicholas Greenwood Onuf, Making Sense, Making Worlds-Constructivism in social theory and international relations, pp.18-19.

鏈條垂直組織起來，這是一種從上而下組織起來的辦事場所被稱為一種「等級制」（hierarchy），此種統治形式源於辦事人員針對指令性規則的執行。當指令性規則具有法定效應，等級制度因而成型化，由此導致所謂「權威」（authority），就是所謂「合法控制權」。另外，非成型化的等級制度可能強化那些已經達到相對較高成型化程度的霸權，例如二次大戰結束以來的「美國治理下的和平」（Pax Americana）：美國以實現自由和繁榮的名義，通過隨心所欲的時候與地方進行干預方式的統治。有關原則地提出，導致美國成為自由世界領導者地位的形成，經由這些原則的實施賦予美國某種非正式全力行使場所。[38]

第三、「承諾性規則」處於明顯優勢時，行為体間具有伙伴關係性質，或是「平等聯盟成員」（associatons）關係。在國際關係的制度範圍內，主權原則與其相互承認的輔助性規則，促使國家相互間在形式上具有平等性。國家在形式上都是平等的，都具有相同的「角色」（role），平等聯盟的功能就是通過存在的承諾性規則，將有關的角色分配給其他行為體。承諾性規則有助於建構出大量基於有限目的，而在形式上平等的行為體。[39]

在此承諾性規則為主體的社會標誌，行為體擔當不同的角色，這些角色是通過參照其他行為體所擔任的角色而被界定出來。沒有一個角色或者制度能夠將特定的行為體演變為統治者，因此，成型化的承諾性規則也強化了成型化的等級制度，由此產生的國家形態為立憲國家，憲法界定了制約國家政府並使其承擔責任的承諾性規則。[40]

行為体所涉及的因素包括地位、辦事場所、角色，依據所處的制度環境不同，每一個行為体都必須具備某種地位、擁有某個辦事處所，或者扮演某種角

---

[38] Nicholas Greenwood Onuf, Making Sense, Making Worlds-Constructivism in social theory and international relations, p.19.

[39] 溫都爾卡·庫芭科娃（Vendulka Kubalkova）、尼古拉斯 奧魯夫（Nicholas Onuf）、保羅 科維特（Paul Kowert）主編，肖鋒譯、張志洲校，建構世界中的國際關係（International Relations in a Constructed World）（北京：北京大學出版社，2006），頁85-86。

[40] Nicholas Greenwood Onuf, Making Sense, Making Worlds-Constructivism in social theory and international relations, p.19.

色。大多數行為體在某種程度上同時具有三個特徵，因為，大多數都是處於多個制度中的行為体，而且，許多制度都綜合具有網絡、組織與和平等聯盟的特性。[41]（參見下表一：言語行為的意涵、特色與統治形態一覽表）

表一：言語行為的意涵、特色與統治形態一覽表

形　　式	內　　涵	統治形態
指導式言語行為	霸權統治	權力平衡
指令式言語行為	等級制度	勢力範圍
承諾式言語行為	夥伴關係	平等結盟

資料來源：筆者整理自製。

以上三種類型的規則相互協作，共同支撐一個制度的原則。由於主權原則的存在，在國際關係體系中沒有存在成型化的指令性規則，但是，還是有一些非成型化的指令性規則的存在。Onuf 認為「無政府狀態」是一種行為體的意圖沒有關聯的統治狀態，國際社會就不是一種無政府狀態，而是存在一種「他治」（heteronomy），由於「主權」原則以及一系列，亦即「行為體」的「自治程度」受到其他行為體的限制，國際社會基本上呈現一種「他治」狀態。[42] 一言之，行為體圍繞著許多「制度」，相關「制度」的安排都有其目的，都受到「行為體」的「意圖」（intention）加以制約，就是如何維護行為體的利益，也包括在相互制約時所呈現的「共同利益」（shared interest in being rules）關係。[43]

是以，根據Onuf提出的「規則構建出行為體、行為體也構建出規則」、「規則形成制度、制度構建社會」，以及「規則產生統治」三項規則建構主義

---

[41] 溫都爾卡、庫芭科娃（Vendulka Kubalkova）、尼古拉斯 奧魯夫（Nicholas Onuf）、保羅 科維特（Paul Kowert）主編，肖鋒譯、張志洲校，建構世界中的國際關係（International Relations in a Constructed World）（北京：北京大學出版社，2006），頁86。

[42] Nicholas Greenwood Onuf, Making Sense, Making Worlds-Constructivism in social theory and international relations, p.20.

[43] Nicholas Greenwood Onuf, Making Sense, Making Worlds-Constructivism in social theory and international relations, p.20.

的邏輯思考，本文提出以下三個命題與推論：

命題 1：美中相互實踐過程，建立以規則為基礎的國際秩序與不干涉內政為原則的戰略對抗情勢。

推論一：影響美中雙方規則互不相容的因素，在於雙方對於無政府狀態下主權的定義與操作的認知差異。

命題 2：美中依據上述雙方對抗性規則，使得第三方以歐盟為主，面臨戰略選擇的矛盾。

推論二：主導美中以外的第三方的規則選擇，主要基於各別的不同國家利益考量

命題 3：依據美中兩國上述規則的對抗下，由於「　」、「指令性」與「承諾性」言語行為相互交雜，呈現國際社會多種「統治形態」的態勢。

推論三：由於非傳統安全（Covid-19）的出現，全球治理能量與決策者特質影響國際社會多元統治形態的生成。

參、美中印太相互規則建構

一、規則為基礎的國際秩序VS.不干涉內政原則

2021年3月18日，美中兩國外交與國安高層激烈交鋒，美國國務卿布林肯直接點出維護國際既有「規則」（rule）的重要性，顯示出美方代表有備而來，他強調：「此種以規則為基礎的國際秩序體系並非一種抽象概念，可以協助國家之間和平解決爭端、有效率協調多邊能量，共同參與全球貿易機制，並確認每一個行為體都能接受同樣的規則。因為，在一個缺乏以規則為基礎的國際制度之中，可能會造成贏者全拿的局面，但是，會形成一個更加暴力與不穩定的世界。」[44]

同時，布林肯直接點出：「我們美國深深關切中國在新疆、香港、台灣，

---

[44] "Secretary Antony J. Blinken, National Security Advisor Jake Sullivan, Director Yang And State Councilor Wang At the Top of Their Meeting, REMARKS ANTONY J. BLINKEN, SECRETARY OF STATE, ANCHORAGE, ALASKA, MARCH 18, 2021", U.S. Department of State, https://www.state.gov/secretary-antony-j-blinken-national-security-advisor-jake-sullivan-chinese-director-of-the-office-of-the-central-commission-for-foreign-affairs-yang-jiechi-and-chinese-state-councilor-wang-yi-at-th/ (2021/04/22)

對美國的網路攻擊，以及運用經濟工具脅迫美國盟邦。這些中國的行為威脅了以規則為基礎的秩序，確保全球穩定。是以，上述事件並非僅僅是內政事務，我們有義務提出上述的議題來討論。」[45]

另外，美國國家安全會議顧問蘇利文也稍後提出：「我們不是要追求對抗，但是我們歡迎激烈競爭，我們將會持續捍衛我們的原則、人民與我們的朋友。」，以及確認與中國的戰略定位是：「我們與中國的關係當它是應該的，會是競爭性的，如果取可的話應該是合作的，如果必要，也是一種對抗性關係。」[46]

上述布林肯與蘇利文的論述表達兩種思維，一方面確認美國對中國的戰略定位是延續前總統川普時代，以中國為主要戰略對手的思考。根據2017年川普公布其任內唯一的「國家安全戰略報告」（National Security Strategy Report），[47] 特別指出當今世界局勢處於「強權競逐」時代，提到「美國將會回應面對世界增長的政治、經濟與軍事的競爭。中國與俄羅斯挑戰美國的權力、影響力與利益，並企圖侵蝕美國的安全與繁榮。」[48]

2021年3月3日，就職不到一百天，美國總統拜登就提出「暫行國家安全戰略指導」（Interim National Security Strategic Guidance），點出中國與俄羅斯對於美國國家安全的威脅，強調「我們必須要瞭解全世界權力分布已經改變了，也帶來新的安全威脅，其中，中國漸漸的表現出更加決斷侵略性。中國是唯一

[45] "Secretary Antony J. Blinken, National Security Advisor Jake Sullivan, Director Yang And State Councilor Wang At the Top of Their Meeting, REMARKS ANTONY J. BLINKEN, SECRETARY OF STATE, ANCHORAGE, ALASKA, MARCH 18, 2021", U.S. Department of State, https://www.state.gov/secretary-antony-j-blinken-national-security-advisor-jake-sullivan-chinese-director-of-the-office-of-the-central-commission-for-foreign-affairs-yang-jiechi-and-chinese-state-councilor-wang-yi-at-th/ (2021/04/22)

[46] "Secretary Antony J. Blinken, National Security Advisor Jake Sullivan, Director Yang And State Councilor Wang At the Top of Their Meeting, REMARKS ANTONY J. BLINKEN, SECRETARY OF STATE, ANCHORAGE, ALASKA, MARCH 18, 2021", U.S. Department of State, https://www.state.gov/secretary-antony-j-blinken-national-security-advisor-jake-sullivan-chinese-director-of-the-office-of-the-central-commission-for-foreign-affairs-yang-jiechi-and-chinese-state-councilor-wang-yi-at-th/ (2021/04/22)

[47] "National Security Strategy of the United States of American", December 2017, https://trumpwhitehouse.archives.gov/wp-content/uploads/2017/12/NSS-Final-12-18-2017-0905.pdf (2021/04/02)

[48] "National Security Strategy of the United States of American", December 2017, p.2, https://trumpwhitehouse.archives.gov/wp-content/uploads/2017/12/NSS-Final-12-18-2017-0905.pdf (2021/04/02)

具有整合其潛在的經濟、外交、軍事與科技能量，能夠挑戰一個固定與開放的國際體系。俄羅斯則是決議強化其全球影響力，以移扮演一個國際社會破壞性角色。」[49] 以上前後兩任美國總統並不因為黨籍不同，就改變對中國「敵對」身份的認知，只是拜登更加具有建構主義三種無政府文化的運用：敵對軍事對抗的的「霍布斯文化」、經濟與科技的競爭的「洛克文化」，以及相對有善在全球事務的朋友為特徵的「康德無政府文化」。

另外，除了定調中國為主要戰略對手之外，從傳統國際關係理論角度言，尋找敵人確認國家安全的威脅來源，才能確定國家利益之所在，以及為達成此利益的目標與其下的各種戰略與政策。華盛頓必須跳脫川普時代的對中戰略，採取「單邊主義」作為，不斷強調「美國優先」（America First）與「美國利益」（America Interest），而是藉由「多邊主義」角度，從「規範」或是「規則」切入，才能進一步號召全世界理念相近國家共同因對中國。

反之，中國早於2011年9月6日，透過「中國的和平發展白皮書」，凸顯出「不干涉內政原則」基本立場，亦即「不允許外部勢力干涉中國內政……不干涉別國內部事務……不把自己的意志強加於人」，並強調：「政治上相互尊重、平等協商，共同推進國際關系民主化。國家不分大小、強弱、貧富，都是國際社會平等成員，都應受到國際社會尊重、各國內部事務應由本國人民自己決定，世界上的事情應由各國平等協商，各國平等參與國際事務的權利應得到尊重和維護。」[50]

2014年6月28日，中國慶祝和平共處五原則之六十週年大會，習近平再度重申，主權平等、共同安全、共同發展、和平共處五項原則，其中「第六，堅持公平正義。我們應該共同推動國際關系民主化，世界上的事情由各國政府和人民共同商量辦。我們應該共同推動國際關系法治化，推動各方遵守國際法和

---

[49] "RENEWING AMERICA'S ADVANTAGES Interim National Security Strategic Guidance", March 2021, The White House, President Joseph R. Biden, Jr. , pp.7-8, https://www.whitehouse.gov/wp-content/uploads/2021/03/NSC-1v2.pdf(2021/04/02)

[50] 「《中國的和平發展》白皮書（全文）」，中華人民共和國國防部，http://www.mod.gov.cn/big5/affair/2011-09/06/content_4295874_3.htm。（檢索日期：2021/05/06）

公認的國際關系基本原則。我們應該共同推動國際關系合理化，推進全球治理體系改革。」[51] 2020年11月10日，習近平在北京以視訊方式，出席「上海合作組織」成員國元首理事會第廿次會議並發表談話表示，要堅決反對外部勢力以任何藉口干涉成員國內政，習近平並強調，要堅定支持有關國家依法平穩推進重大國內政治議程，堅定支持各國維護政治安全和社會穩定，堅決反對外部勢力以任何藉口干涉成員國內政。[52]

2021年4月21日，面對拜登政府透過多邊主義方式批判中國人權議題，習近平在海南博鰲亞洲論壇的開幕視訊演說上，他表示國際上的事應該由大家共同商量著辦，世界前途命運應該由各國共同掌握，不能把一個或幾個國家制定的規則強加於人，也不能由個別國家的單邊主義給整個「世界帶節奏」，並強調：「世界要公道，不要霸道。大國要有大國的樣子，要展現更多責任擔當。要把平等相待、互尊互信挺在前面，「動輒對他國頤指氣使、干涉內政不得人心」。[53] 換言之，北京除了強調「不干涉別國內部事務」，持續主張「國際關係民主化」，並尊重國家主權平等原則，並透過北京主導的區域組織推廣上述理念，以爭取更多友好國家支持。

二、民主自由尊重人權VS獨裁專制迫害人權原則

基本上，民主、自由與人權是美國相當重視的立國基本價值，在美國歷屆總統的「國家安全戰略報告」（National Security Strategy Report）中呈現。2021年2月8日，美國總統拜登接受CBS訪問提到：習近平沒有民主素養，習近平非常聰明，非常強硬。他骨子裡頭沒有一點民主的成分。我跟習近平強調美中兩國不需要衝突，但是，兩國會有很激烈的競爭。基於習近平已經釋放出訊號，我也會如同川普總統一樣，我們將會聚焦於國際秩序的道路。[54]

---

[51] 「習近平出席和平共處五項原則發表60周年紀念大會」，中華人民共和國國防部，http://www.mod.gov.cn/big5/leader/2014-06/29/content_4519197.htm。（檢索日期：2021/05/06）

[52] 「習近平：反對外力干涉內政」，聯合新聞網，https://udn.com/news/story/7331/5005375。（檢索日期：2021/05/06）

[53] 「博鰲論壇開幕演說　習近平句句嗆美：干涉內政不得人心」，蘋果日報，https://tw.appledaily.com/international/20210420/2Q5U2NIN4NGYBJOPDIIBAJWZWI/。（檢索日期：2021/05/06）

[54] "Biden discusses vaccinations, school reopenings, foreign policy in first network interview as president",

2021年3月25日，拜登出席第一場就職以來的記者會，[55] 針對Bloomberg記者提問中國議題，強調儘管美中存在非常嚴峻的競爭，美國堅持中方遵循國際規範，公平競爭、公平行為與公平貿易，[56] 並且說明「我即將邀請各國來美國參加民主聯盟會議，為了順利達成民主推廣，我們必須讓中國遵循國際的規則，不管是攸關南海、東海或是與台灣的協議，或是其它相關與中國有關議題。」[57] 最後，拜登強調：「這是一個民主有效性與獨裁在二十一世紀的戰爭，我們必須證明民主是有效的。」[58] 從上述拜登講話，凸顯出美國要以「民主」（democracy）與「民主制度」成為因應中國的一項戰略工具。

事實上，中國為了因應全球民主發展的潮流，也理解美國會透過「民主」來牽制中國政治發展過程，於2019年11月2日，習近平在上海考察時，針對民主制度的回應，指出：「我們走的是一條中國特色社會主義政治發展道路，人民民主是一種全過程的民主，所有的重大立法決策都是依照程序、經過民主醞釀，通過科學決策、民主決策產生的。」[59]

同時，針對美國對於中國人權議題的批判，早於川普總統時期，於2019年9月22日，中國國務院就發佈「為人民謀幸福：新中國人權事業發展70年」白皮書，強調：「努力維護世界和平與安全。作為和平共處五項原則的積極倡導者和堅定實踐者，中國不僅自身積極奉行和平外交思想、注重與各國和平共處，而且積極倡導共同、綜合、合作、可持續的安全觀，致力於推動南南合作

The Washington Post, https://www.washingtonpost.com/politics/2021/02/07/biden-china-iran-cbs-interview/ (2021/05/07)

[55] "Remarks by President Biden in Press Conference, MARCH 25, 2021• SPEECHES AND REMARKS", The White House, https://www.whitehouse.gov/briefing-room/speeches-remarks/2021/03/25/remarks-by-president-biden-in-press-conference/ (2021/04/29)

[56] 「拜登：習近平是聰明人 但骨子裡沒有民主」，聯合新聞網，https://udn.com/news/story/122051/5344720。（檢索日期：2021/05/09）

[57] "Remarks by President Biden in Press Conference, MARCH 25, 2021 • SPEECHES AND REMARKS", The White House, https://www.whitehouse.gov/briefing-room/speeches-remarks/2021/03/25/remarks-by-president-biden-in-press-conference/ (2021/04/29)

[58] "Biden Defines His Underlying Challenge With China: 'Prove Democracy Works'", The New Yorks Times, https://www.nytimes.com/2021/03/26/us/politics/biden-china-democracy.html (2021/05/09)

[59] 「習近平首提「全過程民主」 新詞匯引發新質疑」，BBC中文網，https://www.bbc.com/zhongwen/trad/chinese-news-50285833。（檢索日期：2021/05/06）

和南北對話，努力縮小南北差距。」[60] 其中：「七、全面參與全球人權治理 中國在大力推進自身人權事業發展的同時，始終堅持平等互信、包容互鑒、合作共贏、共同發展的理念，積極參與聯合國人權事務，認真履行國際人權義務，廣泛開展國際人權合作，積極為全球人權治理提供中國智慧、中國方案，以實際行動推進全球人權治理朝著更加公正合理包容的方向發展。」[61]

2021年1月19日，前國務卿蓬佩奧在拜登總統即將上任的前一天宣布，中國政府在新疆實行了種族滅絕的認定。他指責中國共產黨至少從2017年3月開始針對維吾爾人和其他穆斯林少數民族犯下反人類罪。[62] 同時，美國於拜登上台兩個月之後，於2021年3月20日，旋即公佈第45次年度「國際人權實踐年度地區報告書」（annual Country Reports on Human Rights Practices），報告書中引述拜登總統強調：我們必須開始以民主價值為基礎的外交工作：保衛自由、創造機會、確保全球人權、尊重法治，並且尊重每一個人的尊嚴。[63] 針對中國方面提出：「2020年很多人依舊遭遇殘忍的條件。在中國政府針對新疆的種族淨化，包括勞改、迫害、強制勞動，以及其他宗教與少數族群的破壞。」[64]

兩天之後，3月22日，美國國務卿布林肯協同英國外交大臣拉布（Dominic Raab）、加拿大外交部長賈諾（Marc Garneau）發表聯合聲明指出，美英加3國團結一致，對中國在新疆侵犯人權的行為表示深切和持續的關注，包括嚴格限制宗教自由、強迫勞動、關押再教育營、強迫絕育，以及對維吾爾族資產的破壞。[65] 針對布林肯聯合北約主要國家的反中國人權作為，2021年3月24日，

---

[60] 「為人民謀幸福：新中國人權事業發展70年」，中華人民共和國中央政府，http://big5.www.gov.cn/gate/big5/www.gov.cn/zhengce/2019-09/22/content_5432162.htm（檢索日期：2021/05/07）

[61] 「為人民謀幸福：新中國人權事業發展70年」，中華人民共和國中央政府，http://big5.www.gov.cn/gate/big5/www.gov.cn/zhengce/2019-09/22/content_5432162.htm（檢索日期：2021/05/07）

[62] 「美國務院：維持有關中國政府在新疆實行種族滅絕的認定」，美國之音，https://www.voacantonese.com/a/US-China-xinjiang-genocide-20210309/5808954.html。（檢索日期：2021/04/07）

[63] "2020 Country Reports on Human Rights Practices", U.S. Department of State, https://www.state.gov/reports/2020-country-reports-on-human-rights-practices/ (2021/05/06)

[64] "2020 Country Reports on Human Rights Practices", U.S. Department of State, https://www.state.gov/reports/2020-country-reports-on-human-rights-practices/ (2021/05/06)

[65] 「美英加聯手歐盟制裁中國！指北京侵犯人權證據確鑿」，自由時報，https://news.ltn.com.tw/news/world/breakingnews/3475665。（檢索日期：2021/05/06）

中國國務院發佈「2020年美國侵犯人權報告」其中有關：「六、踐踏國際規則造成人道災難」層面，中國方面強調：「在抗疫需要全球團結的時刻，美國卻執意奉行本國優先，推行孤立主義、單邊主義，揮舞制裁大棒，霸凌威脅國際機構，殘酷對待尋求庇護者，成為全球安全與穩定的最大麻煩製造者。」，包括：「悍然退出世界衛生組織。背信棄義退出《巴黎協定》。霸凌行徑威脅國際機構。單邊制裁加重人道危機。殘酷對待尋求庇護者。疫情期間繼續強制遣返移民。赦免屠殺他國平民的戰爭罪犯。」[66]

此外，2021年4月23日，中國外交部長王毅晚間與美國智庫「外交關係協會」（CFR）視訊交流。針對對於中美之間是否陷入制度之爭，他強調：「民主不是可口可樂，美國生產原漿，全世界一個味道。如果地球上只有一種模式、一種文明，這個世界就會失去了生機，沒有了活力」。[67] 同時「中國實行的社會主義民主政治，「是一種全過程、最廣泛的民主，體現人民意志，符合中國國情，得到人民擁護」，若僅僅因為實行民主的形式跟美方不一樣，「就給中國扣上『威權』、『專制』的帽子，這本身就是不民主的表現。如果打著民主、人權旗號搞價值觀外交，干涉他國內政，人為製造分裂和對抗，只會引發動盪甚至災難。」[68]

三、全球公海自由航行權VS.傳統領海主權原則

從1979年起，美國海軍就開始從事「公海自由航行活動」（Freedom on Navigation Operations, FONOPs），挑戰美國認為不符合國際法的海洋主權申張活動。除了中國之外，還有其它洲與國家成為美國此種活動的對象國。截至2019年以來美國挑戰了全世界22國家的主權過度伸張行動，只是從中國在南沙組建人工島礁之後，就成為美國海軍主要活動的對象。2021年3月16日，美國

---

[66] 「2020年美國侵犯人權報告」，中華人民共和國國防部，http://www.mod.gov.cn/big5/topnews/2021-03/24/content_4881827.htm。（檢索日期：2021/05/06）

[67] 「王毅辯稱民主非可口可樂 全世界跟美國同口味」，中央社，https://www.cna.com.tw/news/aopl/202104240046.aspx。（檢索日期：2021/05/09）

[68] 「王毅辯稱民主非可口可樂 全世界跟美國同口味」，中央社，https://www.cna.com.tw/news/aopl/202104240046.aspx。（檢索日期：2021/05/09）

國務院發表「2020年自由航行年度報告」，2020年一年，美國海軍挑戰不符合國際法的海洋主權伸張的19國家高達28個違法主張，[69]

　　基本上，美國總統拜登會延續川普的公海自由航行活動，也應該修正歐巴馬總統時代的作法，此一行動必須經常與定期性，至少每一季兩次行動。換言之，美國必須透過此種在南海地區進行的「公海自由航行活動」，表達「公海自由航行活動」是沒有妥協的餘地。[70]

　　2015年11月7日，習近平在新加坡國立大學針對南海主權的演講，重申「南海諸島自古以來就是中國領土」的立場，維護自身的領土主權和正當合理的海洋權益，是中國政府必須承擔的責任，而中國也「堅持同直接當事國在尊重歷史事實的基礎上，根據國際法，通過談判和協商解決有關爭議」，他說：「在中國和南海沿岸國共同努力下，南海局勢總體是和平的，航行和飛越自由從來沒有問題，將來也不會有問題，因為首先中國最需要南海航行通暢…我們歡迎域外國家參與亞洲和平與發展事業，為此發揮積極作用。」[71]

　　2016年7月13日，中國國務院新聞辦發表《中國堅持通過談判解決中國與菲律賓在南海的有關爭議》白皮書，強調：中國一貫遵守《聯合國憲章》的宗旨和原則，堅定維護和促進國際法治，尊重和踐行國際法，在堅定維護中國在南海的領土主權和海洋權益的同時，堅持通過談判協商解決爭議，堅持通過規則機制管控分歧，堅持通過互利合作實現共贏，致力於把南海建設成和平之海、友誼之海和合作之海。」[72]

　　2020年7月13日，美國前國務卿龐畢奧提出關於「美國對於南海海權主張

---

[69] "Freedom of Navigation Report Annual Release", MEDIA NOTE, OFFICE OF THE SPOKESPERSON, U.S. Department of State, ,https://www.state.gov/freedom-of-navigation-report-annual-release/ (2021/05/02)

[70] Jeff M. Smith, "Biden Must Keep Challenging China on Freedom of Navigation", Foreign Policy, https://foreignpolicy.com/2021/02/16/biden-south-china-sea-spratlys/ (2021/05/07)

[71] 「習近平新加坡演講重申中國南海領土主權」，BBC中文網，https://www.bbc.com/zhongwen/trad/china/2015/11/151107_china_singapore_xi_speech。（檢索日期：2021/05/06）

[72] 「國務院新聞辦發表《中國堅持通過談判解決中國與菲律賓在南海的有關爭議》白皮書」，中華人民共和國中央人民政府，http://big5.www.gov.cn/gate/big5/www.gov.cn/xinwen/2016-07/13/content_5090813.htm。（檢索日期：2021/05/06）

的立場」（U.S. Position on Maritime Claims in the South China Sea）[73] 聲明，涵蓋三點重要立場宣示，第一、美國主張中國不能合法的要求從黃岩島、南沙島礁延伸的任何專屬經濟區、大陸架的要求。第二、既然北京無法一致性、合法性的主張南海地區的海洋權利，美國方面拒絕任何依據中國所佔有的南沙島礁的12浬領海主權要求。最後、美國認為中國對距離馬來西亞50浬，距離中國南部海岸1000浬的「曾母暗沙」（James Shoal）不具備任何合法的領土或是海洋權利主張，也不是中國最南端的國土。因此，美國認為：「世界不允許北京將南中國海視為其海上帝國。美國與我們的東南亞盟國與合作夥伴站在一起，根據國際法賦予它們的權利和義務，來維護它們對離岸資源的主權。」[74]

2021年1月22日，中國全國人民代表大會常務委員會通過《中華人民共和國海警法》並於2月1日生效，2021年2月19日，美國清楚地表達針對中國推動「海警法」的憂心考量，透過此法允許中國海警局可以使用武力對付其他國家海上活動。[75] 美國國務院發言人普賴斯（Ned Price）進一步表示，中國「海警法」中的條文「強烈暗示了這部法律可能被用來恐嚇（中國的）海上鄰國」，而北京堅稱，該「海警法」內容符合國際慣例和各國實踐。[76]

2021年3月15日，美日外交與國防部長會談時，布林肯指出：「兩國2+2部長會議時，美國表達承諾防衛日本與中國在東海島嶼之爭，以及重複強調反對中國在南海地區非法的主權聲張。」[77] 4月8日，布林肯與菲律賓外長談話重申對於美菲共同防禦條約的重要性，並且強調：「兩國共同關切中國海上民兵於

[73] "Statement by Secretary Michael R. Pompeo, U.S. Position on Maritime Claims in the South China Sea", U.S. Embassy in Laos, https://la.usembassy.gov/statement-by-secretary-michael-r-pompeo-u-s-position-on-maritime-claims-in-the-south-china-sea/(2021/03/29)

[74] "Statement by Secretary Michael R. Pompeo, U.S. Position on Maritime Claims in the South China Sea", U.S. Embassy in Laos, https://la.usembassy.gov/statement-by-secretary-michael-r-pompeo-u-s-position-on-maritime-claims-in-the-south-china-sea/(2021/03/29)

[75] "Biden administration's policy towards Vietnam, and the South China Sea", modern diplomacy, https://moderndiplomacy.eu/2021/03/02/biden-administrations-policy-towards-vietnam-and-the-south-china-sea/ (2021/03/29)

[76] 「中國海警法上路 執法權擴容下釣魚島火藥味漸濃」，BBC中文網，https://www.bbc.com/zhongwen/trad/chinese-news-56249465。（檢索日期：2021/03/29）

[77] "Blinken warns China against 'coercion and aggression' on first Asia trip", Reuters, https://www.reuters.com/article/us-usa-asia-blinken-japan-idUSKBN2B71C9 (2021/05/10)

南沙群島所屬牛軛礁附近大量集結活動，也再度強調要求中國遵守2016依據國際海洋法規定下的南海仲裁案的判決。」[78] 是以，當中國主張在南海地區的傳統海域主張權利，並通過「海警法」維護其海洋權利，並希望通過談判解決爭端，美國則是認為依據國際海洋法與公海自由航行權，否認中國在第一島鏈三海：東海、台海與南海的主張伸張，形成兩種都是根據國際法與國際海洋法伸張權利的戰略競逐態勢。

肆、美中印太相互制度檢證

一、美中相互言語行為檢證：指導、指令與承諾性規則

基於，透過行為體之間的「言語行為」（speech act）：「通過語言的方式，促使其他人採取相應行為的做法」，Onuf提出一個檢視「規則」的途徑，亦即「規則」可以透過三種「言語行為」（speech act）加以觀察：指導性、指令性、承諾性等言語行為，讓行為體之間的互動有了依據準則，從而形成規範社會的「規則」。行為體對於「規則」做出回應時，係透過具體的目標：「制度」，有助於其利益的實現。

是以，根據前述美中兩個行為體相互透過「言語行為」形成不同的「規則」，本文整理出三種規則：1.規則為基礎的國際秩序VS.不干涉內政原則2.民主自由尊重人權VS獨裁專制迫害人權原則3.全球公海自由航行權VS.傳統領海主權原則。上述三種「規則」之爭，都涉及如何看待國際社會基本原則：「無政府狀態」與「國家主權原則」。

Onuf認為「無政府狀態」是一種「統治形態」，屬於受制於他者的狀態，並非完全可以「自治」的形態。美國透過上述三個「規則」的提出，本意其實就是告訴中國必須遵守一定的國際「遊戲規則」，不能僅僅以自身國家利益為思考，不去考慮整體國際公義與國際利益。一定程度，美國運用新自由制度主

---

[78] "Secretary Blinken's Call with Philippine Secretary of Foreign Affairs Locsin", U.S. State of Department, https://www.state.gov/secretary-blinkens-call-with-philippine-secretary-of-foreign-affairs-locsin-2/ (2021/05/10)

義的理念，強調全球國家「複合式相互依存」的統治狀態。

如同2021年5月7日，布林肯在聯合國安理會視訊會議中提出「多邊主義」的重要性，並表達對於強國國家主權行使的克制思考，同時強調：「安理會中最有權力國家同意自我克制，不僅有利於全人類，也對他們有利。我們相信其他國家的成功對我們也是關鍵議題，我們也不希望力量較弱國家感受威脅，進而結合一起對抗我們。」[79]

二、美中印太相互制度建立：四方對話到印太區域機制

2021年1月20日拜登普就職美國總統不久，隨即於3月15日以「視訊會議」方式，在華盛頓舉行首度「四方安全對話」（Quadrilateral Security Dialogue, The Quad），透過四國元首聲明顯示出其主要目的在於「我們重新強化我們的承諾來確保我們所處區域透過國際法加以治理，承諾保護全球價值，免除被脅迫」[80]

事實上，「四方安全對話」構想起源於2004年印尼發生大海嘯之後，四國協助成立的機制，到了2021年之前，並沒有任何具體元首高峰會議的成型來處理攸關印太事務。[81] 2021年拜登一上台就思考運用既有機制，從多邊主義角度整合主要印太國家。此次「四方安全對話」的元首高峰會議之後，發表共同聲明具體針對因應「新冠肺炎」，提出一個「四方疫苗夥伴關係」（The Quad Vaccine Partnership），「四方領袖同意強化2021年疫苗製造，協助印太國家疫苗供應，強化與既存多國衛生機制合作關係包括世界衛生組織與COVAX。」[82]

---

[79] "Secretary Antony J. Blinken Virtual Remarks at the UN Security Council Open Debate on Multilateralism", U.S. Department of State, https://www.state.gov/secretary-antony-j-blinken-virtual-remarks-at-the-un-security-council-open-debate-on-multilateralism/ (2021/05/10)

[80] "Remarks by President Biden, Prime Minister Modi of India, Prime Minister Morrison of Australia, and Prime Minister Suga of Japan in the Virtual Quad Leaders Summit", MARCH 12, 2021 • SPEECHES AND REMARKS", The White House, https://www.whitehouse.gov/briefing-room/speeches-remarks/2021/03/12/remarks-by-president-biden-prime-minister-modi-of-india-prime-minister-morrison-of-australia-and-prime-minister-suga-of-japan-in-virtual-meeting-of-the-quad/ (2021/05/10)

[81] "Biden's meeting with "the Quad," a new alliance to counter China, explained", VOX, https://www.vox.com/22325328/biden-quad-japan-australia-india-vaccine-rare-earth (2021/05/10)

[82] "Fact Sheet: Quad Summit", The White House, https://www.whitehouse.gov/briefing-room/statements-releases/2021/03/12/fact-sheet-quad-summit/ (2021/05/10)

亦即，由印度生產美國研發的武漢肺炎疫苗，並由美國國際開發金融公司（US International Development Finance Corporation，DFC）和日本國際協力銀行（Japan Bank for International Cooperation）出資支持，由澳洲資助培訓，協助疫苗配送。[83]

2021年4月14日，美國國務卿布林肯前往北約總部與秘書長Jens Stoltenberg會面發表聲明，強調與北約合作因應全球反恐與在阿富汗反恐戰爭，透過拜登總統的正式宣布，美國將從阿富汗撤軍，但是，也不會忘掉對阿富汗未來安全保證的承諾。[84] 5月4日，布林肯於倫敦與七大工業國家部長會晤，主要會談議題聚焦於防衛民主價值與開放社會議題，主要防衛對象在於中國與俄羅斯。其他議題還包括疫情後經濟復甦，並邀請七大工業國家與盟邦共同協調，面對一個與莫斯科和北京之間的全球民主與專制競爭。[85]

美國白宮也公佈拜登即將於本年6月前往英國與比利時參訪，這也是拜登第一次國外參訪，表達強化聯盟關係，與跨大西洋夥伴關係，與多邊夥伴共和合作，加以應對全球挑戰與確保美國利益。[86] 拜登將於6月11日至13日，將參加在英國舉辦的七國高峰會議，拜登會再次強調「多邊主義」，強化美國政策優先順序：公共衛生、經濟復甦、與主要民主國家分享價值，並與其他七大工業國家元首舉行雙邊會晤包括英國首相強森。[87]

在中國方面也是透過各種國際場合宣傳多邊主義的必要性，2021年4月22

---

[83] 「四方安全對話抗疫計畫 2022年底前配送10億疫苗至印太地區」，中央社，https://www.cna.com.tw/news/firstnews/202103130086.aspx。（檢索日期：2021/05/10）

[84] "Secretary Antony J. Blinken and NATO Secretary General Jens Stoltenberg Before Their Meeting, REMARKS, ANTONY J. BLINKEN, SECRETARY OF STATE, NATO HEADQUARTERS, BRUSSELS, BELGIUM, U.S. Department of State, https://www.state.gov/secretary-antony-j-blinken-and-nato-secretary-general-jens-stoltenberg-before-their-meeting-2/ (2021/05/10)

[85] "Blinken and G7 Allies Turn Their Focus to 'Democratic Values'", The New York Times, https://www.nytimes.com/2021/05/04/world/europe/blinken-G-7-china-russia.html (2021/05/10)

[86] "Statement by Press Secretary Jen Psaki on the President's Travel to the United Kingdom and Belgium", The White House, https://www.whitehouse.gov/briefing-room/statements-releases/2021/04/23/statement-by-press-secretary-jen-psaki-on-the-presidents-travel-to-the-united-kingdom-and-belgium/ (2021/05/10)

[87] "Statement by Press Secretary Jen Psaki on the President's Travel to the United Kingdom and Belgium", The White House, https://www.whitehouse.gov/briefing-room/statements-releases/2021/04/23/statement-by-press-secretary-jen-psaki-on-the-presidents-travel-to-the-united-kingdom-and-belgium/ (2021/05/10)

日，習近平應邀在「全球氣候領袖峰會」上發表「共同構建人與自然生命共同體——在『領導人氣候峰會』上提出五個原則：「堅持人與自然和諧共生；堅持綠色發展；堅持系統治理；堅持以人為本；堅持多邊主義；堅持共同但有區別的責任原則」，[88] 並強調中美共同發布了「應對氣候危機聯合聲明」，期待同包括美方在內的國際社會一道，共同為推進全球環境治理而努力。[89] 此外，中國已決定接受「蒙特利爾議定書」基加利修正案」，加強非二氧化碳溫室氣體管控，還將啟動全國碳市場上線交易。顯示出中國希望在氣候變遷問題上，擴大與美國的合作利基。

基本上，美中兩國在印太區域的具體作為，呈現出美國的攻勢作為，以多邊主義為基礎，強調民主與人權為外交政策核心，並以多個雙邊、多個多邊機制因應中國崛起的挑戰，當然中國方面也是步步為營，鞏固本身在印太的「影響區域」，另外則是尋求與俄羅斯同盟關係，進行與美國的「權力平衡」關係，整體印太區域呈現「勢力範圍」、「權力平衡」與「平等結盟」的混合統治形態。

伍、結語

一、研究總結與檢討心得

美中印太兩洋戰略的競逐從2011年開起，美國前總統歐巴馬倡議「亞太再平衡戰略」，重返亞太地區，主要是因應一個崛起的中國。同時，2013年中國開始　動「一帶一路」國際合作，範疇從亞洲到歐洲，從太平洋到印度洋，呈現印太兩洋戰略價值的重要性。加上，從2013日本實施「釣魚台列嶼國有化」政策之後，北京公佈「東海防空識別區」，並展開南沙群島的「填砂造陸」工程，改變南海現狀，使得美國加強基於維持以「規則」為基礎的國際秩序維

---

[88] 「共同構建人與自然生命共同體——在『領導人氣候峰會』上的講話（2021年4月22日，北京）中華人民共和國主席習近平」，新華網，http://www.xinhuanet.com/politics/leaders/2021-04/22/c_1127363132.htm。（檢索日期：2021/04/24）

[89] 「共同構建人與自然生命共同體——在『領導人氣候峰會』上的講話（2021年4月22日，北京）中華人民共和國主席習近平」，新華網，http://www.xinhuanet.com/politics/leaders/2021-04/22/c_1127363132.htm。（檢索日期：2021/04/24）

護，白熱化美中兩強印太戰略對峙態勢。

本文根據Onuf提出「規則構建出行為體、行為體也構建出規則」、「規則形成制度、制度構建社會」，以及「規則產生統治」三項規則建構主義的邏輯思考，從美國的角度出發，針對美中在印太區域戰略的競逐過程的三個命題與推論的研究結果：首先，針對命題 1：美中相互實踐過程，建立以規則為基礎的國際秩序與不干涉內政為原則的戰略對抗情勢。推論一：影響美中雙方規則互不相容的因素，在於雙方對於無政府狀態下「主權」的定義與操作的認知差異。北京認為主權平等不容干涉國家內部事務，華盛頓確認為逾越民主、人權，以及影響國際法規則事件，就非單純內部事務。

其次，命題 2：美中依據上述雙方對抗性規則，使得第三方以歐盟為主，面臨戰略選擇的矛盾。推論二：主導美中以外的第三方的規則選擇，主要基於各別的不同國家利益考量。美國透過美日、美韓外交與國防部長2+2會議、四方安全對話（QUAD）、「七大工業國家會議」（G7）、「北約組織」（NATO），從雙邊到多邊機制，未來可能會糾結於安全利益與經濟利益的選擇困境。

第三、命題 3：依據美中兩國上述規則的對抗下，由於「指導性」、「指令性」與「承諾性」言語行為相互交雜，呈現國際社會多種「統治形態」的態勢。推論三：由於非傳統安全（Covid-19）的出現，全球治理能量與決策者特質影響國際社會多元統治形態的生成。美中雙方透過主導的國際機制運作，呈現「權力平衡」、「勢力範圍」與「平等結盟」的世界多元統治形態。

二、整合途徑檢證與發現

Onuf提出行為體之間互動產生「規則」，「規則」再型塑行為體之間的「社會實踐」，如果以「美中關係」為例，兩國於印太區域互動發展出「規則」對立現象：以規則為基礎的國際秩序VS.以主權不干涉內政原則（國際關係民主化），呈現在國際經貿、民主人權、公海自由航行等等，透過近期兩國外交主管在聯合國安理會視訊會議上，再度呈現出「各說各話」、「各取所

需」與「各自表述」對立情勢。

其實，另一位美國學者溫特（Alexander Wendt）提出「行為體」之間透過「有意義互動」（meaningful interaction），產生「共有知識」（shared ideas），形成三種國際無政府文化：霍布斯、洛克、康德文化，從主導行為體之間的「利益」與「政策」。溫特強調透過「武力」、「利益」與「合法性」驅動無政府文化內化的三個客觀決定因素，從而影響行為體之間後續的政策與作為。

因此，美中雙方缺乏彼此之間的「有意義互動」與「共有知識」，是以，雙方建構更深的「敵意」，以及產生後續更多對立政策與作為。雖然美國強調與中國產生主要以「競爭」為主，也不排除「對抗」與有限「合作」的戰略混合態勢，如同溫特提出的三種無政府文化，可以區隔對待加以處理。

另外，美中兩國相互規則建立過程，透過不同的「言語行為」：指導性、指令性與承諾性三種模式，只要涉及「主權」議題，就會牽動北京無法讓步的邏輯思考。主要基於中國實力的增強，當習近平提出可以「平視世界」，就在於要求一個世界地位身份的需求。

三、後續發展與研究課題

最後，本文檢證美中在印太戰略競逐下的實際「統治狀態」，及其未來可能發展趨勢，從而思考台灣在兩強之間可行的戰略與政策作為。目前台灣處於美國以「規則」為基礎的國際秩序，而此種國際秩序系透過「民主」與「人權」為核心的外交政策來加以實踐，進而對抗中國在全球與印太區域逾越上述「價值」的國際作為。基於台灣的民主實踐、防疫表現，以及高科技半導體能量，成為美國戰略競逐印太的重要相關「行為體」。

此外，中國因應美國以規則為基礎的國際秩序戰略，面對需要一個短期穩定內、外環境考量下，必定會增強與美國可能於「氣候變遷」、「防治新冠肺炎」等議題合作空間，消除在經貿、科技方面的摩擦，至於針對民主與人權議題，只能採取「各說各話」、「堅持底線」，毫不退縮的立場。

　　不過，美國針對台海問題透過美日華盛頓高峰會議的聯合聲明，很清楚展現出兩大主軸：「和平與穩定的台海情勢」與「透過對話解決爭端」，也正是美國一個中國政策的三個要項：中國不武、台灣不獨，兩岸人民和平解決爭端。是以，如果中國方面持續軍機船艦繞台、擾台、封台，目標在於「以武逼談」，台北除了持續深化「美台關係」，[90] 避免糾結於美國台海政策的「戰略模糊」或是「戰略清晰」之辯，強化台灣的「能動性」，主導成為一個印太戰略平衡中的「籌碼」者，誠為國際事務與戰略研究學界的當務之急。

---

[90] 根據報載國安局研判，「中共將對美國表明底線以壓縮「台美關係」，中共總書記習近平將以強勢對外及對台作為爭取延任，推動更多「單向促融」統戰分化作為，持續對台施壓發動複合式威脅。」請參見：「壓縮台美關係 國安局：中共將對美表明底線」，自由時報，https://news.ltn.com.tw/news/politics/paper/1448092。（檢索日期：2021/05/11）

日本的「自由開放印太戰略」理念與政策及軍事外交行動

許衍華 *

壹、前言

一、美中阿拉斯加高層對話：兩種規則的對立？

　　2020年1月19日，迎來日美安保條約締結60周年紀念，安倍晉三（Abe Shinzo, 1954-）於紀念會上，大表謝忱！「日美同盟以來是促進與守護世界及亞洲、印度、太平洋和平與繁榮之不動支柱，更須與時俱進的深化與努力」。[1] 1月20日，安倍晉三於第二百一回國會總理大臣施政方針演說：「在日美同盟強固基礎上與歐洲、澳洲、印度、東盟等國家，以共有基本價值，實現「自由開放印度太平洋」為目標。[2] 然新年開春，日本全力奔向奧運主辦之際，1月5日，國際世衛組織發布《疾病風暴新聞》通告全世界在武漢發生的群聚性肺炎病例患者狀況和公共衛生對策。[3] 旋即亞洲及全世界相繼淪陷新冠肺炎（COVID-19）的傳染。疫情不輕的日本，安倍政府不得於3月24日與國際奧運委員會商談後，一致同意將本屆奧運會推遲於明年夏天前舉行。[4] 籠罩在新冠肺炎感染的印太地區，都在戒慎恐懼實施防疫政策。然印太區域的海上熱點—南海島礁，正進行著海上軍事對抗。4月21日，由海軍美國號（The amphibious assault ship USS America, LHA6）協同澳大利亞導彈護衛艦等艦，

---

* 國立彰化師範大學助理教授

[1] https://www.mofa.go.jp/mofaj/na/st/page6_000482.html/日米安全保障条約60周年記念レセプション（安倍總理挨拶）

[2] https://www.kantei.go.jp/jp/98_abe/statement/2020/0120shiseihoushin.html/第二百一回国会における安倍內閣總理大臣施政方針演說：「日米同盟の強固な基盤の上に、歐州、インド、豪州、ASEAAN など、基本的價值を共有する国々と共に、「自由で開かれたインド太平洋」の實現を目指します」。

[3] https://www.who.int/zh/news-room/detail/27-04-2020-who-timeline---covid-19/世衛組織應對COVID-19疫情時間線。

[4] https://www.kantei.go.jp/jp/singi/tokyo2020_suishin_honbu/pdf/200324pressrelease.pdf/安倍總理大臣とバッハ國際オリンック委員會會長電話會談。

自由航行於南海周域；4月22日，美軍升空B1-B遠程戰略轟炸機與F-16戰鬥機並聯合日本航空自衛隊F-15、F-2戰鬥機等聯合編隊飛行，展示威力巡航印太地區。[5] 安倍政府雖忙於「戰疫」，其統合幕僚監部，也馬不停蹄地以軍事外交訪問印太地區各國，討論著印太地區安全保障等議題。[6]

「印度太平洋」作為一個新的地緣政治概念，始於2004年12月26日，印度洋大海嘯造成多個國家重大災難。時任自民黨幹事長安倍晉三協同美日澳印等國進行人道救援為嚆矢；[7] 2006 年擔任首相後倡議價值外交色彩的「自由與繁榮之弧」（the Arc of Freedom and Prosperity）理念。第二任後，2016年8月在「東京國際非洲聯盟發展會議」（TICAD）中正式倡議「俯瞰地球儀外交積極的平和主義」（地球儀を俯瞰（ふかん）する外交と「積極的平和主義」）。[8] 2017年11月，美國川普總統公布「美國國家安全戰略」報告中，也首次使用「印太地區」（Indo-Pacific region）；[9] 2018年5月，將太平洋司令部改為美國印太司令部（U.S. Indo-Pacific Command）。「戰略印太」一方面串接地緣政治與地緣經濟的運作支撐，同時在廣博的地緣空間裡，鑲嵌主要大國的利害關係與利益競合情勢特別是當前的印太戰略論述，主要聚焦在美國、日本、印度、澳洲等應對印度洋與太平洋串接之後的新區域政策與介入布局，這種「利益的競合」不僅彰顯大國之間的角力情結，同時也進一步促成印太區域的安全挑戰。[10] 本文認為，安倍政府形塑印太戰略，有其戰略文化、領導人意象與外交政策等因素作為其戰略理念、政策實踐及軍事外交行動。

---

[5] https://www.pacom.mil/Media/News/News-Article-View/Article/2161338/us-japan-bomber-fighter-integration-demonstrates-dynamic-force-employment/

[6] https://www.mod.go.jp/js/Joint-Staff/js_topics.htm#20200424/山崎統幕長は、EU軍事委員長グラチアーノ陸軍大将と會談を行い、インド太平洋の地域情勢の認識を共有するとともに、「自由で開かれたインド太平洋」の實現に向けた防衛協力・交流の推進について、意見交換を行いました。

[7] 吳定南 著，《新世紀日本對外戰略研究》（北京：時事出版社，2010年），頁177。

[8] https://www.mofa.go.jp/mofaj/gaiko/bluebook/2019/html/chapter1_00_02.html#s10201/地球儀を俯瞰（ふかん）する外交と「積極的平和主義」。

[9] https://www.whitehouse.gov/briefings-statements/president-donald-j-trump-announces-national-security-strategy-advance-americas-interests/"The Strategy in a Region Context", pp.45-47.

[10] 楊昊 著，〈形塑中的印太：動力、論述與戰略布局〉，政大國關中心，《問題與研究》57期2卷（台北：政大國關中心，2018年6月），頁88。

貳、研究理論與文獻分析

　　一國大戰略制定應考慮國內因素的制約，而不能像狹義的結構現實主義所強調的，只考慮國際體系結構的制約、物質實力及外部威脅來檢視大戰略是不夠全面的；相反的，國內各集團壓力、社會觀念、憲法特徵、經濟相互依存等思考因素，在大戰略的決策中有著重要的影響，實際上是關鍵性的作用。因此，這些思考與決策也是決定國際衝突與合作前景的重要因素。[11] 新古典現實主義（Neoclassical Realist Theory of International Politics）不否認國際體系的等級和無政府會影響單元行為者之間互動。然國內領導人的意象、戰略文化、國內制度、國家－社會關係，操作為仲介變量，會改變國際戰略環境的包容性與制約性，以解釋外交政策與大戰略的制定。[12] 戰略家思考的架構是實用的（プラグマティズム, pragmatic）。他不像科學家只以發現最後真理為目的，而是志在幫助政治和軍事領袖來準備他們的心靈和裝備，以便有效的因應其對手。[13] 因此，新古典現實主義嘗試綜合結構現實主義之無政府狀態下國際體系對國家施加的約束和古典現實主義者有關於謀略（statecraft）及外交政策制定的一種實用性的見解。[14] 理論多元化能對行為者或文本敘事的脈絡裏找出關於行為者採取行動「理由」。[15] 自日本印太戰略出台後，有許多關於此議題的發表與論述：安倍《日本の決意》著作中，收錄其施政方針及元首出訪與在國外智庫演說，都清晰論述其對「印度太平洋」打造之決心，頗具參考價值。[16] 學者何思慎〈印太戰略VS.印太構想〉[17]、楊昊〈形塑中的印太：動力、論述與戰

---

[11] Richard Rosecrance and Arthur Stein, *The Domestic Bases of Grand Strategy* (Cornell University Press, 1993). pp.3-5.

[12] 諾林・理普斯曼（Norrin M. Ripsman）、傑佛瑞・托利弗（Jeffery W. Taliaferro）、斯蒂芬・洛貝爾（Steven E. lobell），劉豐、張晨譯、劉豐校，新古典現實主義國際政治理論（*Neoclassical Realist Theory of International Politics*）（上海：上海人民出版社，2017年），頁9-10。

[13] 鈕先鍾 著，《戰略研究入門》（台北：麥田文化，1998年月），頁116-117。

[14] 諾林・理普斯曼（Norrin M. Ripsman）等，同前註，頁5-6。

[15] Tim Dunne、Milja Kurki、Steve Smith, 主編、葉宗顯 譯、國家教育研究院 主譯，《國際關係理論：學科的面貌》（International Relations Theories: Discipline and Diversity）（台北：韋伯文化，2013年），頁3。

[16] 安倍晉三 著，《日本の決意》（東京：新潮社，2014年）。

[17] 何思慎，〈印太戰略VS.印太構想〉2018年12月21日發表於台北論壇。作者依日本產經新聞報導認為安倍首相將「印太戰略」修正為「印太構想」，呼應日中關係「從競爭到協調」的思路。

略布局〉、高珮珊〈川普政府亞洲政策分析：印太戰略意涵〉[18]、大陸學者盧昊〈日本外交與印太構想－基於國際公共產品角度的分析〉[19] 等文章，從不同視角研究，亦值得參考。

參、日本形塑印太戰略之理念內涵

一、歷史意識

　　1904年，麥金德（H.J. Mackinder, 1861-1947）於《歷史的地理樞紐》（The Geographical Pivot of History）文中，最後結論曾有一段警語：「如果日本和中國統一起來，征服了蘇俄的領土，推翻了俄羅斯帝國，則可以使海權與陸權結合，這種利益是樞紐地帶的俄國人從未享有過的」。這也正是日本發動侵華戰爭的理論依據。而影響更為顯著的，則是德國地緣政治學家豪斯霍夫（Karl Haushofer）的「泛地區論」。由於豪斯霍夫在1909-1911年，曾任駐日武官，也曾到過中國東北及華北，其生存空間與自給自足論點，不僅影響了日本接受了他的地略概念，並且依「泛亞地區—亞洲由北至南統一起來，以日本為區域統治者泛亞地區」。使日本具體的發展成為「大東亞共榮圈」的思想。[20] 1890年，明治維新時代，山縣有朋（Yamagata Arimoto, 1838-1922）在日本群島周圍劃了一條「主權線」，在亞洲地區則劃了一條「利益線」。這一戰略描繪出圍繞日本國土的主權線，以及深入到鄰國區域如朝鮮與清朝滿洲的利益線與海上安全線。[21] 自1907年，山縣有朋的「國家防禦基本對策」至2013年的「國家安全保障戰略」的制定，均將自身於亞（印）太區域，視其至為重要的國家安全與利益範圍。[22]

---

[18] 高珮珊，〈川普政府亞洲政策分析：印太戰略意涵〉，李大中主編，《川習時期美中霸權競逐新關係》（台北：淡江大學出版中心，2019年）。

[19] 盧昊，〈日本外交與印太構想－基於國際公共產品角度的分析〉，中國社會科學研究院日本研究所。http://ijs.cssn.cn/xsyj/xslw/rbwj/

[20] Zoppo、Zorgibibe, 鈕先鍾 譯、張式琦 校訂，《古典及核子時代：地略學》（ON GEOPOLITICS Classical and Nuclear），頁5-6。

[21] 藤村道生 著，山縣有朋（東京：吉川弘文館，1997年），頁156-157。

[22] http://www.mod.go.jp/j/approach/agenda/guideline/pdf/security_strategy.pdf/2016/0707.国家安全保障戰略について。

二、國家意識與保守主義

（一）國家意識

　　日本學者船曳建夫（Funabiki Takeo, 1948-）《是右又是左，我的祖國日本》（右であれ左であれ，わが祖國日本）著作中，依據國家主義與民族意識，把織田信長、豐臣秀吉、德川家康的時期，以文化人類學對明治維新的日本國家主義，重新定調與解讀。他認為，及至現在日本只有三種國家類型：即織田信長時代的「國際日本類型」、豐臣秀吉時代的「大日本類型」以及德川家康時代的「小日本類型」。他強調指出，明治維新以來日本所有的國家形式，無論有何種變化，都大抵不離其左右。「國際日本類型」特徵是實行國際貿易，因此稱為「國際日本主義」；「大日本類型」特徵是彈壓歐洲轉向亞洲，發動戰爭成為區域強權，因此稱為「大日本主義」。船曳強調，在後來的昭和時代外交上所展現的模式就有著秀吉時代的明顯軌跡。即與歐美保持競爭與合作，擴張對中國與東亞的侵略。[23] 德川家康的「小日本類型」：對外封閉、對內開發，就成為了德川幕府260餘年的基本國策。因此也稱「小日本類型」。[24] 以上分析日本國家意識的話語中，這些不一致的路線－國際主義、自由主義與亞洲主義匯聚成了日本國家完整性的共同信念，會隨著權力自主與威望而變體。

（二）保守主義

　　日本山口縣古稱長州藩，遠古曾與朝鮮半島相連。與其有著密切文化交流與人員往來。明治維新前，長州藩曾長期秘密積蓄實力，素有維新孕育地之稱。幕末思想家吉田松陰，曾興辦私學主持「松下村塾」培養出木戶孝允、高杉晉作、伊藤博文、山縣有朋等維新領袖。吉田譽有明治維新精神導師，被梁啟超稱，居維新之首功。這些人思想中具有強烈的民族擴張主義成分，對日本軍國主義思想的形成也有一定影響。山口縣的政治家們，血脈代代相傳，在日

---

[23] 船曳建夫 著，《右であれ左であれ，わが祖國日本》（東京：PHP研究所，2007年）頁45-51。
[24] 同前註 著，頁53-58。

本政治史上占據重要地位。其中伊藤博文、山縣有朋、桂太郎、寺內正毅、田中義一、岸信介、佐藤榮作及安倍晉三等先後出任首相。人數如此集中，居全國之首。此外，安倍的父親晉太郎，也曾被視為距首相寶座最近的人。山口縣出身的政治家被稱為「長州幫」。他們的共同點之一，就是多數持國家主義立場。因此，在安倍身上有著長州藩的精神滋養。[25]

自民黨成立後，於1960年發表了《保守主義政治哲學綱要》……《綱要》對「保守主義精神」，做了明確界定。保守主義是保持優良的傳統與秩序，去惡揚善，在傳統上創造、在秩序中發展進步。這意識著保守主義的政治觀，既排除破壞性的激進主義，也不與固守過去與現狀的消極保守主義同流合污。[26] 這與日本的社會民族性是相迎合的……愛美又黷武、喜新又守舊的保守傳統社會底蘊是相聯繫的。[27] 然自民黨保守主義的理念，也有著天皇制的崇敬、強調國家主義、獨佔的資本主義與務實的現實主義的內容。[28] 安倍重新執政後加強了官邸主導的能動性，這些理念的閃爍，就熔鑄成冷戰後的新自由主義與新保守主義的金塊。[29] 自民黨的新保守主義，相對提出了國家發展方向它一方面要保守日本傳統，另一方面又主張對戰後秩序做出重大改變。換言之，對內而言，實行「國家主義、經濟復興、自主外交、憲法正名、富國強兵」之復興大業；[30] 在對外關係，自民黨2010年《綱領》強調，要打造能履行世界和平義務，為全人類共同價值做出貢獻及有德的日本。這與戰後日本國內要求提升國際地位的政治訴求相契合。[31] 在尊重繼承歷史文化傳統，自民黨掌握了日本自主意識的話語權，更迎合了社會心理的訴求。以安倍的話語來註解：「開放的

---

[25] 張勇 著，〈韜晦之"鷙"：安倍晉三人格特質與對外政策偏好〉，中國社科院日本研究所《外交評論2017年第6期，2017年》，頁113。

[26] 林尚立 著，《日本政黨政治》（上海：上海人民出版社，2016年），頁206。

[27] Ruth Benedict, *The Chrysanthemum and the Sword: Patterns of Japanese Culture*. (New York: New American Library, 1946), pp.1-20.

[28] 林尚立 著，同前註，頁206-214。

[29] 鄭子真 著，《重返榮耀：解構21是繼日本政治的新進化》（台北：五南文化，2018年），頁68-72。

[30] Tsai His-Husn, Abe Verison of "Rich Nation, Strong Army", *TAMKANG JOURNAL of INTERNATIONAL AFFAIRS*. Volume22. Number3. January 2019., pp.164-165.

[31] 楊伯江，〈日本政壇為何形成一黨優位格局〉中國社會科學院日本研究院，2018年《日本外交》。http://ijs.cssn.cn/chuban/lanpishu/qk/201908/t20190807_4952738.shtml.

保守主義是我的立場⋯⋯保守主義對我而言,是對日本長達百年、千年的歷史中所累積、編織而成的傳統,時時持有更慎重、更小心、更熟慮的理解之心,正是所謂保守精神」。[32]

(三)安倍晉三的領導

安倍自傳《邁向美麗之國》書中所說:「『開放保守主義』是我的立場⋯⋯自衛隊的行動基準充滿限制,由於過於嚴苛的憲法限制,使得日本幾乎沒有政治判斷的餘地」。書中論述:「日中關係應政經分離為原則、日印澳美合作、集體自衛權、自衛隊交戰權力與行動等議題到安倍親臨觀艦式的談話,將自衛隊入憲正名,實現正常國家,帶領日本走向美麗家園」。[33] 2013年2月22日(平成25年)安倍晉三在美國CSIS發表演說:「日本再起⋯⋯日本絕不會成為二等國家⋯⋯眺望地球世界,心中感覺著,過去半個世紀,美國盟友長期地與日本並肩合作緊密的夥伴關係裨益了太平洋的和平與繁榮⋯⋯維繫著這樣的榮景,是兩國同盟並具有普世價值而增進的⋯⋯描繪現今的世界,期待著日本在人權伸張、反恐威脅、地球暖化、貧病疾苦等諸多問題,能和貴國一起並肩作戰這是我重任總理的決意與責任」。同年9月25日,在美國哈德遜智庫接受頒獎後演說:「決心推動積極平和主義是我的歷史使命,也很自豪地擔當舉起積極和平主義大旗的推手,鼓舞著日本人民向前邁進。」在這篇演說文本裡,更對積極和平主義有清晰的說明:「當前日本首要課題,是經濟重建⋯⋯健全財經政策、對山姆大叔及外國更加開放投資政策和TPP(Trans-Pacific Partnership)推動是三支箭重要組成內容。有關自衛隊集體自衛權和武力的使用,與現行日本憲法違背必須嚴肅檢討的⋯⋯對我國安全保障的架構,也必須重新啟動新的模式設立國家安全保障會議為重要開端。這些都是積極和平主義重要的支柱」。 也是這位自命不凡「戰鬥政治家」的鏗鏘領導與表現。

現今日本大戰略,可以形象地被概括為「安倍路線」,即由「安倍政治學」、「安倍經濟學」、「安倍外交學」、「安倍戰略學」組成的綜合治國方

---

[32] 安倍晉三 著,《新しい国へ・美しい国へ》(東京:株式會社文藝春秋,2013年),頁18-28。

[33] https://www.kantei.go.jp/jp/98_abe/statement/2018/1014kunji.html/平成30年度自衛隊記念日觀閱式安倍內閣總理大臣訓示。

針，圖謀在全方位、多領域、全要素出發改革和振興日本的大政方略。從這個大戰略的行動看，日本的印太戰略及其政策突顯著自主與威望，即安倍推行「印度太平洋」理念的閃爍金塊：「國家主義、經濟復興、自主外交、富國強兵」之明治維新（Meiji Ishin）再造。[34]

肆、日本「印太」戰略（構想）的政策實踐

第二任上台的安倍晉三即積極布局印太戰略。先於 2007 年8 月 22 日，安倍訪問印度國會的公開演說便強調日本基於國際協調主義以「積極的平和主義」立場，推動俯瞰地球儀外交、掌握印度洋與太平洋的「兩洋合流（二つの海の交わり）」（confluence of the Two Sea）關鍵，期待能具體促成印度、美國、澳洲與日本等民主國家在亞洲架構中成為具體夥伴關係。[35] 該次演講深具戰略意義，它不僅呼應「自由與繁榮之弧」的理念，框架出新的亞洲權力政治輪廓，同時也轉換成日後安倍政府推動自由開放印太戰略的理念與價值基礎。安倍在 2012 年延續「自由與繁榮之弧」提出「亞洲民主安全鑽石同盟」（Asia's Democratic Security Diamond）的倡議，將執政初期的廣泛亞洲架構中的夥伴關係轉換成捍衛區域和平穩定與航行自由繁榮的民主國家同盟，明確對應包括海上爭端在內的海事安全挑戰。在 2013 年，安倍於美國華府戰略與國際研究中心（Center for Strategic and International Studies, CSIS）的公開演講，再次確認日本在推進「以印太為框架」、「以積極外交政策為基調」的新角色，也就是成為亞太或印太區域中的規則推動者、國際公共財的捍衛者、以及推動理念相近國家合作的促成者。[36]

2017 年 4 月，日本外務省國際合作局在「年度開發重點協力方針」中揭示將以更具戰略性與效能考量的ODA政策，更加具體落實「自由開放印太戰略」。2018年2月，安倍在國會發表施政演說：「從廣闊太平洋到印度洋……

---

[34] Tsai His-Husn, Abe Verison of "Rich Nation, Strong Army", *TAMKANG JOURNAL of INTERNATIONAL AFFAIRS*. Volume22. Number3. January 2019., pp.164-165.

[35] 安倍晉三 著，《日本の決意》（東京：新潮社，2014年），頁154-155。

[36] https://warp.ndl.go.jp/info:ndljp/pid/11236451/www.kantei.go.jp/jp/96_abe/statement/2013/0223speech.html/日本は戻ってきました。

平等地給世界人民帶來和平與繁榮的公共產品。基於航行自由、法的基礎上……推進自由開放印度太平洋戰略」。2019年1月，安倍在達沃斯論壇年會上表示：「日本堅決維護並努力增強自由、開放、有規則的國際秩序」，將觸角延伸至歐盟。[37] 2018年底，日本公布新《防衛計畫大綱》首次引入「印太區域」的概念。日本將與澳洲、印度共同合作，進行演習、部隊共同訓練、武器裝備改進等，強化推進美澳印戰略聯繫。[38] 在戰略理念的設定上，日本的印太戰略是安倍政府的「俯瞰地球儀外交」與「積極和平」倡議的延伸。特別在地緣設定上，它所涉及的範疇更由印度洋與太平洋的兩洋合流，架構出亞洲與非洲兩個大陸的連動關係。日本希望透過此一戰略來推動區域的穩定與繁榮，以推動法治（rule of law）海事安全以及優質基礎建設輸出等價值的落實轉換成對南太平洋湄公河流域、東協、南亞以及中亞等區域國家的實際援助計畫。[39] 同時也將軍事外交議程提上台面，更加了一層領域的援助。

　　安倍自推動印太戰略至今，到達國家81國。誠如其自許的「積極的政治家」戮力以公的精神，殊值敬佩。在其表現及政府的宣傳，推進印太戰略的政策實踐，歸納如下：

一、明確印太戰略的政策及規則：印太戰略由三大支柱：（一）基於法的支配、固定航行自由、自由貿易等；（二）追求經濟繁榮；（三）確保和平與穩定。[40] 為實現以上採取三大基本手段：1.維護國際秩序的基本原則；2.改善提升地區內部網絡化；3.開展「能力構築」與人道救援、海上救難；日本構想打造一套系統性標準即規則，作為提供公共產品向印太輸出。具體涵括：透明可持續的高質量基礎設施標準即合作規則；高度開放共享的自由貿易體制及商務規則；健康的、可持續性的經濟社會與環境及商務規則；以民主化與尊重人權為前提治理規則；以國際法與和平協商為

---

[37] https://www.kantei.go.jp/jp/98_abe/statement/2019/0123wef.html/2019.01.30/世界経済フォーラム年次総會安倍總理スピーチ。

[38] https://www.mod.go.jp/j/approach/agenda/guideline/2019/pdf/20181218.pdf/平成31年度以降に係る防衛計画の大綱について。

[39] https://www.mofa.go.jp/mofaj/gaiko/bluebook/2017/html/chapter1_02.html#s10201.

[40] https://www.mofa.go.jp/mofaj/files/000430631.pdf/自由で開かれたインド太平洋の具体化。

前提的國際行為規則。

二、制定「印太規則」作為國際秩序基礎：日本推行印太戰略高度符合西方主
流價值觀與世界發展潮流，力圖主張區域性規則標準嵌入至國際層面，作
為兩洋交流的印太地區的秩序原則；同時，更積極推進「雙高」即「高質
量基礎設施」及「高標準貿易體制」的國際合作。藉由場域如G20峰會、
ASEAN、APEC、TICAD等組織，宣揚理念與政策，營造主導優勢與國際
貢獻。[41]

三、突顯對印太地區連結性重點的建設：日本在印太主導建設連接性，突顯自
身公共產品貢獻者身分的關鍵。這一連接性有三層次即通過建設港口、鐵
路公路、能源設施、通信技術聯絡網等高質量基礎建設，實現「物理的連
結性」；經由提供人才技能培訓，實現「人的連接性」；通過簽署自貿
協定，實現「制度的連接性」。如此緊密的連結性將印太地區的各節點
構成網絡化進而相互聯繫。[42] 2017年5月，所宣示推動的「亞－非成長走
廊」（Asia-Africa Growth Corridor），即為日印之間的指標性戰略布局成
果。同年9月，安倍再度造訪印度，並且與穆迪總理簽署了《日本－印度
聯合聲明：朝向自由、開放與繁榮的印太》（Japan-India Joint Statement:
Toward a Free, Open and Prosperous Indo-Pacific），這份聲明涵蓋日本與印
度的全面合作領域，將兩國的國家利益、發展利益、經貿利益以及人民之
間的休戚與共網絡關係緊密相繫。[43]

四、拉攏更多合作者，強化印太同盟：2018年11月，日本將「印太戰略」修正
為「印太構想」。順應日中關係「從競爭到協調」的思考。[44] 日本強調，
在滿足特定前提下，期望與中國深化第三方合作市場。同時，日本維持與
美、印、澳等合作的前提下，重點尋求將非洲與南太平洋島國打造為在印

---

[41] 盧昊，〈日本外交與印太構想－基於國際公共產品角度的分析〉，中國社會科學研究院日本研究
所，《日本學刊》，（2019年第6期），頁12。http://ijs.cssn.cn/xsyj/xslw/rbwj/。

[42] 盧昊，〈日本外交與印太構想－基於國際公共產品角度的分析〉，同前註，頁13-14。

[43] 楊昊 著，〈形塑中的印太：動力、論述與戰略布局〉，同前註，頁92-93。

[44] 何思慎，〈印太戰略VS.印太構想〉，同前註。

太開展合作的新增長點。又以開放地區名義，將作為外在勢力的歐洲國家，特別是英、法拉入印太戰略的框架中。2019年10月，安倍在國會施政演說中表示：「俯瞰地球儀外交以日美同盟為基軸與英國、法國、澳大利亞、印度等共享基本價值觀的國家攜手，實現自由開放印度太平洋而努力。[45]

伍、軍事外交行動

2020年1月19日，安倍在慶祝日美同盟60周年紀念演說：「日美同盟以來是促進與守護世界及亞洲、印度、太平洋和平與繁榮之不動支柱，更須與時俱進的深化與努力」。2019年1月20月，安倍國會演說：「在強固日美同盟基礎上，與英國、法國、澳大利亞、印度等共享基本價值觀的國家，實現自由開放印度太平洋為目標」；同時亦強調今年設立宇宙太空部隊，並強化電磁波、網路空間等優勢能力與體制建設，以應對印太區域之嚴峻環境：[46]

一、自2008年至2016年，日印兩國領導人多次互訪並在安全合作領域，達成多項協議、多次進行戰略對 ，實施聯合軍事演習等。2016年9月，安倍與莫迪舉行峰會時，安倍向莫迪說明了「自由開放印度洋太平洋戰略」……印度處於亞洲和非洲大陸間「最重要的位置」，應在區域發 更重要的作用。希望能實現其日本「印太戰略」與印度「東向戰略」的無縫接軌。2016年 11 月 11 日，印度總理莫迪在訪日期間與安倍首相簽署了日本向印度出口核能相關技術的「日印核能協定」；另外，日印兩國還達成軍售協議，日本向印度出售12架 US－2反潛機。此或許成 日本突破「武器出口三原 」之後的首次對外軍售。日本政府在莫迪上台後進一步強化了對印度的外交力道。2017年9月14日，安倍與印度總理莫迪舉行會談，雙方簽署了 為「面向自由開放與繁榮的印度太平洋」的聯合聲明，強調了要強化防衛與安全保障合作，擴大日美印3國海上聯合訓練，以及日本陸上自

---

[45] https://www.kantei.go.jp/jp/98_abe/statement/2019/1004shoshinhyomei.html第二百回国會における安倍內閣總理大臣所信表明演說/外交・安全保障－（地球儀を俯瞰（ふかん）する外交）。

[46] https://www.kantei.go.jp/jp/98_abe/statement/2020/0120shiseihoushin.html/第二百一回国會における安倍內閣總理大臣施政方針演說/積極的平和主義、安全保障。

衛隊與印度陸軍聯合訓練的可能性等。安倍表示「要攜手實現日印新時代」的巨大飛耀，主導亞洲太平洋地區及世界的和平與繁榮。

二、日本與澳洲軍戰略安全合作：澳洲是日本構築「亞洲民主安全菱形」同盟是遏制中國崛起的重要一環。在亞太格局變動及美國「重返亞太」戰略影響之下，澳洲也在延伸其戰略安全空間，傳統視域中的「亞太」向西延伸至印度洋區域，試圖構建印太「戰略弧」。2007年3月13日，時任日本相安倍晉三和澳洲總理霍華德簽署〈日澳安全保障聯合宣言〉這是日本首次與美國之外的國家建立安全保障體制。根據宣言：日澳兩國將在反恐、邊界防衛、海事安全、救災和打擊毒品走私等傳統安全領域開展合作，雙方決定建立部長 （外交部長和國防部長「2+2」）定期會晤機制，以加強安全與防務對話，更進一步增強双方的互信與協調。2010年5月先後達成《獲得與交互支援協議》和《日澳資訊安全協定》；前者規定：除食物、燃料外日本自衛隊與澳洲軍隊將「相互提供交通工具、住所、保養 護和醫療衛生服務」；後者主要是為兩國共享軍事機密和反恐情報提供法律依據深化日澳安全合作。2017年川普就任美國總統後，美國相繼退出TPP巴黎氣候協定等國際組織，川普政府對外政策的不可預測性引發了亞太盟國的不安。在日澳的牽頭下，日本、澳洲、印度和越南等國家加大相關對話及合作力度以抵消或緩衝美國在該地區安保力量的弱化。近年來，日本以海洋安全為切入點，加強了與越南、菲律賓、新加坡馬來西亞和印尼等國的軍事合作並構建國家安全與戰略夥伴關係。在東盟國家軍事能力建設方面，提供全面協助包括人員交流與培訓、軍用裝備（如巡邏艇）出口與聯合演習等。日本與東盟國家展開海洋安全合作把和中國大陸存在南海主權争端的菲律賓和越南為重點對象大力展開軍事安全外交，意圖加強防務合作，以應變突發狀況。同時在國內武器出口政策「武器輸出原則」限制的背景下日本正努力推進與東盟國家在武器裝備及軍事技術方面的深度合作。[47] 2018年12月，日本《新防衛計畫大綱》中首次將印度太平洋寫入

---

[47] 張耀之，〈日本的印太戰略理念與實踐〉，中國社會科學院日本研究所，《日本問題研究》，2018年第2期第32卷，頁12-14。

《大綱》。要構築我國總合防衛作戰能力；強化安全保障協力，前瞻自由開放印度太平洋區域，須以多角度、多層次，開展軍事交流、共同防衛演習、技術合作。謀求共同價值觀與戰略利益。[48]

陸、對日本印太戰略的反思

日本論述著印太區域掌握著兩洋合流—兩陸連結的關鍵之鑰，殊不知日本自古以來就處於兩大洲（亞洲與美洲）之中，其戰略文化雙重性就牽引著日本國家性格。安倍政府實行「俯瞰地球儀外交之積極平和主義」，就有著兩手準備：一是對美國印太戰略採取追隨戰略；二是在其國家利益線與安全線中，謀求自主性的戰略。安倍政府的印太戰略，「以地緣上的印太、經貿上的印太、戰略上的印太」，[49] 謀求多議題組合、多重聯盟方式，制定國際規則，達到其國際貢獻。最為表現的就是：「海洋民主國家聯盟」及「亞非增長走廊」政策的實踐。彷彿也有著「一帶一路」的暗勁來與中國大陸的「一帶一路」互別苗頭。然今非昔比，地緣政治上所孕育的形態權力、人口權力、經濟權力、組織權力、軍事權力、外交權力等實質與形式，[50] 隨著工業互聯網、AI、大數據、5G等高技術而革新。緣此，中共的北斗系統預於今年覆蓋全球，提供海陸通訊導航的服務，試問印太區域的公共產品是否會改朝換代，日本的經濟體量是否足以支撐印大地區的擴張；再者，中共的高速鐵路已在亞非幾各國興建，是否也會在「亞非增長走廊」形成角力造成日本退縮。在戰略上的印太，中共高超音速滑翔導彈，是否讓在印太區域，讓防範反恐、打擊海上盜賊的軍事行動變得欲蓋彌彰而使日本進行軍備競賽，連續增長軍費。最後，高唱著「地球規模課題深刻化」以難民保護、氣候變遷、自然災害、傳染疾病等人類永續生存與發展問題，最為關注的議題，卻被此刻新冠病毒的侵襲而蒼白無力。因此，印太地區有著豐盛的資源是必須要提升更高的層次，讓這地區的人們「雨露同

---

[48] https://www.mod.go.jp/j/approach/agenda/guideline/2019/pdf/20181218.pdf/平成 31年度以降に係る防衛計画の大綱について。

[49] 楊昊 著，〈形塑中的印太：動力、論述與戰略布局〉，同前註，頁88。

[50] 鈕先鍾 著《戰略研究入門》（台北：麥田出版，1998年），頁165-166。

沾、共榮共存」切勿跌入「修昔底德陷阱」的困境。[51] 印太區域絕非由單一國家所主導，它的發展議程更不會由特定國家群所獨享。換言之，如何突顯印太區域或印太戰略的包容性與可行性，是日本積極平和主義必須奉獻的而不是惠澤的。

柒、結語

印太戰略已經成為當前日本外交戰略的重要組成部分。雖然受新冠肺炎影響，但是安倍政府仍一生懸命（勤勉）的推行軍事外交活動。面向未來，今年美國總統大選，安倍推遲主辦奧運到明年，明年的國會選舉關係著日本「憲法改正」成功與否，疫情防控等重大情事，都會成為安倍政府施政的重點。尤以，新冠肺炎的疫情正影響著全世界，也困住了日本。未來的印度太平洋區域，將會面臨著權力格局的挑戰。

---

[51] 格雷厄姆・艾利森 著，包淳亮 譯《注定一戰？中美能否避免修昔底德陷阱》（Can American and China Escape Thucydides's Trip?）（台北：大塊文化，2018年）。

澳洲在中美南海競逐的戰略抉擇

湯智凱 *

壹、前言

國際南海仲裁案判決菲律賓勝訴後，新政權上台改採合作大轉向，[1] 緊接著，中越南海主權爭端再起，美國自由巡航加劇，中國大陸南海填土造島經營日固，中美南海競逐逐日趨擴及印太周邊區域，澳洲介於印太區域又鄰近南海，不僅對美國行動亦步亦趨，擴大結盟抗衡中國的跡象更日趨顯著，其立國文化屬性海陸兼具，國家利益戰略權重分配究竟偏向大陸或海洋選項，攸關澳洲國運發展至鉅！國際關係理論有關國家利益權重分配主要區分三種觀點，[2] 澳洲在南海既不需現實主義高階與低階政治利益判準，亦難辨出自由制度主義核心與周邊利益或主要與次要利益，社會建構主義主觀與客觀利益權重認知觀點，或可為澳洲在南海戰略行動提供最佳指引！蓋澳洲戰略資源分配有其必然的優先次序，宜從南海文化結構角度切入，慎思其南海身分與角色，並以此尋求其南海國家利益最佳戰略選項。本文先從中美南海競逐探求其發展趨勢，接著，循社會建構主義文化認知觀點，逐步探求澳洲最佳南海行動戰略，俾提供澳洲在南海競逐獲取最佳戰略利益之參考。

貳、中美南海競逐發展趨勢

一、國際仲裁

（一）法理爭議擴大

中國大陸早在海牙國際仲裁菲律賓獲勝前就聲稱，仲裁結果「不過是一張廢紙」！仲裁出爐後，中國大陸外交部更以一千五百字嚴正聲明裁決無效，

* 淡江大學國際事務與戰略研究所博士
[1] 〈杜特蒂南海仲裁態度轉向　或為緩和國內批評與壓力〉，中廣新聞網，https://www.rti.org.tw/news/view/id/2080431。2020-9-24（檢索日期：2021.3.14）
[2] 鄭端耀，〈國際關係新古典現實主義理論〉，《問題與研究》，44卷1期，（民94年，1.2月）。

沒有拘束力，中國不接受、不承認。判決結果也把中華民國「太平島」判成「礁」，迫使蔡英文政府發表嚴正聲明，絕不接受，沒有法律拘束力！

在《聯合國海洋法公約》談判時，有關爭端解決條款中國大陸就建議單列成文，不應由法院或法庭以裁定解決，南海仲裁案最終裁決不僅損及國家自主選擇爭議解決方法的權利，[3] 也使屬於領土主權和海洋邊界爭端，被曲解成有關海洋地形地貌和海洋權益來源的爭端，[4] 此外，菲國投訴選的是常設仲裁法庭不是聯合國國際法院，合法性與正當性皆備受質疑！[5] 聯合國海洋法公約尤其無法解決1982年以前就已存在有主權爭議的歷史聲稱海域的自證與公證問題，也無法有效釐清經濟海域重疊、領海基線與大陸棚劃分基準等屬於規範本身的衝突問題，相關爭議遂又不得不回歸國際現實權力的老規範框架。[6]

傳統公海自由航行起自傳統陸地海外3海浬，1945年美國率先打破傳統，宣布領海管轄延伸至大陸棚，國際對公海自由航行即呈現各自表述現象。1982年尼加拉瓜控訴美國，美國不接受國際法院判決，2011年，澳大利亞控訴日本南極捕鯨案，日本政府不接受國際法庭判決，基此預判，南海爭議的國際司法介入不會因南海仲裁案止息，反而將更形興起與擴大。[7]

（二）軍備防務激化

2016年1月27日美軍太平洋艦隊司令哈里斯（Harry B. Harris Jr.）表示，南海是有爭議區域，美軍最新戰具如F-35、P-8A反潛機等都將部署到太平洋地

---

[3] 〈金永明：南海仲裁的影響及中國的若干對策〉，中國評論新聞網。http://hk.crntt.com/doc/1043/1/2/2/104312258_4.html?coluid=7&kindid=0&docid=104312258&mdate=0721002216。2016-07-21。（檢索日期：2016.7.21）

[4] 〈德學者：南海仲裁若侵權 中國可退出海洋法〉，中國評論新聞，http://hk.crntt.com/doc/1041/4/9/7/104149793_2.html?coluid=7&kindid=0&docid=104149793&mdate=0307103905。2016-3-7（檢索日期：2016.6.16）

[5] 劉國興，〈名家觀點－看菲律賓打外交戰〉，中時電子報。http://www.chinatimes.com/newspapers/20160718000376-260109。2016.7.18。（檢索日期：2016.7.18）

[6] 〈日本未公開消息：中國戰機五月曾逼近釣魚台〉，中時電子報，http://www.chinatimes.com/realtimenews/20160616006833-260409。2016-6-16（檢索日期：2016.6.17）

[7] 〈別指望用仲裁結果逼中國就範〉，中國評論新聞網，http://hk.crntt.com/doc/1042/4/0/0/104240031.html?coluid=93&kindid=2910&docid=104240031。2016-5-21（檢索日期：2016.7.11）

區，[8] 同年4月G7外長會議發表抗中意涵鮮明的《海洋安保聲明》，[9] 中共軍機降落永暑礁執行後送任務，美國立即表示將向菲國提供感測與通訊裝備，提高南海監控能力，[10] 第三與第七艦隊亦將一起開展任務巡航，[11] 日本同步表示將協助東南亞國家提高軍事安全防衛能力，[12] 開赴菲律賓蘇比克灣參加美菲2016年度「肩並肩」聯合軍演的日本兩艘護衛艦與潛艦，更首度停靠越南軍港。[13] 2017年1月，日本與澳洲雙方簽署《軍需相互支援協定》，2018年1月18日澳洲總理莫里森（Scott Morrison）展開日本訪問，[14] 日美澳合作不僅大步向前邁進，更積極探索日美澳印共同塑造印太地區多國安全合作機制的可行性，[15] 太平洋在日本協管與澳洲積極參與下，美日澳聯合與中國競逐南海的防務對抗似已無可避免。[16]

二、競逐消長

（一）美國軍力虛巡演

美軍於2015年出動「拉森號」（Lassen）驅逐艦駛入南海中國大陸島礁12浬內開啟首次巡航行動，美日《新安保法》生效後，更正式邀請日本自衛隊協

---

[8] 〈制中 美最大航母戰鬥群挺進南海〉，中時電子報。http://www.chinatimes.com/newspapers/20160129000876-260301。2016-1-29。（檢索日期：2016.1.29）

[9] 〈陸強化海上民兵 宣示南海主權〉，中時電子報。http://www.chinatimes.com/newspapers/20160309000858-260309。2016-3-9。（檢索日期：2016.3.12）

[10] 〈美提供飛船 助菲監控南海〉，青年日報。http://news.gpwb.gov.tw/news.aspx?ydn=026dTHGgTRNpmRFEgxcbfVEV3cQibTDk%2f3zFY4u8tBcDjtwEJe%2bqPQC0WkMxKZqsE4JL3JMoYfkr%2bco3SmrtFaQgtUB2XPUOBztr9%2bu%2fnfQ%3d。2016-04-20。（檢索日期：2016.4.20）

[11] 〈美軍西移抗中 第三艦隊協防東亞〉，中時電子報。http://www.chinatimes.com/newspapers/20160616000849-260301。2016-06-16。（檢索日期：2016.06.16）

[12] 〈日挺美艦 稱要合作保護自由海洋〉，中時電子報。http://www.chinatimes.com/newspapers/20160202000882-260309。2016-2-2。（檢索日期：2016.2.2）

[13] 〈日2船艦 首次停靠越南金蘭灣〉，青年日報。http://news.gpwb.gov.tw/news.aspx?ydn=026dTHGgTRNpmRFEgxcbfVEV3cQibTDk%2f3zFY4u8tBcDjtwEJe%2bqPQC0WkMxKZqsmEpqhgnctxpCqoqnIJwuE1Z5F5%2bdgWnIpRWJpNakuQM%3d。2016-4-13。（檢索日期：2016.4.13）

[14] 〈抗中 安倍構海洋「鑽石」戰略〉，中時電子報。http://www.chinatimes.com/newspapers/20151223000865-260301。2015-12-23。（檢索日期：2016.1.4）

[15] 〈軍事合作為抓手 日澳打造特殊關係〉，中國評論新聞網。http://hk.crntt.com/doc/1049/4/8/5/104948564_3.html?coluid=91&kindid=2710&docid=104948564&mdate=0120112635。2018-1-20。（檢索日期：2018.1.21）

[16] 〈美日澳菲潛艦 編織中國包圍網〉，中時電子報。http://www.chinatimes.com/newspapers/20160518000857-260309。2016-5-18。（檢索日期：2016.5.18）

同巡航南海，[17] 2016年美軍飛彈驅逐艦又駛進西沙群島中建島12海里內，[18] 並刻意一視同仁點名中國大陸、中華民國、越南三方，扮演幕後操縱者的意圖明顯。[19] 2016年4月美菲「肩並肩」聯合軍演、6月美「海上戰備暨訓練聯合演習」，美國「史塔森號」（USS Stethem）等三艘驅逐艦與航空母艦雷根號密集現身南海，第三艦隊太平洋水面作戰群（Pacific Surface Action Group）更於4月間即派駐協防東亞。[20] 美國在南海先後扮演旁觀者、中立者、制衡者與直接介入爭端的挑撥者等四種角色，[21] 美國直指外國機艦行經島礁附近，應「事前獲得批准」或「事前通報」的要求違反《國際法》航行自由權，將繼續飛航、航行、行動，[22] 不過預判其所得成效也將僅此虛名而已，畢竟實質意義不大。

（二）中國島礁實經略

南海水域目前有100多個島礁，實際島礁控佔者有中華民國、中國大陸、越南、菲律賓、馬來西亞、汶萊等6個聲索國，[23] 1996年越南僅占有24個島礁，如今倍增為48個，顯示越南積極經略南沙海域事實不假，不過中國大陸雖實際控制7個，卻經營成效顯著。[24] 菲律賓於2013年提出南海仲裁案後，2014年中國大陸即將價值10億美元的鑽井平台拖至中越有爭議海域鑽探，同年，中國大陸在三沙市永興島修建第一所學校，更把1999年開始實施的年度禁漁令，

---

[17] 〈大國角力 南海成亞洲新火藥庫〉，中時電子報。http://www.chinatimes.com/newspapers/20151227000660-260301。2015-12-27。（檢索日期：2016.1.4）

[18] 〈美艦駛近中建島 陸批美挑釁 危險操作〉，中國評論新聞。http://www.chinatimes.com/newspapers/20160131001080-260102。2016-01-31。（檢索日期：2016.2.1）

[19] 〈旺報觀點－擔心兩岸聯手護南海 美動作多〉，中國評論新聞。http://www.chinatimes.com/newspapers/20160131001507-260301。2016-01-31。（檢索日期：2016.2.1）

[20] 〈美軍西移抗中 第三艦隊協防東亞〉，中時電子報。http://www.chinatimes.com/newspapers/20160616000849-260301。2016-06-16。（檢索日期：2016.06.16）

[21] 李曦，〈海峽兩岸共衛東海與南海邊疆的幾點思考〉，中國評論新聞。http://www.zhgpl.com/doc/1032/1/3/3/103213392.html?coluid=7&kindid=0&docid=103213392&mdate=0717103402。2014-07-09。（檢索日期：2015.08.11）

[22] 〈美公開承認 台為南海聲索國〉，中時電子報。http://www.chinatimes.com/newspapers/20160109000877-260301。2016-1-9。（檢索日期：2016.2.1）

[23] 〈旺報觀點－擔心兩岸聯手護南海 美動作多〉，中國評論新聞。http://www.chinatimes.com/newspapers/20160131001507-260301。2016-01-31。（檢索日期：2016.2.1）

[24] 〈南海仲裁將出爐 越南聲索占最多〉，青年日報。http://news.gpwb.gov.tw/news.aspx?ydn=026dTHGgTRNpmRFEgxcbfVEV3cQibTDk%2f3zFY4u8tBfYCEcIyHtuvpZ3MVYBPSwD1pvGm4Z5NXqKJFe7GkBnYtpQK20DHFirLYSYQyZ4mvU%3d。2016-5-9。（檢索日期：2016.5.9）

擴大適用於外國漁民，但同時開放黃岩島供菲律賓漁民使用，[25] 2015年5月，中國大陸正式在禁漁區內實施休漁期，並在南海相關島礁與海域進行大規模填海造陸，[26] 新造的永暑礁超越太平島躍居南沙第一大島，[27] 對照美軍的巡演，顯示中國大陸在南海島礁的經營日趨實化與固化。[28]

三、協商發展

（一）政治和平倡議

1995年東協外長會議，中國大陸宣示，願意依照公約精神與相關國家和平解決爭端，2002年東協十加一高峰會，中國大陸與東協十國簽署「南海各方行為宣言（DOC）」，重申以《聯合國憲章》、《聯合國海洋法公約》、《東南亞友好合作條約》、和平共處五項原則與公認的國際法原則作為處理國家間關係的基本準則，美國同步呼各方遵守南海各方行為宣言。[29] 2005年中國大陸擴大慶祝鄭和下西洋600周年紀念，將7月11日定為中國「航海日」，鄭和巡航三大洋展現的東方儒教風範，為當道西方文明體系提供另類選擇參考。[30] 中國大陸外長王毅強調，南海九段線的歷史性權利並不來自於《聯合國海洋法公約》，而是源自一般國際法，中方有權維護自己的主權和正當海洋權益，[31] 南海非軍事化應通過對話管控分歧，通過談判解決爭議，以維護各國航行自由。[32]

---

[25] 〈設國家級研究院 陸積極經營南海〉，中時電子報。http://www.chinatimes.com/newspapers/20151101000655-260301。2015-11-01。（檢索日期：2015.11.1）

[26] 〈陸填海擴主權 南海6礁變島〉，中國評論新聞。http://www.chinatimes.com/newspapers/20140901001572-260301。2014-09-01。（檢索日期：2015.08.15）

[27] 〈渚碧礁 月底將成南沙第一大島〉，中時電子報。http://www.chinatimes.com/realtimenews/20150318003065-260409。2015-3-18。（檢索日期：2015.01.04）

[28] 〈設國家級研究院 陸積極經營南海〉，中時電子報。http://www.chinatimes.com/newspapers/20151101000655-260301。2015-11-01。（檢索日期：2015.11.1）

[29] 〈中評智庫：南海 21世紀地緣政治主戰場〉，中國評論新聞網。http://hk.crntt.com/doc/1049/3/4/5/104934563_9.html?coluid=7&kindid=0&docid=104934563&mdate=0221002242。2018-02-21。（檢索日期：2018.2.22）

[30] 白德華，〈莫忘來時路／7月11日－海洋夢！大國夢！〉，中時電子報。http://www.chinatimes.com/newspapers/20150711000377-260109。2015-07-11。（檢索日期：2015.07.17）

[31] 〈高之國答中評：聯合國海洋法公約非萬能〉，中國評論新聞網。http://hk.crntt.com/doc/1043/1/1/1/104311187.html?coluid=266&kindid=0&docid=104311187&mdate=0716010337。2016-7-16。（檢索日期：2016.7.17）

[32] 〈王毅：驅逐艦轟炸機出沒南海何以視而不見？〉，美麗島電子報。http://www.my-formosa.com/DOC_96189.htm。2016-2-24。（檢索日期：2016.4.26）

2015年5月，中華民國馬總統發表南海和平倡議，[33] 蔡英文總統表態「擱置爭議，共同開發」，[34] 越南尊重領海無害通行權並爭取合作發展經濟，[35] 中菲2017年5月建立的營運持續管理（BCM）機制，被認為是建立信任措施和促進海上合作與海上安全的重要平台，[36] 印尼發起的「科莫多海上聯合軍演」（MNEK）著重於非軍事領域的合作，尤以海上災害救援為演訓主軸，[37] 據此，協商管控分歧，力促爭議擱置，共同尋求資源淵開發、共享解決途徑的政治協商與倡議，將為南海行為準則簽署增添更多助力。[38]

（二）島礁和平轉化

永興島成立海南省三沙市，涵蓋西沙、中沙和南沙群島，循序漸進規畫開放觀光，將不需軍事存在的島礁供遊客旅遊，[39] 自郵輪旅遊航線開通後，島間形成的環南海郵輪旅遊航線交通條件大為改善，[40] 現代漁業與海上旅遊兩大產業發展獲得充分保障。[41] 郵輪業是海南近期大力推動的旅遊產業，中國大陸國防部網站不僅配合發布《西沙旅遊攻略》和最佳線路圖，[42] 更提供半潛船作為人道救援的海上基地。[43] 中國大陸批量建造海上浮動核電站將近20座，除力促

---

[33] 〈中評智庫：郭震遠指南海問題成美台關係熱點〉，中國評論新聞網。http://news.gpwb.gov.tw/news.aspx?ydn=026dTHGgTRNpmRFEgxcbfcCSN9Fhd8KFbqLRgMWauV8DzofFKLgOXFbWKC%2fW0VQd0x2TSlzlg%2bpqIZuaN691VMdoyiNKk5TjHUruGj6b8J0%3d。2016-4-10。（檢索日期：2016.4.14）

[34] 〈社評－堅守南海11段線主權〉，中時電子報。http://www.chinatimes.com/newspapers/20160611000722-260310。2016-6-11。（檢索日期：2016.6.11）

[35] 郭正亮，〈越菲拒當美國棋子，只有台灣例外？〉，美麗島電子報。http://www.chinatimes.com/newspapers/20160413000793-260301。2016-3-11。（檢索日期：2016.4.26）

[36] 〈中菲南海問題磋商"以合作為中心"〉，中國平論新聞網。http://hk.crntt.com/doc/1049/7/7/4/104977485.html?coluid=202&kindid=11690&docid=104977485&mdate=0214152859。2018-02-14。（檢索日期：2018.2.14）

[37] 〈印尼「科莫多」軍演揭幕37國海軍大會師〉，中國評論新聞網。

[38] 〈馬登島陸媒鼓掌不避諱中華民國〉，中時電子報。http://www.chinatimes.com/newspapers/20160130001013-260301。2016-01-30。（檢索日期：2016.02.02）

[39] 〈開發三沙島礁成陸版馬爾地夫〉，中時電子報。http://www.chinatimes.com/newspapers/20160530000633-260301。2016-5-30。（檢索日期：2016.5.30）

[40] 〈通信交通、島礁綠化、民生事業……中國全力開發西沙、中沙、南沙〉，風傳媒。http://www.storm.mg/article/88820。2016-3-18。（檢索日期：2016.4.18）

[41] 〈反制美艦進逼 陸推西沙旅遊線〉，中時電子報。http://www.chinatimes.com/newspapers/20160202000879-260309。2016-2-2。（檢索日期：2016.2.2）

[42] 〈維權 陸發布西沙旅遊攻略〉，中時電子報。http://www.chinatimes.com/newspapers/20160414000800-260301。2016-4-14。（檢索日期：2016.4.14）

[43] 〈陸半潛船出塢 海上戰力大增〉，中時電子報。http://www.chinatimes.com/newspapers/

南沙建成國際航運中心外，並藉此推動中國大陸從陸地經濟轉向海洋經濟。[44]
南海上空是國際繁忙空域，島礁機場可大幅提升南海地區海空交通服務與國際公共服務能力。[45]

中國大陸大型水陸兩棲飛機「蛟龍－600」（AG-600），除軍事用途外，亦應用於森林滅火、水上應急救援等任務。越南在南威島（越方稱長沙島）修建的基礎設施完備，[46] 除邀集海外僑民遊覽外，並向其他島礁贈送民生設備與系統。[47] 中華民國控有的太平島，曾提供國際海難救援，海岸警衛隊和海軍也在附近舉行搜救演習，2016年馬總統登上太平島，中國大陸外交部發言表示，兩岸中國人都有責任維護中華民族祖產，中國政府致力於將南海建設成和平、友誼、合作之海，持續擴充並轉化島礁建設的和平用途，值得各方支持與擴大推動。

參、澳洲南海戰略行動分析

一、澳洲南海文化結構

（一）四面環海

澳洲大陸也是南太平洋島國，海洋和海灘是澳洲國家標誌，跟澳洲文化密不可分，海洋是澳洲人生命中不可或缺的組成，更象徵健康、活力、熱愛戶外與運動休憩的展示。澳洲統計資料顯示，85％的澳洲人住在離海50公里內的沿岸區域，海濱度假是澳洲文化傳統，澳洲原住民靠近海濱、殖民澳洲的西方人也來自航海文明，不管是原住民還是殖民者，澳洲重視海洋文化，海洋休閒觀

20160607000855-260301。2016-6-7。（檢索日期：2016.6.7）

[44] 〈鞏固南海 陸將建20座浮動核電站〉，中時電子報。http://www.chinatimes.com/realtimenews/20160422005763-260417。2016-04-22。（檢索日期：2016.4.22）

[45] 〈美濟礁、渚碧礁新建機場 陸試飛成功〉，中時電子報。http://www.chinatimes.com/realtimenews/20160713005796-260409。2016-07-13。（檢索日期：2016.07.14）

[46] 〈越南蘇－30飛越南威島 PO照示威〉，中時電子報。http://www.chinatimes.com/newspapers/20160413000793-260301。2016-4-13。（檢索日期：2016.4.13）

[47] 〈遊覽宣示主權 越僑登南沙島礁〉，中時電子報。http://www.chinatimes.com/newspapers/20160529000598-260301。2016-5-29。（檢索日期：2016.5.29）

光發展始終是大洋洲包含澳洲最重要的文化象徵。[48]

（二）特殊濱海廊道

澳洲獨占一塊大陸，面積幾乎涵蓋整個大洋洲陸地，東臨南太平洋，西臨印度洋，內陸沙漠和半沙漠占據澳洲全國面積35%，約70%國土屬於乾旱或半乾旱地帶，[49] 海岸線長達37,521千公尺是全球最長國家，沿海地帶適合畜牧及耕種，廣闊的海灘緩坡和蔥鬱草木，緩緩向西傾斜逐漸過渡到平原，東部沿海擁有世界上最大的珊瑚礁群—大堡礁，[50] 整個沿海除南海岸外，形成了一條環繞大陸的「綠帶」，成為澳洲不可或缺的生命臍帶。[51]

（三）歸屬大洋洲島鏈

澳洲最著名的世界遺產烏魯魯（Uluru）號稱「地球的肚臍」，傳統持有者為中澳原住民阿南古人（Anangu），烏魯魯是原民創世神話的起點，也是祖靈記憶的歷史核心，英國殖民後以南澳洲首相亨利・艾爾斯（Henry Ayers）的名字冠稱為「艾爾斯岩」（Ayers Rock），是澳洲內地觀光的熱門景點，1985年10月26日澳洲政府正式將烏魯魯「返還」（Uluru handback）阿南古人，1987年雙方在共同成立「烏魯魯－卡達族塔國家公園管委會」，阿南古人將烏魯魯「租借」給澳洲政府，[52] 象徵澳洲已充分融入大洋洲島鏈文化群。

二、澳洲南海身份認同與角色定位

（一）大洋洲區域合作領導角色

維護南太地區穩定、促進島國經濟發展符合澳洲利益，2003年7月澳洲與

---

[48] 姜茉安，〈自以為是大陸國家的美麗島？澳洲人談台澳兩個島國看待「海洋」的文化差異〉，The News Lens關鍵評論，https://www.thenewslens.com/article/38292。2020-06-13。（檢索日期：2020.7.23）

[49] 〈大洋洲的澳洲和新西蘭，先後遭到英國殖民統治，原住民皆受到欺壓〉，大洋洲殖民地簡史。https://sites.google.com/site/historyofcolony/home/oceania。（檢索日期：2020.8.7）

[50] 〈四面環海，獨占一洲，這個國家，從何而來？〉，每日頭條，https://kknews.cc/news/zrer9rl.html。2020-06-13。（檢索日期：2020.7.23）

[51] 〈全球海岸線最長的國家是位於南半球的澳大利亞〉，每日頭條。https://kknews.cc/travel/x3y3bvo.html。2018-03-13。（檢索日期：2020.7.24）

[52] 〈澳洲禁爬「烏魯魯」：尊重原民先祖傳說，2019年10月封閉〉，轉角國際，https://global.udn.com/global_vision/story/8662/2791532。2017-11-1。（檢索日期：2020.7.23）

紐西蘭及部分南太島國應所羅門群島政府要求，對其進行聯合軍事干預，同年12月澳洲再與巴布亞紐幾內亞簽署一系列援助方案，協助其整治經濟、治安等，2004年2月又與諾魯簽署諒解備忘錄，協助其擺脫危機，並聯合紐西蘭推出實現地區和平、和諧、安全與繁榮的太平洋計劃，更提出聯合地區管理主張，推動建立南太地區聯合航線和設立地區警察培訓中心。[53] 2007年底澳洲加大對島國援助力度，投入10億澳元實施南太夥伴計劃，設立南太民事和軍事夥伴中心，並積極修復與索羅門、巴布亞紐幾內亞等島國關係，重申加強地區後援團（RAMSI）和對東帝汶的安全承諾。2010年8月，澳總理首度出席太平洋島國論壇首腦會議，並宣布撥款幫助太平洋島國促進男女平等。2013年3月，總理吉拉德（Julia Gillard）開啟訪問巴布亞新幾內亞之旅，兩國發表《巴布亞新幾內亞澳洲新夥伴關係聯合聲明》，[54] 顯見澳洲對大洋洲積極展現合作領導的角色。

（二）國際海洋公約倡導

《聯合國海洋法公約》，是當今國際解決海域爭端的主要規範，但卻也常是形成爭端的主要緣由，如何解讀與維護的確需要精心設計與共同協商努力。[55] 南海仲裁結果出爐後，美日等國先後呼籲中國大陸遵守，2016年7月13日澳洲外交部長畢曉普（Julie Isabel Bishop）呼應表示，各方都應尊重南海仲裁結果，並指名中國大陸若不理會，將付出沉重的聲譽成本，中國大陸外交部回應表示，希望澳洲不要把非法仲裁庭的非法結論當成國際法，而應努力維護國際法的嚴肅性、合法性和代表性，[56] 澳洲前駐中大使芮捷銳（Geoff Raby）表示，南海緊張局勢升級不符合區域內任何國家利益，緩和南海緊張局勢的唯

---

[53] 〈澳大利亞国家概况〉，中華人民共和國外交部，https://www.fmprc.gov.cn/web/gjhdq_676201/gj_676203/dyz_681240/1206_681242/1206x0_681244/。2020-05。（檢索日期：2020.7.23）

[54] 〈澳大利亞国家概况〉，中華人民共和國外交部，https://www.fmprc.gov.cn/web/gjhdq_676201/gj_676203/dyz_681240/1206_681242/1206x0_681244/。2020-05。（檢索日期：2020.7.23）

[55] 〈《南海仲裁案》台大國際法教授姜皇池：仲裁庭擴權，有損國際法治〉，風傳媒，http://www.storm.mg/article/141388。2016-7-13。（檢索日期：2016.7.20）

[56] 〈中方狠批澳就南海仲裁表態：看輕國際法〉，中時電子報，2016-7-15。http://hk.crntt.com/doc/1043/1/0/7/104310792.html?coluid=93&kindid=10094&docid=104310792。（檢索日期：2016.7.17）

一方式是讓所有聲索國進行談判，澳洲兩大政黨都有與中國大陸發展良好關係的意願，澳洲應在南海有更適切的國際立場與主張。[57]

（三）南海競合平衡角色

1974年澳洲與東盟正式建立對話夥伴關係，並先後與馬來西亞、新加坡和印尼等國簽訂雙邊或多邊安全防務條約，2005年，澳洲加入《東南亞友好合作條約》，2009年，澳洲總理陸克文（Kevin Rudd）出席第四屆東亞峰會，2010年1月，東盟－澳洲－紐西蘭自貿協定正式生效，2015年10月美、澳國防及外交部長進行會談，會後發表聯合聲明，雙方在南海爭議上「立場一致」，並對中國大陸在南海填海造陸和興建設施表達「強烈關切」，同時呼籲各方克制，不應採取片面行動升高區域緊張關係，對於有關主權的法律爭議雙方沒有意見，但警告會繼續支持航行自由權，[58] 澳洲國防部長安德魯斯（Kevin Andrews）更就南海問題強硬表態稱，澳洲將與美國和其他國家一起反擊中國的南海造島計畫，繼任的國防部長佩恩（Marise Payne）亦發表聲明強調，根據國際法，所有國家的船舶或飛行器都有自由航行的權利，澳洲60%出口商品運輸必須經過南海，澳洲有維持區域和平及穩定的合法利益，[59] 澳洲外長畢曉普（Julie Bishop）接著表示，南海存在許多民航飛機及商船航行路線，南海聲索國應該停止填海造島行動，2016年，澳越簽署雙邊行動計劃深化全面夥伴關係，2017年，澳洲總理和外交部長聯合出席環印度洋聯盟峰會和外長會，2018年3月，澳洲—東盟特別峰會首度在雪梨舉行，[60] 澳洲日益呈現的《外交政策白皮書》與政壇評論及媒體似亦逐漸導向冷戰思維和意識形態，[61] 顯見澳洲政

---

[57] 〈澳前駐華大使：南海局勢不符合地區國家利益〉，中國評論新聞網，2016-7-15。http://hk.crntt.com/doc/1043/1/0/7/104310764_2.html?coluid=7&kindid=0&docid=104310764&mdate=0715144523。（檢索日期：2016.7.17）

[58] 〈美澳嗆中國：捍衛南海航行自由〉，自由時報。http://news.ltn.com.tw/news/world/paper/923780。2015-10-15。（檢索日期：2016.03.19）

[59] 〈強調自由航行權利 澳挺美艦駛入南海 擬跟進航行 支持美軍「依法行動」〉，中時電子報，http://www.chinatimes.com/newspapers/20160201000707-260309。2016-2-1。（檢索日期：2016.2.2）

[60] 〈澳大利亞國家概況〉，中華人民共和國外交部，https://www.fmprc.gov.cn/web/gjhdq_676201/gj_676203/dyz_681240/1206_681242/1206x0_681244/。2020-05。（檢索日期：2020.7.23）

[61] 〈時評：抹黑中國凸顯澳外交焦慮〉，中國評論新聞網。

府與媒體輿論「不選邊站」的平衡態度有逐漸傾斜的跡象。[62]

　　不過澳洲也不乏有識之士，如前外交部長羅伯特‧約翰‧卡爾（Robert John Carr）與著名國防專家麥格雷戈（Richard McGregor）質疑澳洲參與美國南海巡航，是一項冒險行動。[63] 中國大陸是澳洲礦石最主要的出口國，[64] 中國大陸若暫停進口澳洲礦產並轉向非洲尋求能源進口，澳洲除面臨失去中國大陸主要市場與國內大量礦產資源面臨滯銷的風險，經濟亦將受到沉重打擊，[65] 澳洲一項調查顯示，澳洲民眾認為美國在亞太影響力正在下降，對同盟支持也不堅定，與中國大陸維持緊密關係更甚於美國，澳洲人不一定認同日本人對中國大陸的擔憂，也不會在歷史問題、南海爭議上與日本同調，[66] 顯見澳洲南海競合平衡正站在政府與民間分歧的十字路口上。

三、澳洲南海利益分析

（一）經貿發展利益

　　澳洲是亞洲地區唯一的西方國家，與美國雖有緊密的傳統聯盟關係，但澳洲畢竟礦產豐富，有先進科技也有廣大市場的需求，地緣上又臨近東協國家與中國大陸，與南海有密切地緣關係。2015年中國大陸山東嵐橋集團與澳洲北領地簽署達爾文港九十九年租賃合約，同年底中澳自由貿易協定（FTA）正式生效，兩國85.4％的產品立即零關稅，澳洲承諾對中國大陸的服務貿易以負面清單方式開放，大陸則以正面清單向澳洲開放。[67] 然而，美國強烈抗議達爾文港租約後，澳洲政府隨即宣布緊縮外國投資基礎設施規定，並自2016年3月31日

---

[62] 〈澳配合美制中 打南海消耗戰〉，中時電子報。http://www.chinatimes.com/newspape rs/20150603000925-260309。2015-06-03。（檢索日期：2015.06.03）

[63] 〈南海風雲－參加美軍南海行動 澳前外長：險棋〉，中時電子報。http://www.chinatimes.com/newspa pers/20160305000980-260301。2016-3-5。（檢索日期：2016.03.12）

[64] 〈澳大利亞這次火了！眼巴巴羨慕中非百億訂單合作：中國你不能這樣〉，每日頭條。https:// kknews.cc/world/m6ggl46.html。2018-06-23。（檢索日期：2020.8.3）

[65] 〈澳大利亞慌了，中國開始從非洲進口這些資源，網友表示：活該！〉，每日頭條。https://kknews. cc/world/g5e8m4m.html。2018-06-16。（檢索日期：2020.8.3）

[66] 〈過半韓澳民調 陸將超美成強權〉，中時電子報。http://www.chinatimes.com/newspape rs/20160611000599-260309。2016-6-11。（檢索日期：2016.6.11）

[67] 〈FTA生效滿月 韓國超日本成為大陸第2大貿易夥伴〉，中時電子報。http://www.chinatimes.com/new spapers/20160130000442-260108。2016-1-30。（檢索日期：2016.2.11）

起生效，只要超過2.5億澳元（約63億台幣），都須通過澳洲相關委員會審查，澳洲財政部擁有否決權。[68] 2014年，澳美雙邊貿易400億澳幣時，中澳兩國已達1420億澳幣，達爾文港在2014年至2015年間約有340公噸的貨物出口量，其中近半是與中國大陸貿易，主要是礦物出口，[69] 誠如2015年10月16日美國華爾街日報的報導顯示，達爾文港租賃合約的憂慮凸顯澳洲在與其最大貿易國中國大陸的經貿利益正面臨美國傳統盟邦政治安全利益的侵蝕挑戰。

（二）合作開發利益

1989年澳洲前總理霍克（Robert Hawke）主動倡議成立「亞太經濟合作」（Asia-Pacific Economic Cooperation, APEC）會議，1993年美國前總統柯林頓（Bill Clinton）基此倡議成立最高層級經濟領袖會議（APEC Economic Leaders' Meeting, AELM），[70] APEC成為亞太區域主要經濟諮商論壇，也是亞太地區最重要的多邊官方經濟合作論壇之一，[71] 澳洲又是東協加六的主要成員國，藉由澳洲、紐西蘭和印度的資源及市場，東協加六主要形塑一個更強大而廣泛的經濟合作架構－「東亞綜合經濟夥伴」（Comprehensive Economic Partnership in East Asia, CEPEA），2013年5月，東協與中國大陸、日本、韓國、紐西蘭、澳洲、印度等16個國家進一步展開「區域全面經濟夥伴協定」（RCEP）談判，2019年完成簽署，成為亞太區域最大經濟組織。[72]

2016年，澳洲等12國代表即在紐西蘭簽署《跨太平洋戰略經濟夥伴協定》（The Trans-Pacific Partnership, TPP），原本可打造出經濟產值占全球經濟40％全球規模最大的自由貿易區，因美國退出出現中挫，後由日本於2018年繼起

---

[68] 〈美抗議後 澳洲宣布基礎建設防堵中資〉，自由時報。http://news.ltn.com.tw/news/focus/paper/969929。2016-3-19。（檢索日期：2016.03.19）

[69] 〈中企租下達爾文港 澳軍方憂國安〉，自由時報。2015-10-17。http://news.ltn.com.tw/news/world/paper/924375。（檢索日期：2016.03.19）

[70] 〈亞太經濟合作，APEC〉，中華民國外交部。https://www.mofa.gov.tw/igo/News_Content.aspx?n=6D4F15F9D9692A0D&sms=EC05196BBCD002A6&s=E6395E7D94B7952F。（檢索日期：2020.07.23）

[71] 〈亞太經濟合作，APEC〉，中華民國外交部。https://www.mofa.gov.tw/igo/News_Content.aspx?n=6D4F15F9D9692A0D&sms=EC05196BBCD002A6&s=E6395E7D94B7952F。（檢索日期：2020.07.23）

[72] 〈東協加六〉，中華經濟研究院東協研究中心。http://www.aseancenter.org.tw/ASEAN6.aspx。（檢索日期：2016.1.16）

完成跨太平洋夥伴全面進步協定（Comprehensive and Progressive Agreement for Trans-Pacific Partnership, CPTPP）簽署，[73] CPTPP號稱亞太地區最大的區域經濟整合體，經濟規模達28兆美元，約占全球生產總值的36%，遠高於歐盟、北美自由貿易區（NAFTA）的23%跟26%，[74] 澳洲積極參與各型經濟合作組織的發起與形成，顯見其經濟合作發展的角色與份量不可分割亦不可或缺。

（三）體育觀光休閒活動發展利益

澳洲除礦產豐富並積極參與國際各項經濟合作發展組織外，其旅游資源與國際海、空運輸業更形發達，雪梨是南太平洋主要交通運輸樞紐，港口多達100個，墨爾本為全國第一大港，各類機場和跑道約2000個，其中國際機場12個，對擴大發展體育觀光休閒甚有助益。[75] 聯合國世界觀光組織（UNWTO）將2017年訂為「國際永續觀光發展年（International Yearof Sustainable Tourism Development）」，強調環境面、社會面及經濟面之永續發展，並訂有17項永續發展目標（Sustainable Development Goals, SDGs），包含海洋生態、陸地生態、就業與經濟成長、永續城市及全球夥伴等面向的均衡發展。

據聯合國世界觀光組織統計（Tourism Highlights 2017 Edition），2016年澳洲觀光外匯收入排名世界第10位，[76]「達沃斯論壇」（Davos Convention）世界經濟論壇（World Economic Forum, WEF）發布之《旅遊觀光競爭力報告書》（The Travel & Tourism Competitiveness Report）「2017 年全球觀光競爭力指數」顯示，澳洲旅遊觀光競爭力排名全球第7，[77] 2017年APEC體育政策網

---

[73] 徐遵慈，〈民意趨向是TPP的必修課〉，中時電子報，2016-2-5。http://www.chinatimes.com/newspapers/20160205001164-260109。（檢索日期：2016.3.1）

[74]〈TPP簽了 美爭經濟霸權勝中一籌 2018年生效 台積極搶第2輪門票〉，中實電子報，http://www.chinatimes.com/newspapers/20160205001527-260301。2016-2-5。（檢索日期：2016.2.5）

[75]〈澳大利亞國家概況〉，中華人民共和國外交部，https://www.fmprc.gov.cn/web/gjhdq_676201/gj_676203/dyz_681240/1206_681242/1206x0_681244/。2020-05。（檢索日期：2020.7.23）

[76]〈Tourism Highlights 2017 Edition〉，World Tourism Organization,UNWTO。https://www.e-unwto.org/doi/epdf/10.18111/9789284419029。（檢索日期：2020.8.04）

[77]〈世界經濟論壇報告書 旅遊觀光競爭力日本名列第4〉，朝日新聞中文版，2019-4-9。https://www.msn.com/zh-hk/news/world/%E4%B8%96%E7%95%8C%E7%B6%93%E6%BF%9F%E8%AB%96%E5%A3%87%E5%A0%B1%E5%91%8A%E6%9B%B8-%E6%97%85%E9%81%8A%E8%A7%80%E5%85%89%E7%AB%B6%E7%88%AD%E5%8A%9B%E6%97%A5%E6%9C%AC%E5%90%8D%E5%88%97%E7%AC%AC4/ar-AAGOtpN。（檢索日期：2020.8.04）

絡（APEC Sports Policy Network, ASPN）的「2017 APEC國際婦女與運動研討會」，澳洲等10個APEC會員體產官學代表及運動員熱烈與會，[78] 1996年澳洲旅遊收入即已成為最大宗出口商品，旅游資源扮演澳洲經濟頭號角色指日可見，2000年雪梨奧運會已標誌出這一轉變的端倪。澳洲政治家們費盡心機計劃和推進初級產品導向技術產業經濟轉型，始終受阻於國內市場的狹小和來自國外的強大競爭，旅遊業的異軍突起將使澳洲經濟再次擁有柳暗花明的驚喜。[79]

四、澳洲南海戰略行動取向

（一）戰略能力指導

美國與北約國防規畫從「威脅導向」轉為「能力導向」（capabilities-based approach），[80] 即要求部隊必須具備多層次安全防護與整合軍民資源的廣泛能力，[81] 澳洲應轉化陸海空軍軍事行動為陸海空域體驗休閒活動的戰略思維，發揮其體育觀光休閒資產優勢，為南海和平休閒觀光之海做出主導貢獻，以支撐澳洲海洋國家發展的願景。文化教育體育休閒要素在國際戰略環境的比重日漸上升，[82] 戰略思維應更聚焦於行動體驗與學習能力的探索與發展。[83] 澳洲本土多年未受戰火波及，又具四面環海與濱海廊道戰略環境特質，應積極轉化威脅導向為能力導向，修正與美國聯防中國大陸威脅的政策作為與機制運作，並優先在爭議的南海體現，為大洋洲的領導角色起帶頭示範作用。

---

[78] 〈2017 APEC 國際婦女與運動研討會〉，APEC 體育電子報 2017 10月第3期。2017-10-31。https://www.sa.gov.tw/PageContent?n=2065。（檢索日期：2020.7.31）

[79] 〈大洋洲的澳洲和新西蘭，先後遭到英國殖民統治，原住民皆受到欺壓〉，大洋洲殖民地簡史。https://sites.google.com/site/historyofcolony/home/oceania。（檢索日期：2020.8.7）

[80] 史考特・賈斯柏（Scott Jasper）編，劉慶順譯，《國防能力轉型：國際安全新策略》（北市：國防部史政編譯室，民國101年6月），頁13- 21。

[81] 史考特・賈斯柏（Scott Jasper）編，劉慶順譯，《國防能力轉型：國際安全新策略》，頁33- 35。

[82] 羅建波，〈中國崛起的對外文化戰略－一種軟權力的視角〉，《中共中央黨校學報》，第10卷第3期，2006年6月，頁98-99。

[83] 參見謝智謀、王怡婷譯，《體驗教育：帶領內省指導手冊》（北市：幼獅初版，民92年）。吳兆田，《探索學習的第一本書：企業培訓實務》（北市：五南出版，2006年10月初版）。蔡居澤、廖炳煌，《探索教育與活動學校》（桃園：社團法人中華探索教育發展協會，2007年1月初版）。蔡居澤、廖炳煌，《探索教育引導技巧培訓手冊》（桃園：社團法人中華探索教育發展協會，2007年1月初版）。蔡居澤、廖炳煌，《將探索教育帶回學校》（桃園：社團法人中華探索教育發展協會，2007年1月初版）。

（二）海域獨力巡航

澳洲《2013年國防白皮書》（2013 Defence White Paper）顯示，澳洲與美國及中國大陸三邊關係傾向崛起中的中國，同時呈現東南亞區域的和平與穩定，是澳洲國防戰略的關注焦點，2016年的國防白皮書宣布增加軍事預算，以因應日益崛起的解放軍軍力，國防白皮書除凸顯美澳之間緊密的戰略合作關係，同時也對南海島礁爭議發出聲明，要中國大陸與聲索國儘速達成共識，以免造成區域秩序不穩，影響海域巡航的安全。[84]

澳洲國防部長佩恩（Marise Payne）希望中國大陸推出更透明的國防策略，並停止在南海地區造島與駐軍，共同為區域穩定盡一份心力，澳洲總理特恩布爾（Malcolm Turnbull）考慮對爭議島礁進行單獨自由航行，澳洲新一代潛艇原由法國奪標，國防部長佩恩曾指出，這個決定反映澳洲是一個海洋貿易國家，[85] 堅持獨力巡航並兼顧美國傳統戰略友誼與中國感受才是澳洲應有的戰略選項。

（三）多元休閒活動倡導

澳洲地理位置獨立又具眾多國際組織成員國身份，軍隊傳統主要用於聯合國維和，與紐西蘭同被評為世界上和平的國家。[86] 國際軍事圈在2013年出現「軍事體育」新運動，目前已發展成「世界軍人運動會」，[87] 澳洲的體育休閒活動興盛，海域活動更具競爭優勢，倡導海洋國防體育休閒活動，將更能彰顯澳洲的地緣價值與活動利基優勢。[88] 隨著國際戶外活動開展「和平建立中心」

---

[84] 劉致廷，〈防範中國獨霸亞太 澳洲宣布7100億擴軍計劃〉，風傳媒。http://www.storm.mg/article/83246。2016-2-26。（檢索日期：2016.1.16）

[85] 〈澳洲潛艇項目：法國奪得巨額合同在澳洲建造〉，BBC中文網。2016-4-26。http://www.bbc.com/zhongwen/trad/world/2016/04/160426_australia_submarines_france_contract。（檢索日期：2016.4.27）

[86] 〈驚人的真相！紐西蘭究竟有多強大？〉，中時電子報，2015-8-26。http://www.chinatimes.com/photo-app/20150826003558-260802。（檢索日期：2015.10.10）

[87] 〈軍事體育多元化 提升戰技能力〉，青年日報，http://news.gpwb.gov.tw/news.aspx?ydn=026dTHGgTRNpmRFEgxcbfb0e%2b%2beJP7D3HGsxhDrckoVxfU8y7EZYfKgg%2bDOodmcZ4%2ftwN8%2fqC0gci7OdJbnO%2fE7E8LMdrdrU42ESKnxsiHU%3d。2015-08-04。（檢索日期：2015.08.04）

[88] 〈綠推改良式募兵制 執政後提國防總檢討〉，中國評論新聞。http://www.zhgpl.com/doc/1039/1/4/1/103914184_3.html?coluid=93&kindid=2931&docid=103914184&mdate=0903005757。2015-09-03。（檢索日期：2015.09.07）

（Center for Peacebuilding）訓練體驗活動，關懷觸角與服務對象已延伸至「衝突預防」、「衝突管理」、「後衝突調解」等全球議題，[89] 訓練活動體育化、體育活動休閒化與聚焦海域活動需求的發展趨勢已日益彰顯，[90] 澳洲南海戰略行動取向應奠基於國家四面環海與濱海廊道的海洋文化為基石，在海域獨力巡航規劃的同時發揮體育觀光休閒的優勢，將開創南海合作開發與體育休閒觀光的發展利基。

肆、結語

　　面對國際仲裁加速南海地緣競逐的激化，域內外國家對和平發展倡議的提升，與島礁和平多元用途轉化的高度期許，澳洲在加大獨力海域巡航的力度時，充分發揮其四面環海、特殊濱海廊道的休憩生活型態，與大洋洲完整海域島鏈的資產優勢，將讓澳洲在南海的身份認同與角色發展朝大洋洲區域合作領導者、國際海洋公約倡導者與南海競合平衡者的方向努力形塑，並在澳洲倡導海域巡航自由時，引導南海區域的經貿發展、合作開發成效向海域體育觀光休閒活動發展利益共同努力開發與創造，以避免澳洲深陷傳統地緣戰略兩強選邊的困窘，並成為全球海洋活動相關議題的倡導者與奉行者，開創澳洲在南海海域活動發展獨一優勢。

---

[89] 〈什麼是外展〉，外展教育基金會。http://www.obtaiwan.org/?FID=15&CID=154。（檢索日期：104.5.4）

[90] 教育部體育署組織法。2013-12-09。http://www.sa.gov.tw/wSite/ct?xItem=3113&ctNode=274&mp=11。（檢索日期：106.1.21）

美國南海自由航行任務對台灣主權的挑戰

郁瑞麟 *

壹、前言

中華民國（台灣）由於國際處境艱難、中共威脅增強與國內政治氛圍的驅使，政治方面越來越依賴美國。美國前總統川普外顯的「聯合抗中」模式，讓台灣方面受到鼓舞，台美關係也因此進入新的里程碑。2020年底的美國總統大選，川普雖不願言敗，但仍於今（2021）年初離開了白宮，將政權移交給現任總統拜登。美國總統拜登上任後，台灣政界即湧現政府押錯寶的聲浪，再加上拜登所屬的民主黨與川普所屬的共和黨相較，傳統上顯得較為親中，且喜以圓融的外交方式處理爭議，使台灣朝野皆緊盯拜登團隊對中國大陸與台灣方面的一舉一動，深怕台美關係會自高點滑落。

2021年1月20日拜登總統上任，在2月4日於國務院發表首次外交政策演說，強調「美國回來了、外交回來了」，隨然重申中國是美國最嚴峻的競爭者，但仍表示願在符合美國利益的前提下與北京合作。不久後的2月16日，美海軍飛彈驅逐艦羅素號（USS Russell, DDG 59）即在南海進行拜登上任後的首次「自由航行任務」，美國印太戰區第七艦隊即發布新聞稿表示該任務符合國際法，強調南沙群島附近航行的權利與自由，美國堅持在國際法認可下對於海洋權利、自由與合法的使用；這次任務主要是挑戰中國、越南與台灣對「無害通過」的非法限制。[1] 該聲明將台灣與中國大陸同列為被挑戰的對象，自然成為台灣政治的敏感話題，國民黨包括前馬總統辦公室皆對此提出質疑，提醒政府要注意台灣在南海的主權並質疑美方非《聯合國海洋法公約》（文後簡稱《公約》）的簽約國，何以反過來質疑其他國家的主張不符合國際法？

---

* 國防大學戰略研究所上校副教授兼所長
[1] 7th Fleet Public Affairs, "7th Fleet destroyer conducts Freedom of Navigation Operation in South China Sea," U.S. Pacific fleet, Feb. 16, 2021, https://www.cpf.navy.mil/news.aspx/130811 (accessed by 2021. 05. 07).

據此，本文認為有幾個問題有必要進一步釐清：

一、美國非《公約》簽約國是否有權主張國際法的權利？

二、美國自由航行任務挑戰我南海主張是否象徵著台美關係改變？

三、美國此舉是否有損我南海主權？

四、如果台灣不想再被列為被挑戰的對象，該如何調整？而此調整又會對我主權帶來什麼影響？

在此必須先說明的是，文中所涉及的有關外國軍艦通過我國領海的「無害通過」規定皆不涉及中國大陸軍艦。「無害通過」係為外國船舶通過沿海國領海的所應遵守的相關規範，惟依據我國現行憲法增修條文第十一條：「自由地區與大陸地區間人民權利義務關係及其他事務之處理，得以法律為特別之規定。」我方遂以《臺灣地區與大陸地區人民關係條例》（以下簡稱《兩岸人民關係條例》）規範之，表明大陸船舶有異於外國船舶。又依據特別法優先於普通法的原則，大陸船舶應優先適用於《兩岸人民關係條例》而非我國的《領海及鄰接區法》，目前依據我國國防部的作法仍以不讓中國大陸軍艦越過中線為原則，更遑論進入我國領海。

本文後續將依上述所列問題，首先討論美國與國際海洋法之間的權利義務關係，接著透過美國「自由航行任務」報告檢視台灣被挑戰的歷程，以釐清川普時期與拜登時期對台方式有無改變，後續再檢視我國的南海主張與被美國挑戰原因之間的關聯性，從中思考修改我國海洋主張以規避成為美國「自由航行任務」挑戰對象的方式，但作者提醒亦須反思修改後對我造成的可能影響。

貳、美國與國際海洋法

美國迄今尚未簽署《聯合國海洋法公約》，卻又以國際法的維護者自居，因此常遭受非議。但若認定美國未簽署公約，故不必遵守國際海洋法，也不能主張維護國際法，此認知亦有所差池。

美國自1979年迄今由美軍所執行的「自由航行任務」，旨在挑戰沿海國過

度的海洋主張，而這些所謂的過度主張就是根據美國所詮釋的《公約》規範。在1991年公布的第一版「自由航行報告」中就表明，依據美國總統在1983年3月10日的「海洋政策宣言」（Oceans Policy Statement）：「美國會實踐與堅持在全球領域進行符合1982年《公約》中基於利益平衡的航行與飛越的權利與自由（The United States will exercise and assert its navigation and overflight rights and freedoms in a worldwide basis in a manner that is consistent with the balance of interests reflected in the 1982 Law of the Sea Convention.），[2] 後續的報告亦將此聲明臚列其中。至於美國是如何詮釋1982年《公約》的相關規範，可參考美國國務院下轄的「海洋、國際環境和科學事務局」（Bureau of Oceans and International Environmental and Scientific Affairs）針對沿海國的海洋主張所提出的《海洋界限》（Limits in the Seas）系列報告。

　　另一方面，我國亦為1958第一屆與1960年第二屆「聯合國海洋法會議」的參與國，也同樣簽定了第一屆會議後所產生的四項公約。[3] 其中1958年《領海及鄰接區公約》，在1964年9月10日正式生效，其主要法律規範後來皆被整合於1982年《公約》之中，其中包括領海無害通過規定。故若我國與美國等非1982年《公約》簽約國，在1958年《領海及鄰接區公約》未違背後續1982年《公約》條文或現行國際習慣時，理應繼續遵守1958年《領海及鄰接區公約》的規範。另我國雖非《公約》的簽訂國，但同樣依賴《公約》主張自己的海洋權益，也將《公約》條文整合到我國的海域二法之中，將其國際法國內法化，這與美國的作法類似，兩者並無軒輊。故我國若指責美國沒有簽訂《公約》，卻用《公約》主張權利，反而顯得有些不當。

---

[2] US Department of Defense, "DoD Annual Freedom of Navigation (FON) Reports," https://policy.defense.gov/OUSDP-Offices/FON/. (accessed by 2021. 05. 08).

[3] 在1960年第二屆聯合國海洋法會議於3月17日召開的第1次的「全員會議」中，當時蘇聯代表Mr. TUNKIN針對中國的代表權問題提出質疑，認為中國的合法代表應為「中華人民共和國」而非「中華民國」，且「德意志民主共和國」（German Democratic Republic）、「朝鮮民主主義人民共和國」（Democratic People's Republic of Korea）、「越南民主共和國」（Democratic Republic of Vietnam）與「蒙古人民共和國」（Mongolian People's Republic）也應該有權派代表參加，該提議獲得了波蘭、捷克、阿爾巴尼亞等國家代表的發言支持；惟美國、中華民國與南韓代表發言主張現有代表的正當性，執行主席也認為該提議超出了原來聯合國1307號（XIII）的決議內容，故未成案。參見：United Nations. *Conferences on the Law of the Sea, official Records, 2nd Conference, Vol. 1, 1960* (New York: William S. Hein & Co., 2000), 2-3.

此外，國際法院規約（Statute of International Court of Justice）第三十八條規定：「1.不論普通或特別國際條約，確立訴訟當事國明白承認之規條者。2.國際習慣，作為通例之證明而經接受為法律者。3.一般法律原則（general principles of law）為文明各國所承認者。4.在五十九條規定下，司法判例及各國權威最高之公法學家學院，作為確定法律原則之補助資料者。」亦已表明當事國應循國際條約，若無簽訂相關條約，則應依序遵守國際習慣、國際判例、一般法律原則或法學家學說逐一釐清法律效果。任何受國際法院審視的當事者皆無法逃避此一規範的制約。在《維也納條約法公約》（The Vienna Convention on the Law of Treaties）第38條也表明：「條約所載規則由於國際習慣而成為對第3國有拘束力」，換言之，國際法的理論與實踐皆認為，無論是雙邊或多邊條約，一旦成為國際習慣法規則後，即可約束非簽約國，特別是那些國際多邊立法性條約，如《公約》，因締約國已成為全球國家的大多數，極易形成國際習慣法規則。故本文認為美國無法辯稱因未簽署《公約》而置身於國際海洋法規範之外，當然它也有權主張國際海洋法所賦予的權利與義務。

參、美國自由航行任務

始於1979年，由美國總統指導行政部門所推動的「自由航行計畫」（Freedom of Navigation Program），其目的就是在挑戰那些超出國際海洋法規範的沿海國主張，以維護長久以來符合美國國家利益的海洋自由原則；該計畫強調美國不會容許某些國家單方面地限制國際社會所應享有的海洋權益與航行自由。[4] 換言之，該計畫即是為美國政府擔心某些不利於美國的法律主張會逐漸形成國際習慣，而特別採取行動以凸顯該等沿海國的海洋主張並非現今國際法所承認之習慣。基此，由該計畫延續至今所推動的「自由航行任務」便成為美國海軍用來挑戰沿海國海洋權益主張的重要工具。

美海軍在執行自由航行任務時，為落實國際法中的「非歧視原則」，本

---

[4] US Department of Defense, "U.S. Department of Defense Freedom of Navigation (FON) Program," (DoD, February 28, 2017), accessed July 13, 2020, https://policy.defense.gov/Portals/11/DoD%20FON%20Program%20Summary% 2016.pdf?ver=2017-03-03-141350-380.

不應針對某些特定國家進行旨在挑戰其法律主張的行動，故在執行該項任務時本應宣稱挑戰所有具管轄權的沿海國的相關海洋主張。2020年1月25日，由美軍近岸戰鬥艦「蒙哥馬利號」（USS Montgomery）在南沙群島所執行的「自由航行任務」，美國第七艦隊的官方聲明即指出，該任務是挑戰中國、越南與台灣等聲索方要求軍艦進入領海進行無害通過前，須獲得許可或事先通知的規定。[5] 2020年4月29日，由美巡洋艦邦克山號（USS Bunker Hill）執行類似任務時，第七艦隊也同樣發出挑戰中國、越南與台灣主張的類似聲明。[6] 唯有趣的是，在這些被挑戰的沿海國中，除中國大陸發表了嚴正申明抗議外，越南與台灣兩方似乎都未對此挑戰做出明顯回應。

　　本文針對美國國防部自1991年所公布的自由航行任務報告進行檢視（如表1），可發現在2000財政年度（Fiscal Year，簡稱FY，FY 2000代表自1999年10月1日到2000年9月30日為止的這段時間，以下報告所指年份皆為財政年度）報告中，首次挑戰台灣，更將台灣的海洋主張（過度直線領海基線）列在國家──「中國」之下。之後FY 2000年至2003年的綜合報告中，又將台灣與中國大陸分開，單獨列為聲索方（claimant）；在FY 2006年的報告中更將台灣列為國家（country）且當年未將中國列為挑戰對象；在FY 2011年的報告中不但將台灣列為國家，同時與中國並列成為當年的挑戰國，且自該年開始，台灣即被列為每年挑戰的對象；FY 2012年開始，報告又改回聲索方的稱呼至今（FY 2020年）。另該報告自FY 2017年起列出挑戰地點，自FY 2018年起，增加列出沿海國法律主張的法律來源。

5　Ben Werner, "USS Montgomery Conducts First 2020 FONOP in South China Sea," (USNI, 2020, 01, 08), accessed July 13, 2020, https://news.usni.org/2020/01/28/uss-montgomery-conducts-first-2020-fonop-in-south-china-sea.主要爭議是美方認為依據現行國際海洋法，軍艦等同於其他船舶可在沿海國領海享有無害通過的權利，且無需事先經過沿海國的許可或通知沿海國。中國與越南主張需事先報備許可後軍艦方可進入領海無害通過，越南於2012年頒布越南海洋法後，改為與我國相同的事先報備制。

6　Sam LaGrone, "USS Bunker Hill Conducts 2nd South China Sea Freedom of Navigation Operation This Week," (USNI, 2020, 04, 29), accessed July 4, 2020, https://news.usni.org/2020/04/29/uss-bunker-hill-conducts-2nd-south-china-sea-freedom-of-navigation-operation-this-week.

表1：台灣與其他國家共同納入自由航行任務報告紀錄（US FY1991 to 2020）

Fiscal Year	Country/ Claimant	Excessive Maritime Claims	Geo. Location
FY1991- FY1999	X	X	X
FY2000	**Country: China**	**Taiwan's excessive straight baselines**	X
FY2000- FY2003	Claimant: Taiwan	Excessive straight baselines; 24 nm security zone	X
FY2006	**Country: Taiwan**	Restriction on right of innocent passage through territorial sea; requirement of prior notice of warships transiting territorial sea.	X
FY2011	**Country: Taiwan**	Excessive straight baselines; prior notification required for foreign military or government vessels to enter territorial sea.	X
	Country: China	Excessive straight baselines; prior permission required for innocent passage of foreign military ships through territorial sea.	X
	Country: Vietnam	Excessive straight baselines; authorization required for foreign warships to enter territorial sea and contiguous zone;	X
FY2012	Claimant: Taiwan	Excessive straight baselines; prior notification required for foreign military or government vessels to enter territorial sea	X
	Claimant: China	prior permission required for innocent passage of foreign military ships through territorial sea	X
	Claimant: Vietnam	authorization required for foreign warships to enter territorial sea and contiguous zone	X

Fiscal Year	Country/ Claimant	Excessive Maritime Claims	Geo. Location
FY2013	Claimant: Taiwan	Excessive straight baselines; prior notification required for foreign military or government vessels to enter territorial sea	X
	Claimant: China	prior permission required for innocent passage of foreign military ships through territorial sea	X
	Claimant: Vietnam	prior notification required for foreign warships to enter territorial sea	X
FY2014	Claimant: Taiwan	Excessive straight baselines; prior notification required for foreign military or government vessels to enter the territorial sea.	X
FY2015	Claimant: Taiwan	Prior notification required for foreign military or government vessels to enter the TTS.	X
	Claimant: China	prior permission required for innocent passage of foreign military ships through the TTS.	X
	Claimant: Vietnam	Excessive straight baselines; prior notification required for foreign warships to enter the TTS.	X
FY2016	Claimant: Taiwan	Prior notification required for foreign military or government vessels to enter the TTS.	X
	Claimant: China	Prior permission required for innocent passage of foreign military ships through the TTS.	X
	Claimant: Vietnam	Prior notification required for foreign warships to enter the TTS.	X
FY2017	Claimant: Taiwan	Prior notification required for foreign military or government vessels to enter the TTS	Paracel Islands

Fiscal Year	Country/ Claimant	Excessive Maritime Claims	Geo. Location
	Claimant: China	Excessive straight baselines	Paracel Islands
		Prior permission required for innocent passage of foreign military ships through the TTS	
	Claimant: Vietnam	Prior notification required for foreign warships to enter the TTS	Paracel Islands
FY2018	Claimant: Taiwan	Prior notification required for foreign military or government vessels to enter the territorial sea. [Law on the Territorial Sea and the Contiguous Zone, art. 7, Jan. 21, 1998.]	Paracel Islands, Spratly Islands
	Claimant: China	Prior permission required for innocent passage of foreign military ships through the territorial sea. [Declaration upon Ratification of 1982 Law of the Sea Convention, June 7, 1996.]	Paracel Islands, Spratly Islands
		Straight baselines not drawn in accordance with the law of the sea. [Declaration of the Government of the People's Republic of China on the Baselines of the Territorial Sea of the People's Republic of China, May 15, 1996.]	Paracel Islands
	Claimant: Vietnam	Prior notification required for foreign warships to enter the territorial sea. [Law of the Sea of Vietnam, Law No. 18/2012/QH13, art. 12, June 21, 2012.]	Paracel Islands, Spratly Islands
FY2019	Claimant: Taiwan	Prior notification required for foreign military or government vessels to enter the territorial sea. [Law on the Territorial Sea and the Contiguous Zone, Art. 7, Jan. 21, 1998.]	Paracel Islands, Spratly Islands

Fiscal Year	Country/ Claimant	Excessive Maritime Claims	Geo. Location
		Straight baseline claims. [Law on the Territorial Sea and the Contiguous Zone, Art. 4, Jan. 21, 1998; Decree No. Tai 88 Nei Tze #06161, Feb. 10, 1999.]	Philippine Sea
	Claimant: China	Prior permission required for innocent passage of foreign military ships through the territorial sea. [Law on the Territorial Sea and Contiguous Zone, Feb. 25, 1992.]	Paracel Islands, Spratly Islands
		* Straight baseline claims. [Declaration of the Government of the People's Republic of China on the Baselines of the Territorial Sea of the People's Republic of China, May 15, 1996.]	Paracel Islands
	Claimant: Vietnam	Prior notification required for foreign warships to enter the territorial sea. [Law of the Sea of Vietnam, Law No. 18/2012/ QH13, Art. 12, June 21, 2012.]	Paracel Islands, Spratly Islands
FY2020	Claimant: Taiwan	Prior notification required for foreign military or government vessels to enter the territorial sea. [Law on the Territorial Sea and the Contiguous Zone, article 7, Jan. 21, 1998.]	South China Sea
	Claimant: China	Prior permission required for innocent passage of foreign military ships through the territorial sea. [Law on the Territorial Sea and Contiguous Zone, Feb. 25, 1992.]	South China Sea
		* Straight baseline claims. [Declaration of the Government of the People's Republic of China on the Baselines of the Territorial Sea of the People's Republic of China, May 15, 1996.	South China Sea

Fiscal Year	Country/ Claimant	Excessive Maritime Claims	Geo. Location
	Claimant: Vietnam	Prior notification required for foreign warships to enter the territorial sea. [Law of the Sea of Vietnam, Law No. 18/2012/QH13, article 12, June 21, 2012.]	South China Sea
	Claimant: Malaysia	Prior consent required for military exercises or maneuvers in the exclusive economic zone. [Declaration upon Ratification of the 1982 Law of the Sea Convention, Oct. 14, 1996.]	South China Sea

資料來源：Under Secretary of Defense for Policy, accessed July 4, 2020, https://policy.defense. gov/OUSDP-Offices/FON/.

　　該報告自2017年起開始公布自由航行任務所挑戰的海域位置，我們可發現自該年起，該任務挑戰台灣的海域位置皆為在南海海域的西沙群島（Paracel Islands）或南沙群島（Spratly Islands）。由此反推，自2011年起，該任務便將台灣與中國大陸及越南並列，成為挑戰的聲索方，故其地點應該也同樣是位於南海的這些島礁。

　　回顧2009年歐巴馬上任後即宣示「重返亞洲」（pivot to Asia），至2010年5月，發佈《美國國家安全戰略報告》（The National Security Strategy of the United States of America），將中國大陸、印度、俄羅斯並列為未來世界「影響力中心」。[7] 2010年7月23日美國國務卿希拉蕊在越南河內召開的第十七屆東協區域論壇部長會議中表示，美國是南海情勢的利害關係國，在南海的航行自由、亞洲共有海洋的開放通道（open access to Asia's maritime commons）、及對國際法的尊重皆涉及美國國家利益，對於南海主權爭議，美國雖不會直接支

---

[7] U.S. Department of Defense, "The Quadrennial Defense Review Report 2010," accessed May 6, 2021, https://archive.defense.gov/qdr/QDR%20as%20of%2029JAN10%201600.pdf.

持任何一方之主張，但反對任何一國單邊宣稱主權，或以任何形式之威脅作為解決爭議的手段，並希望南海主權爭議能以《公約》循多邊外交方式解決。[8] 由於歐巴馬外交政策團隊和國務院之間對這項新策略的歧異，使得正式的宣佈延遲到2011年10月，當時是希拉蕊在「外交政策」（Foreign Policy）上以「美國的太平洋世紀」為題，闡述了這項新策略。[9] 希拉蕊在2011年7月24日的ARF部長會議中又再提聲明，希望各南海聲索方遵守國際法，針對南海主權之主張提出有效的法律證據，不該僅以歷史水域作為佐證依據。[10]

2016年的「南海仲裁案」為美國自由航行任務更增添了許多法律支撐，該裁決的主要重點如下：

1. 中國大陸為《公約》的締約國，不應主張超過《公約》以外的權利，而中共的南海「九段線」歷史水域主張，並不符合現今的國際法規範。[11]

2. 美濟礁、渚碧礁、南薰礁、西門礁（包括東門礁）及仁愛礁等5處皆為低潮高地不能主張領海。[12]

3. 南沙群島中包括太平島、中業島、西月島、南威島、北子島、南子島等較大的地物，均為無法產生專屬經濟海域或大陸礁層的「岩礁」。[13]

4. 《公約》並未規定如南沙群島的一系列島嶼可以作為一個整體共同產生海洋區域。[14]

美國於2016年南海仲裁案頒布後的10月21日，由美國飛彈驅逐艦「迪凱特

---

[8] Hillary Clinton, "Remarks at Press Availability," http://www.state.gov/secretary/rm/2010/07/145095.htm. 轉引自：李瓊莉，〈美國「重返亞洲」對區域主義之意涵〉，《全球政治評論》39期（2012年）：93。

[9] 鄧中堅，〈美國在南海面對中國的理論與政策作為：國際政治與國際法的對話〉，全球政治評論，第61期（2018年1月）：39。

[10] Hillary Clinton, "Sovereignty of South China Sea," http://www.state.gov/secretary/rm/2011/07/169010.htm. 轉引自：李瓊莉，〈美國「重返亞洲」對區域主義之意涵〉，《全球政治評論》39期（2012年）：93。

[11] Permanent Court of Arbitration, *THE SOUTH CHINA SEA ARBITRATION (Arbitration between the Republic of the Philippines and the People's Republic of China)*, The Tribunal Renders Its Award, (The Hague, 12 July 2016), 8-9.

[12] Ibid, p. 9.

[13] Ibid, pp. 9-10.

[14] Ibid, p. 10.

號」（USS Decatur, DDG-73）於西沙群島執行自由航行任務，該任務直接進入中國大陸所繪製的領海直線基線範圍內，但就美國的角度因《南海仲裁案》表明該類島嶼不宜作為一整體產生海洋區域，故美艦只要未進入島礁的12浬內，應屬自由航行的範疇。[15] 但中方認為美國不但沒有尊重中國對於外國軍艦須事先報准的規定，更不遵守國際法有關無害「通過」的規定，擅自進入中國的內水繞行，而非國際法認定的「通過」（如圖1）。

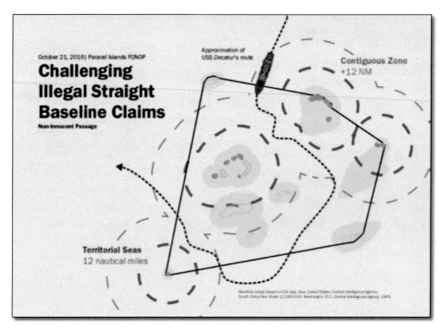

圖1：2016年10月21日美軍艦「迪凱特號」於西沙群島執行自由航行任務示意圖

資料來源：Eleanor Freund, *Freedom of Navigation in the South China Sea: A Practical Guide* (Harvard Kennedy School, Belfer Center for Science and International Affairs, 2017), 38.

---

[15] 另可參照美國國務院海洋、國際環境和科學事務局於2014年針對中國的南海海洋主張公布的第143號海洋界限報告。US Department of State, *Limits in the Seas: No. 143 China: Maritime Claims in the South China Sea* (US Department of State, Bureau of Oceans and International Environmental and Scientific Affairs, 2014).

我們可對照表1的紀錄，此任務歸為FY 2017年（2016年10月1日至2017年9月30日），該報告基於西沙群島的主權考量，當年在挑戰領海無害通過規定時，將中國、越南與台灣皆一併納入，但僅將中國納為西沙群島過度直線基線的挑戰對象，未包含台、越，其原因係因台、越兩方皆未繪製西沙群島的直線領海基線。後至FY 2019年，該任務在隔5年後再次挑戰我國過度領海直線基線，報告顯示挑戰地點為菲律賓海，而非西沙群島；故可推知該任務先前所指挑戰我國過度領海直線主張，應指我台海周邊繪有直線領海基線的海域，且是位於我國東南方的菲律賓海海域部分。

美海軍「自由航行任務」本是其國家政策工具的一環，在美國國家戰略重心調整至亞洲後，作為亞洲航行通道重中之重的南海，自然成為執行該任務的重要場域。既然該行動的目的係在挑戰沿海國的海洋主張，同時以不造成歧視為原則，而中華民國（台灣）身為南海島礁主權的聲索方，而某部分海洋主張又不被美國所接受，自然理當將台灣列為挑戰對象之一；此情形在川普執政時期與拜登執政時期並無二致，故該等事件呈現前後兩任政府政策的一致性，並無表現出台美關係的變化。

至於我國南海主權方面，本文認為將台灣納入反而是彰顯我國在南海島礁的主權主張，至於挑戰我國有關領海「無害通過」的規定，確實也在挑戰我主權所賦予的立法權利。至於台灣在報告中所呈現的名稱與分類，與美國總統任期對照可發現在美國小布希（George Walker Bush）總統時期（2001年1月20日至2009年1月20日）將台灣從中國類別下分離出去，後在「國家」（country）與「聲索方」（Claimant）的名稱間猶疑不定，至2012年歐巴馬（Barack Hussein Obama）時期（2009年1月20日至2017年1月20日）方才定調為聲索方，迄今未再有變化。

肆、我國的海洋主張

前揭美國「自由航行任務」挑戰我國海洋主張的原因不外乎為「過度的直線領海基線主張」與「外國軍艦（公務船舶）通過我國領海需事先報備」等兩

項。依據我國1998年公布的《中華民國領海及鄰接區法》第七條第三款：「外國軍用或公務船舶通過中華民國領海應先行告知。」及該條第五款：「外國船舶無害通過中華民國領海之管理辦法，由行政院定之。」再依據行政院交通部於2002年1月30日訂定的《外國船舶無害通過中華民國領海管理辦法》第十四條、第十五條規定，外國軍艦或公務船舶在通過我國領海時，需於抵達前10日告知我國外交部，即可行使無害通過。這也就是美方所指的外國軍艦（公務船舶）通過我國領海需事先報備的條文。

美國國務院海洋、國際環境和科學事務局於2005年也公布第127號《海洋界限：台灣的海洋主張》（Limits in the Seas: No. 127 Taiwan's Maritime Claims）報告，以檢視我國所頒布的海洋主張。針對1998年所頒布的《中華民國領海及鄰接區法》，該報告認為大致符合《公約》所規範的習慣國際法（customary international law）內容，惟領海基線與無害通過規定明顯與《公約》規範有所偏差，其中我國《領海及鄰接區法》第7條第三款：「外國軍用或公務船舶通過中華民國領海應先行告知」的規定，報告表示在《公約》第二部份第三節，包含其適用於軍艦的C部分，並沒有此要求（No such requirement appears in section 3 of Part II of the LOS Convention, including subsection C on rules applicable to warships.）[16] 至於我國在1999年2月10日由行政院所公布的領海基線（2009年修正，但仍保持原直線基線的部分），美方報告指出，雖《公約》及之前的《領海及鄰接區公約》皆沒有明確規範直線基線的長度限制，但美方透過法理的分析認為領海直線基線長度不應超過24浬。[17] 而我國現行直線基線多處超過24浬，因而美方認為此為過度的領海直線基線繪製方式。

依2009年11月18日行政院院臺建字第0980097355號令修正的中華民國第一批領海基線、領海及鄰接區外界線（如圖2）：

---

[16] US Department of State, *Limits in the Seas: No. 127 Taiwan's Maritime Claims* (US Department of State, Bureau of Oceans and International Environmental and Scientific Affairs, 2005), 4.

[17] Ibid., p. 9.

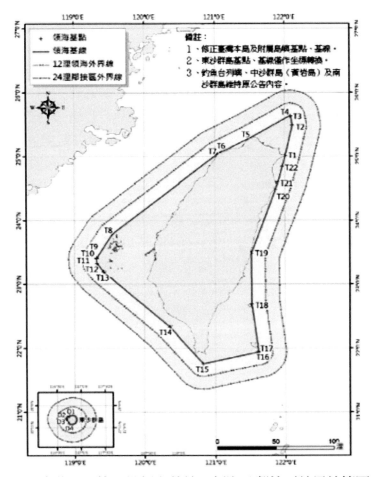

圖2：中華民國第一批領海基線、領海及鄰接區外界線簡圖
（2009年11月18日修正版）

資料來源：內政部網站，http://www.files/20091118第1批領海基線領海及鄰接區外界線修正
發布版(pdf).pdf。

　　有關上述美國挑戰我國海洋主張部分，必須回歸「自由航行任務」的目的，即挑戰美國所認定的沿海國過度海洋主張，美國想藉此行動彰顯該等主張不符合《公約》與「習慣國際法」，換言之，美國的這些主張也尚未形成國際習慣，否則美國也不用大張旗鼓來挑戰這些沿海國的主張。

　　首先，美國國務院《海洋界限報告》對我國以直線基線方式連接本島及澎湖群島，其中尤以T7到T8段（見圖2）長達109浬及T13至T14段超過70浬提出質疑，認為不僅將原本未被劃為內水的部分劃為內水，該直線基線本身也過長，超出了《公約》的規範。[18] 如前所述，美國也體認到《公約》（包括《領海及鄰接區公約》）中，並未對直線基線的長度進行規範。從國際實踐觀之，仍有許多國家採較為寬鬆的直線基線認定方式，如越南也選定了離岸74浬的Hon Hai Islet作為直線基線的基點，故將離岸超過24浬的島礁劃為直線基線的基點，並非我國所獨有。[19]

　　再者，長久以來，中美兩國就事先須經中華人民共和國政府批准存有爭議，國家實踐上對軍艦無害通過領海與沿海國規定事先取得許可或通報的要求也不盡一致。[20] 就國家實踐而言，各國立法略分為兩大流派：一為主張軍艦享有無害通過權，勿須經過沿海國授權或事先報備，此類國家大多為海洋大國，如美國、英國等；另一派則否認軍艦可享有無害通過權，主張外國軍艦通過沿海國領海時，必須取得沿海國同意或事先通知沿海國，此類國家大多為發展中國家，包過中國大陸在內，總計超過40多個國家。[21]

　　最後，美方認為中國在西沙群島採用直線基線劃法以測算領海寬度的作法違反《公約》相關規定。美國認為，中國大陸並非「群島國」，因此，無權採用直線基線劃定西沙群島的領海範圍。[22] 前揭的《南海仲裁案》判決亦持類似意見。[23] 中國大陸在1996年公布《中華人民共和國政府關於中華人民共和國領海基線的聲明》，包含中國大陸領海的部分基線和西沙群島共77個領海基點的名稱和地理坐標；2012年9月10日公布《中華人民共和國關於釣魚島及其附屬

---

[18] Ibid., pp. 10-13.

[19] 姜皇池，《國際海洋法總論》（台北市：新學林，2001年），147-172。

[20] 宋燕輝，〈美軍艦「杜威號」在南海執行FONOPs的政策意圖〉，中國南海研究院，檢索於2018年5月26日，http://www.nanhai.org.cn/paper_c/210.html。

[21] 姜皇池，《國際海洋法（上冊）》（台北市：新學林，2018年4月，2版），342-343。

[22] Eleanor Freund, *Freedom of Navigation in the South China Sea: A Practical Guide* (Harvard Kennedy School, Belfer Center for Science and International Affairs, 2017), 37.

[23] Permanent Court of Arbitration, *THE SOUTH CHINA SEA ARBITRATION (Arbitration between the Republic of the Philippines and the People's Republic of China), The Tribunal Renders Its Award,* (The Hague, 12 July 2016), 10.

島嶼領海基線的聲明》，宣佈釣魚島（我國稱釣魚台）及其附屬島嶼的領海基線和17個領海基點的名稱和地理坐標。在中國大陸的這兩次領海基線公布中，西沙群島與釣魚島（按：釣魚台）皆以直線基線法劃設領海基線。事實上，國際實踐除了中國大陸外，尚有丹麥、西班牙、澳大利亞、葡萄牙、聯合王國、衣索比亞、厄瓜多爾等國，係同樣以直線基線法劃設所屬洋中群島之領海基線。[24]

我們現在可以確定的是，在美國「自由航行任務」挑戰我國的海洋主張方面，確實是依據我國所頒布的國內法條文；換言之，是對我主權所賦予的立法權提出挑戰；但另一方面此舉也象徵著認同我國在南海島礁的主權主張。故維持我國現行的法律規定並加以譴責美方，或修改我國的相關法律規定使台灣不再成為美國自由航行任務挑戰的對象，確實兩難。

伍、代結論：抗議或改變

綜上，美國「自由航行任務」將台灣納為被挑戰的對象，係為彰顯法律的不歧視原則，同時認同我國為南海島礁主權聲索國之一，並非象徵台美關係產生變化。據報告所示，挑戰台灣海洋主張的原因為「過度直線領海基線」與「外國軍艦通過領海需事先報備的規定」兩項，係因美國對此兩項有關國際法方面的詮釋與我國有別，而美國挑戰的是我國現行的海洋立法，若要牽涉主權則為挑戰我國依主權所賦予的立法權。如果我國不想被美方列為挑戰對象，修改現行法律也是一種選項，同樣也是行使國家主權的表徵。

現今我國有關外國軍艦無害通過領海的規定也曾修改過，我國原先於1980年12月9日由外交部及國防部所公布的《外國軍艦請求駛入中華民國領海、港口管制辦法》第1條表明外國軍艦駛入我國領海需於10日前文達外交部請求同意，也就是所謂的「核准制」；然在1998年1月所通過的《中華民國領海及鄰接區法》第7條第3款，卻僅表明「外國軍用或公務船舶通過中華民國領海應先行告知」轉變為所謂的「報備通行制」。2002年1月30日所訂定的《外國船舶

---

[24] 姜皇池，《國際海洋法（上冊）》（台北市，新學林，2018年4月，2版），857。

無害通過中華民國領海管理辦法》中的相關條文，亦已不見「文達外交部請求同意」等字樣。[25]

況且既使是英美等大國，有時也必須配合國際情勢的轉變而更改其海洋主張。例如英國在第一屆聯合國海洋法會議時仍堅持領海寬度3浬的主張並提案表決；而美國在當時有鑑於越來越多的國家支持超過3浬的領海寬度，故改變原有3浬領海寬度主張而改提案6浬的領海寬度，惟兩國提案皆未通過。在第二屆聯合國海洋法會議時，加拿大與美國聯合提案主張6＋6浬（領海6浬加上鄰接領海外部的漁區6浬）寬度，而英國也不再提案3浬領海寬度，惟該提案差一票仍未通過；終至第三屆聯合國海洋法會議，領海寬度以至多不超過12浬成為最終決議條文。[26]

如果我國修法將《領海與鄰接區法》第7條第3款「外國軍用或公務船舶通過中華民國領海應先行告知」對外公告刪除，並調整相對應的法律如《外國船舶無害通過中華民國領海管理辦法》等，這僅代表外國軍艦或公務船舶在通過我國領海時無須先行告知，但當然仍須遵守無害通過的相關規定，且國防部依《外國船舶無害通過中華民國領海管理辦法》第14條：「外國軍用船舶通過中華民國領海時，國防部應通飭所屬，全程監控。」也仍需派遣適當兵力全程監控。而如同前述，這裡的外國軍艦規定不涉及中國大陸軍艦的相關規定與做法。但這樣的修改仍會面臨許多問題，例如有人會認為被美國列在報告中代表美方承認我國的南海島礁主權，這不是很好嗎？

再舉另外一個例子——菲律賓。我們可以發現前揭美國「自由航行任務」報告挑戰南海島礁的海洋主張時，並未將菲國納入。但之所以沒有將菲國納入，絕不會是因為美菲關係較好，所以有差別待遇，或美國不承認菲國為南海島礁的聲索國之一；而是因為菲國至今尚未立法公布相關的海洋主張，如外國

---

[25] 郁瑞麟，〈從國際法及兩岸關係面向思考我國臺灣海峽的航行制度規範〉，《中正大學法學集刊》第62期（2019年1月）：26。相關立法過程可參照：姜皇池，《海洋事務統合法制之研究》，（行政院研考會，2009年），XLVI-XLVII。

[26] 相關立法流程可參考：郁瑞麟，〈國際會議決策制定的建構觀點：以領海寬度的形成為例〉，《問題與研究》第59卷第4期（2020年12月）：1-50。

軍艦通過菲國領海需事先報備等,自然不會被美國列為挑戰的對象。但反思菲國沒有被美國「自由航行任務」納為挑戰對象,是否代表美國不承認菲國為南海島礁的聲索國?或菲國有必要藉著美國的報告,才能主張南海島礁的主權?似乎也非盡然如此。

　　或許是因為中華民國(台灣)的國際處境艱難,導致對所有涉及主權的議題皆十分敏感。既然自詡為海洋國家,我們就應對自己的海洋立法更有自信,在考量國際環境變化、國際法發展趨勢與維護國家利益的前提下,逐漸完善我國的海洋立法。本文就美國「自由航行任務」挑戰我國海洋主張的部分,提供了幾個思考方向,同時釐清了幾個疑慮,但不代表本文主張修改我國現行法律,這部分仍需透過法學先進的討論與台灣社會的溝通方能定奪。

「中國夢」與世界的交流及衝突：
以「一帶一路」為例

施正權 *

壹、問題的緣起

2012年11月29日，甫接任中共中央總書記，以及中央軍委的習近平，在參觀國家博物館之際，首次以官方身分公開提出「實現中華民族偉大復興，就是中華民族近代以來最偉大的夢想。」[1] 2013年3月17日，習近平在第十二屆「全國人民代表大會」第一次會議中當選為中共國家主席，立即在閉幕會議中表示「實現全面建成小康社會、建成富強民主文明和諧的社會主義現代化國家的奮鬥目標，實現中華民族偉大復興的中國夢，就是要實現國家富強、民族振興、人民幸福」。[2] 爾後，習近平在不同的場合陸續提及「中國夢」，[3] 以及將其與中共的國家政策連結，例如在2012年12月8日於廣州戰區考察時提出的「強軍夢」。[4]「中國夢」一詞成為習近平的治國理念，扮演著指導中共國家政策的重要角色。

為了實現「中華民族偉大復興」的「中國夢」，習近平提出「兩個一百年」目標，分別為在中國共產黨建黨百年的2021年「全面建成小康社會」，以及在中共建政百年的2049年「建成社會主義現代化強國」。[5] 因此，為達成上述的目標，習近平執政以來在對內層次上戮力進行政治、經濟、軍事與社會等

* 淡江大學國際事務與戰略研究所副教授

[1] 〈習近平：承前啟後繼往開來繼續朝著中華民族偉大復興目標奮勇前進〉，《新華網》，2012年11月29日，http://www.xinhuanet.com//politics/2012-11/29/c_113852724.htm。

[2] 〈習近平在第十二屆全國人民代表大會第一次會議上的講話〉，《人民網》，2013年3月18日，http://theory.people.com.cn/BIG5/n/2013/0318/c40531-20819774.html。

[3] 〈習近平總書記15篇講話系統闡述「中國夢」〉，《人民網》，2013年6月19日，http://theory.people.com.cn/n/2013/0619/c40531-21891787.html。

[4] 〈「摘編」習近平關於實現中華民族偉大復興的中國夢論述〉，《中國共產黨新聞網》，2013年12月5日，http://theory.people.com.cn/n/2013/0619/c40531-21891787.html。

[5] 〈習近平：承前啟後繼往開來繼續朝著中華民族偉大復興目標奮勇前進〉，《新華網》。

各領域的改革,以求增進國家整體實力的發展;而在對外的層次上,其在「中國夢」的基礎上提出「亞太夢」與「世界夢」,[6] 強調「中國夢」與區域,乃至世界的連結,[7] 以及中共的發展對於世界整體發展的貢獻,[8] 進而達到吸引其他國家支持的目的。

在前述具體實現「中國夢」的行動中,尤其以習近平在2013年下旬提出的「絲綢之路經濟帶」和「21世紀海上絲綢之路」(以下簡稱一帶一路)最受各界關注。在其規劃中,參與各方「共商、共建、共享」,並且逐步將歐亞大陸,乃至世界各國連接為一體。

弔詭的是,各界對於「中國夢」和「一帶一路」的認知呈現兩極化的發展。支持者認為,「一帶一路」不僅能解決中共內部問題與實踐「中國夢」,[9] 同時能為「世界夢」提供更多的動力,協助全世界因應各種公共問題的挑戰;[10] 質疑者則指出,前述兩者為中共透過輸出優勢的國力以擴展區域的政治、安全和文化影響力,[11] 其中近期最受矚目的莫過於有關「債務陷阱」—透過提供巨額的貸款作為建設資金,進而在東道國無力償還貸款時掌握關鍵基礎設施或藉此影響該國的外交、經濟等國家政策—的討論。[12]

面對上述弔詭現象,本文旨在探討「中國夢」在實踐的過程中,究竟能與

---

[6] 〈謀求持久發展共築亞太夢想〉,《人民網》,2014年11月10日,http://politics.people.com.cn/n/2014/1110/c1024-26000531.html。

[7] 〈中國夢與美國夢相通〉,《新華網》,2013年6月9日,http://www.xinhuanet.com//world/2013-06/09/c_124836150_2.htm。

[8] 中華人民共和國國務院新聞辦公室,〈《新時代的中國國際發展合作》白皮書(全文)〉,《中華人民共和國國務院新聞辦公室》,2021年1月10日,http://www.scio.gov.cn/zfbps/32832/document/1618203/1618203.htm。

[9] 趙周賢、劉光明,〈「一帶一路」:中國夢與世界夢的交匯橋梁〉,《人民網》,2014年12月24日,http://theory.people.com.cn/BIG5/n/2014/1224/c40531-26265185.html。

[10] 王義桅,〈「一帶一路」內熱外冷?先減少種種認知誤區〉,《中國評論新聞網》,2015年12月16日,http://hk.crntt.com/doc/1040/4/5/5/104045501.html?coluid=7&kindid=0&docid=104045501&mdate=1216100935。

[11] Satoshi Amako, "China's Diplomatic Philosophy and View of the International Order in the 21st Century," *The Journal of Contemporary China Studies*, Vol. 3, No. 2, 2014, pp. 30-31.

[12] Brahma Chellaney, "China's Debt-Trap Diplomacy," *Project Syndicate*, January 23, 2017, https://www.project-syndicate.org/commentary/china-one-belt-one-road-loans-debt-by-brahma-chellaney-2017-01?barrier=accesspaylog.

世界相通，力求共同發展，又或者將會對現有的國際制度造成衝擊。因此，首先在問題緣起的部分說明本研究的問題意識。接著簡介「一帶一路」自2013年至2020年的在全世界發展的概況。再來以「一帶一路」作為分析的切入點，探討其導致實踐的過程窒礙難行之問題。最後，本文將進行研究回顧，以及討論未來「中國夢」和「一帶一路」發展的可能性。

貳、走向世界的「中國夢」與「一帶一路」

2013年9月，習近平在拜訪哈薩克時，提出「絲綢之路經濟帶」；同年10月，在印尼國會的演講中，他再提出「21世紀海上絲綢之路」的論點，表示與各界共建的意願。2015年3月，中共國務院授權國家發展和改革委員會、外交部、商務部聯合發布《推動共建絲綢之路經濟帶和21世紀海上絲綢之路的願景與行動》白皮書，[13] 強調將透過「五通」—政策溝通、設施聯通、貿易暢通、資金融通、民心相通，以及「五路」—和平、繁榮、開放、創新、文明之路，將「中國夢」與世界共同連結。2017年10月召開的中國共產黨第十九次全國代表大會通過了《中國共產黨章程（修正案）》的決議，將「一帶一路」的推動寫入黨章之中。[14] 換句話說，「一帶一路」不僅僅是國家政策，更成為黨的核心價值。

根據「美國企業公共政策研究所」（The American Enterprise Institute）的「中國全球投資追蹤」（China Global Investment Tracker）的統計資料顯示，中共在2013年宣布推動，一直到2020年年底為止，對於各國在「一帶一路」相關項目的投資總金額超過7,600億美元。其中以2015年度的投資金額最高，約1,250億美元；2013年度的投資金額最低，約298億美元。其他年度的投資則在1,000億美元至1,200億美元之間波動，唯有2020年度受到新冠肺炎疫情影響，投資金額驟降至約465億美元，詳如圖1：

---

13 《推動共建絲綢之路經濟帶和21世紀海上絲綢之路的願景與行動》，《中國政府網》，2015年3月28日，http://www.gov.cn/xinwen/2015-03/28/content_2839723.htm。
14 〈外交部發言人談「一帶一路」建設寫入黨章：體現決心和信心〉，《人民網》，2014年11月10日，http://politics.people.com.cn/n/2014/1110/c1024-26000531.html。

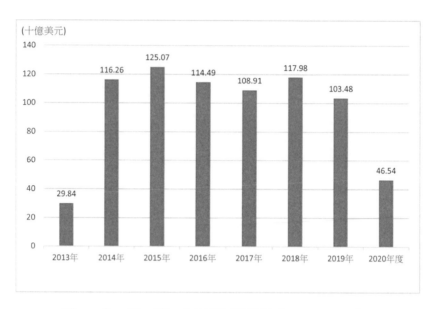

圖1：「一帶一路」年度投資額度（2013-2020）

資料來源：作者整理自American Enterprise Institute, "China Global Investment Tracker".

　　而在整體投資項目上，「一帶一路」所投入的項目並非一般認知的集中在交通運輸等基礎設施上，而是呈現多元的發展。根據「中國全球投資追蹤」的資料顯示，其中包括了農業、化學、能源、娛樂、金融、衛生、物流、金屬、房地產、科技、觀光、運輸、公用設施以及其他等14個項目。有關投資的項目、金額等詳如下表：

表1：「一帶一路」投資項目與金額（2013-2020）

單位：百萬美元

投資項目	子項目	投資項目數量	投資國家數量	投資金額	投資總金額	總投資金額百分比
農業		36	24	13,400	13,400	1.8%
化學		33	18	17,400	17,400	2.3%
能源	可替代性能源	92	43	25,790	296,700	38.9%

投資項目	子項目	投資項目數量	投資國家數量	投資金額	投資總金額	總投資金額百分比
	煤	74	23	55,440		
	天然氣	68	34	42,230		
	水電	86	36	56,920		
	石油	85	34	54,690		
	其他	70	39	61,630		
娛樂		15	12	10,150	10,150	1.3%
金融	銀行	11	7	5,410	13,330	1.8%
	投資	8	6	3,030		1.7
	其他	6	6	4,890		
衛生		13	10	3,160	3,160	0.4%
物流		23	18	18,790	18,790	2.5%
金屬	鋁	16	7	8,550	58,230	7.6%
	銅	9	5	13,900		
	鋼鐵	43	20	21,730		
	其他	25	18	14,050		
房地產	建設	179	52	55,820	74,090	9.7%
	資產	33	14	18,270		
科技	電子通信	30	23	10,620	1,5130	2%
	其他	10	8	4,510		
觀光		21	11	9,060	9,060	1.2%
運輸	陸運	191	72	83,010	187,060	24.5%
	航空	31	24	10,100		
	鐵路	88	32	63,090		
	海運	52	35	29,670		
	其他	1	1	1,190		
公用設施		64	35	23,200	22,960	3%
其他	日常消費	17	12	7,510	22,870	3%
	教育	8	5	2,560		
	工業	21	15	5,850		

投資項目	子項目	投資項目數量	投資國家數量	投資金額	投資總金額	總投資金額百分比
	紡織	9	7	2,470		
	木材	8	5	3,710		
	其他	4	4	770		

資料來源：作者整理自American Enterprise Institute, "China Global Investment Tracker".

由表1可見，在「一帶一路」的各個投資項目中，能源、金屬、房地產、以及運輸為主要的投資項目，占總投資金額比例的八成以上。其中，尤其以能源項目的投資金額比例最高，高達總投資金額的38.9%，次者為運輸項目的24.5%，再次乃房地產項目的9.7%，最後則是金屬項目的7.6%。

再者，在投資的地區與國家上，目前「一帶一路」的範圍已不限於習近平在2013年年底所提的「絲綢之路經濟帶」和「21世紀海上絲綢之路」，而是在其基礎上進一步擴展至全球各地。在2017年6月，中共國家發展和改革委員會與國家海洋局聯合發布《「一帶一路」建設海上合作設想》，計畫和沿線國家加強海上合作，共同設立「中國—印度洋—非洲—地中海」、「中國—大洋洲—南太平洋」和「中國—北冰洋—歐洲」三條藍色經濟通道。[15] 隔年1月，中共和拉美國家在智利召開的「中國-拉共體論壇」共同通過《「一帶一路」特別聲明》，同意一起推動「中拉太平洋海上絲綢之路」；[16] 同時，中共外交部也公布《中國的北極政策》白皮書，呼籲和各國合作建立「冰上絲綢之路」。[17] 根據統計，目前接受「一帶一路」投資和參與建設的國家共有114國，有關「一帶一路」在各國及各地區所投入的資金分布詳見如下：

---

[15] 〈「一帶一路」建設海上合作設想〉，《人民網》，2017年6月20日，http://politics.people.com.cn/n/2014/1110/c1024-26000531.html。

[16] 〈「一帶一路」開闢中拉合作新境界〉，《新華網》，2018年1月23日，http://www.xinhuanet.com/world/2018-01/23/c_1122303910.htm。

[17] 中華人民共和國國務院新聞辦公室，〈《中國的北極政策》白皮書〉，《中華人民共和國國務院新聞辦公室》，2018年1月26日，http://www.scio.gov.cn/zfbps/32832/document/1618203/1618203.htm。

表2：「一帶一路」投資各地區與國家之金額（2013-2020）

單位：百萬美元

投資地區	投家	投資國家金額	投資地區總金額
中東與北非	阿爾及利亞	8,390	109,690
	巴林	1,420	
	埃及	14,300	
	伊拉克	10,020	
	約旦	4,400	
	科威特	8,240	
	摩洛哥	1,260	
	阿曼	5,590	
	卡達	3,380	
	沙烏地阿拉伯	23,210	
	蘇丹	1,450	
	阿拉伯聯合大公國	27,070	
	葉門	510	
東亞	汶萊	3,970	199,360
	柬埔寨	12,010	
	斐濟	250	
	印尼	36,050	
	寮國	21,040	
	馬來西亞	30,100	
	蒙古	5,410	
	緬甸	5,700	
	紐西蘭	2,620	
	巴布亞紐幾內亞	3,040	
	菲律賓	9,830	
	薩摩亞	110	
	新加坡	37,330	

投資地區	投家	投資國家金額	投資地區總金額
	索羅門群島	200	
	南韓	10,460	
	泰國	7,830	
	東帝汶	780	
	越南	12,630	
歐洲	奧地利	230	76,990
	白俄羅斯	1,890	
	波士尼亞與赫塞哥維納	1,960	
	保加利亞	130	
	克羅埃西亞	690	
	賽普勒斯	170	
	捷克	860	
	希臘	4,500	
	匈牙利	2,400	
	以色列	10,850	
	義大利	23,200	
	拉脫維亞	110	
	盧森堡	4,680	
	馬其頓	650	
	馬爾他	440	
	摩多瓦	560	
	蒙特內哥羅	1,220	
	波蘭	2,150	
	葡萄牙	4,430	
	羅馬尼亞	810	
	塞爾維亞	9,600	
	斯洛維尼亞	2,180	

投資地區	投家	投資國家金額	投資地區總金額
	烏克蘭	3,280	
北美洲	安地卡及巴布達	1,000	8,240
	巴貝多	490	
	哥斯大黎加	470	
	古巴	240	
	牙買加	2,140	
	巴拿馬	3,090	
	千里達及托巴哥	810	
南美洲	玻利維亞	4,890	52,870
	智利	12,150	
	厄瓜多	5,160	
	蓋亞那	1,570	
	秘魯	20,960	
	委內瑞拉	8,140	
撒哈拉以南的非洲	安哥拉	9,780	153,820
	貝南	2,180	
	喀麥隆	3,990	
	維德角	130	
	查德	550	
	剛果	6,710	
	吉布地	1,020	
	赤道幾內亞	1,280	
	衣索比亞	11,010	
	加彭	300	
	迦納	7,960	
	幾內亞	10,760	
	象牙海岸	3,250	
	肯亞	8,400	

投資地區	投家	投資國家金額	投資地區總金額
	賴索托	100	
	賴比瑞亞	410	
	馬達加斯加	1,790	
	馬利	1,690	
	莫三比克	3,050	
	納米比亞	3,250	
	尼日	3,040	
	奈及利亞	26,480	
	盧安達	1,050	
	塞內加爾	4,180	
	獅子山	1,940	
	南非	3,910	
	南蘇丹	2,790	
	坦尚尼亞	10,210	
	烏干達	2,310	
	尚比亞	10,370	
	辛巴威	8,760	
西亞	阿富汗	210	161,600
	亞塞拜然	1,440	
	孟加拉	24,960	
	喬治亞	1,020	
	伊朗	11,880	
	哈薩克	11,930	
	吉爾吉斯	4,340	
	馬爾地夫	1,230	
	尼泊爾	2,100	
	巴基斯坦	49,120	

投資地區	投家	投資國家金額	投資地區總金額
	俄羅斯	35,060	
	斯里蘭卡	8,840	
	塔吉克	710	
	土耳其	5,960	
	土庫曼	600	
	烏茲別克	2,200	

資料來源：作者整理自American Enterprise Institute, "China Global Investment Tracker".

　　由上表可見，首先，在投資的地區上，東亞所獲得的投資金額最高，將近2,000億美元，占整體投資金額的26%；其次為西亞獲得約1,600億美元，占整體比的21%；再次者是撒哈拉以南的非洲地區，獲得約1,500億美元的投資金額，占整體比的20%；接著是中東與北非地區，獲得將近1100億美元的投資，占整體比例的14%。而歐洲、南美洲與北美洲三地區所獲得的投資相對較少，皆低於1,000億美元以下，分別獲得約769億美元、524億美元和82億美元的投資。

　　首先，上述的資料顯示出東亞、西亞與撒哈拉以南的非洲三個地區為「一帶一路」投資的重要區域，中共在這三個地區所投入的資金便接近整體投資比例的七成之高。相形之下，歐洲、南美洲與北美洲的吸引的投資總和僅約1,380億美元，甚至不及撒哈拉以南的非洲地區所獲得的投資資金。

　　其次，在各個國家所獲得的投資金額上，前十名國家分別為巴基斯坦、印尼、俄羅斯、馬來西亞、阿拉伯聯合大公國、孟加拉、沙烏地阿拉伯、義大利、秘魯與寮國，占投資總金額將近四成的比例。而在這十個國家中，尤其以東亞與西亞所占的國家最多（兩者各三個國家），所占的金額也最為龐大（前者約870億美元；後者則約1,090億美元），證明了中共的「一帶一路」對此二區域的高度重視。

　　事實上，北京自2013年開始推動「一帶一路」以來，便承襲過去「魅力

91

攻勢」（Charm Offensive）的模式，18 將目標聚焦在消除外界的質疑，並且爭取國際對「一帶一路」更多的支持。因此，中共積極地在國內外的場合中持續傳達「一帶一路」將為全球提供公共財，共同參與其中建設的成員都將成為贏家。19 同時將為充滿不確定、不公正與不合理的西方模式，以及秩序失靈之際，為全球治理作出貢獻，以及解決世界發展所遇到的難題。20

　　換句話說，北京認為「一帶一路」的推行並非挑戰、推翻現有的制度，而是補充其中不足之處。以國際經濟領域為例，論者主張「一帶一路」的推動有助於因應目前國際失序的狀況，21 並且完善與補充國際金融秩序，促進體系的變革。22 特別是「一帶一路」具有包容性的特點將有助於全球化持續的深化與發展，23 並且可以透過「擴散型互惠」補充現有制度的缺陷，24 此外，更能協助非民主與非自由市場經濟體制的國家融入全球經濟體系，分享國際合作的成果。25 最後，由於「一帶一路」所提供的沿線地區公共財之合作網絡，各方得以互相協調與合作，26 解決以往邊陲國家在國際金融與貿易關係中遭到核心國家支配與剝削的問題，成為全球經濟合作的新形態。27

---

18 有關中共如何透過「魅力攻勢」─以文化、經濟以及參與多邊合作與區域組織的方式─消除他國對中共國力快速增長而產生的疑慮，以及爭取支持的行動，詳請參閱Joshua Kurlantzick, *Charm offensive : How China's soft power is transforming the world* (New Haven, Conn.: Yale University, 2007).

19 〈習近平在「一帶一路」國際合作高峰論壇開幕式上的演講（全文）〉，《「一帶一路」國際合作高峰論壇》，2017年5月14日，http://www.beltandroadforum.org/BIG5/n100/2017/0514/c24-407. html。〈參與一帶一路建設的都是贏家〉，2017年5月20日，http://ydyl.people.com.cn/ n1/2017/0520/c411837-29288267.html。〈習近平出席第二屆「一帶一路」國際合作高峰論壇開幕式並發表主旨演講〉，《新華網》，2019年4月26日，http://www.xinhuanet.com/politics/leaders/2019-04/26/ c_1124420373.htm。

20 〈習近平：推動全球治理體制更加公正更加合理〉，《新華網》，2015年10月13日，http://cpc. people.com.cn/n/2015/1013/c64094-27693518.html。〈一帶一路 改變世界的兩條弧線〉，2017年5月11日，《新華網》，http://www.xinhuanet.com//world/2017-05/11/c_1120957146.htm。

21 田文林，〈「一帶一路」：全球發展的中國構想〉，《現代國際關係》，第5期，2017年，頁42-49。

22 靳諾等著，《大國擔當：全球治理的中國主張》（香港：開明出版社，2018年），頁181-182。

23 劉衛東，〈「一帶一路」：引領包容性全球化〉，《中國科學院刊》，第4期，2017年，頁331-339。

24 孫伊然，〈亞投行、「一帶一路」與中國的國際秩序觀〉，《外交評論》，第1期，2016年，頁1-30。

25 徐晏桌，〈2017年「一帶一路」倡議：成果與進展〉，張宇燕編，《全球政治與安全報告2（2018）》（北京：社會科學文獻出版社，2018年），頁178。

26 黃河，〈公共產品視角下的「一帶一路」〉，《世界經濟與政治》，第6期，2015年，頁138-155。

27 曾向紅，〈「一帶一路」的地緣政治想像與合作〉，《世界經濟與政治》，第1期，2016年，頁46-71。

而根據國際貨幣基金組織（International Monetary Fund）對於已開發經濟體（Advanced Economies）、新興市場經濟體（Emerging Market Economies）和發展中經濟體（Developing Economies）的分類，[28] 前述接受「一帶一路」投資的114個國家之中，除了以色列、西班牙、希臘、拉脫維亞、南韓、紐西蘭、馬爾他、捷克、斯洛維尼亞、奧地利、新加坡、義大利、葡萄牙、盧森堡、賽普勒斯等15個國家為已開發經濟體外，其餘的99個國家皆屬於新興市場經濟體和發展中經濟體。相較之下，後者所吸引的投資金額將近「一帶一路」整體投資資金的九成，因此，論者主張在推動「一帶一路」的過程中，中共確實從中獲得更大的利益；然而，在整體的政策作為上，中共也積極的創造平等互惠的雙贏局面，表現出「南南合作」的精神。此外，目前也並無直接的證據顯示中共企圖主導東道國政治與經濟。[29] 而在「南南合作」的基礎上，「一帶一路」亦有望補充全球在和平、發展與治理上的不足，並且推動一個更加平衡與有效的全球經濟治理體系。[30]

參、「一帶一路」與世界的衝突

自提出「中國夢」與「一帶一路」以來，北京一直主張該戰略對全世界的貢獻；然而，誠如前述，各界的質疑也從未消止。「中國夢」以及擔任實踐的主要推手的「一帶一路」似乎呈現與世界格格不入的樣貌。事實上，「一帶一路」在過去八年的投資也存在不少陷入困境而中止的爭議性投資項目。自2013年至2020年年底為止，根據統計，共有涉及47個國家共80筆的投資遭到拒絕、暫時或永久中止，總投資金額為817.7億美元，約占整體投資的十分之一，有關內容詳見表3所示：

---

[28] International Monetary Fund, "*World Economic Outlook: Managing Divergent Recoveries*," March 23, 2021, pp. 107-108. https://www.imf.org/en/Publications/WEO/Issues/2021/03/23/world-economic-outlook-april-2021. 有關IMF對於三者的界定標準，詳請參閱International Monetary Fund, "Frequently Asked Questions World Economic Outlook (WEO)," April 6, 2021, https://www.imf.org/external/pubs/ft/weo/faq.htm#q4b.

[29] 鄧中堅，〈中國對拉丁美洲的資源外交：新殖民主義與南南合作之爭辯〉，《遠景基金會季刊》，第16卷第3期，2015年7月，頁131-180。

[30] Zhou Taidong and Zhang Haibing, "China's Belt and Road Initiative: An Opportunity to Re-energize South-South Cooperation," *China Quarterly of International Strategic Studies*, Vol. 4, No. 4, 2018, pp. 559-576.

表3：「一帶一路」具爭議性的投資項目與國家（2013-2020）

單位：百萬美元

投資地區	投資國家	爭議項目	爭議項目數目	爭議項目金額	國家爭議項目總金額	地區的爭議項目總金額
中東與北非	伊拉克	能源	2	290	1,470	2,460
	沙烏地阿拉伯	農業	1	120	310	
		公用設施	1	190		
	蘇丹	運輸	1	680	680	
東亞	柬埔寨	房地產	1	290	290	16,560
	印尼	能源	2	960	2,460	
		運輸	1	1,500		
	馬來西亞	房地產	1	680	7,350	
		能源	2	4,760		
		運輸	1	1,910		
	蒙古	能源	1	250	250	
	緬甸	運輸	1	760	910	
		房地產	1	150		
	紐西蘭	金融	1	460	460	
	巴布亞紐幾內亞	金屬	1	1,800	1,800	
	南韓	衛生	1	200	200	
	泰國	運輸	1	300	300	
	越南	金屬	1	890	1,160	
		能源	1	270		
歐洲	捷克	金融	1	910	910	11,860
	希臘	旅遊	1	1,200	1,200	
	以色列	金融	5	2660	7,350	
		科技	1	290		
		娛樂	1	4,400		

投資地區	投資國家	爭議項目	爭議項目數目	爭議項目金額	國家爭議項目總金額	地區的爭議項目總金額
	義大利	娛樂	1	280	870	
		房地產	1	590		
	波蘭	能源	1	930	930	
	羅馬尼亞	能源	1	350	350	
	烏克蘭	運輸	1	250	250	
北美洲	千里達及托巴哥	物流	1	490	490	8,240
南美洲	玻利維亞	運輸	1	250	250	9,330
	智利	金屬	1	310	310	
	秘魯	金屬	1	530	530	
	烏拉圭	運輸	1	250	250	
	委內瑞拉	運輸	1	7,500	7,500	
撒哈拉以南的非洲	查德	能源	1	400	400	14,390
	剛果	運輸	1	680	810	
		化學	1	130		
	衣索比亞	運輸	1	990	3,250	
		能源	1	1,460		
		科技	1	800		
	加彭	能源	1	400	400	
	迦納	能源	1	1,000	1,000	
	肯亞	運輸	3	1,460	2,800	
		能源	2	1,340		
	尼日	能源	1	220	220	
	奈及利亞	運輸	1	1,850	1,850	
	獅子山	金屬	1	990	990	
	南蘇丹	能源	1	260	260	
	尚比亞	其他	1	210	210	

投資地區	投資國家	爭議項目	爭議項目數目	爭議項目金額	國家爭議項目總金額	地區的爭議項目總金額
西亞	辛巴威	能源	3	2,200	2,200	27,170
	孟加拉	運輸	3	3430	3430	
	喬治亞	能源	1	630	630	
	伊朗	能源	2	3,250	3,250	
	哈薩克	化學	1	1,850	2,200	
		交通	1	350		
	巴基斯坦	能源	2	4,100	4,100	
	俄羅斯	能源	3	10,880	11,770	
		金屬	1	890		
	斯里蘭卡	運輸	1	1,430	1,430	
	土耳其	能源	1	380	380	

資料來源：作者整理自American Enterprise Institute, "China Global Investment Tracker".

　　而產生這種弔詭發展的主要原因，一方面在於「中國夢」與「一帶一路」在走向世界的過程中，中共與其他國家，尤其是美國、俄羅斯與日本等大國的交流產生了利益上的衝突，以及權力的競爭。[31] 換句話說，在中共的國力快速增加，並且在國際場域投射相應的影響力與權力的同時，既有的霸權將中共的行動視為對其地位，以及建立的規則與秩序的挑戰。另一方面，則是中共在推動「一帶一路」過程中，由於自身問題所導致的結果，例如對於東道國政治與投資環境的認知不足等。以下分別從政治、經濟、文化，以及地緣戰略的角度進行分析。

一、政治

　　2013年6月，習近平在訪美之際提出「新型大國關係」的概念，象徵著中

---

[31] 鄒磊著，《中國「一帶一路」戰略的政治經濟學》（上海：上海人民出版社，2015年），頁278-269。

共戰略思考的轉換，從「韜光養晦」到「有所作為」；而「一帶一路」的推行則是此思維指導下的具體實現手段。[32] 不過，這並不意味著中共意圖與美國等既有強權進行全方位的競爭，雙方在經濟、安全、政治與文化等領域仍然存在著合作的空間。[33]

在過去約二十年，美國深陷反恐戰爭的泥淖，以及歷經2008年金融危機的打擊，由第二次世界大戰後高舉自由與多邊主義重建國際秩序的領導者，轉而逐漸擁抱保守與孤立主義。雖然歐巴馬在執政期間曾企圖改變此趨勢，但是在川普上任之後，強調「美國優先」，只考量自身國家利益的思維達到新一波的高峰。

誠如包道格（Douglas H. Paal）所言：「中國多年來對於承擔全球責任的呼聲一直採取迴避態度，但隨著美國放棄領導全球化的角色」，中共「接過了美國的衣缽」，「扮演世界領導者的角色」。[34] 在這段美國缺席參與國際事務的空檔中，中共以充沛的經濟實力作為後盾，在既有的國際規則下，補充了權力的真空，成為崛起的平行競爭者。

因此，隨著中共在國際與區域的影響力增加，尤其是靈活運用經濟手段對目標國進行強制、說服，或是吸引目標國，使其採取中共偏好的政策，[35] 美國對中共的認知開始轉變，將其視為強而有力的新興競爭者，並採取了針對性的動作，例如歐巴馬（Barack Obama）時期提出的「亞太再平衡」（Asia-Pacific Rebalance），以及川普（Donald John Trump）時期提出的「印太戰略」（Indo-Asia-Pacific Strategy）。而「中國夢」與「一帶一路」的推動，毫無疑問便成為美國對中共認知改變之後的打擊的首要目標。

---

[32] 趙可金，〈「一帶一路」戰略：外交「哥白尼式革命」〉，薛力主編，《「一帶一路」：中外學者的剖析》（北京：中國社會科學出版社，2017年），頁70。

[33] 薛力，〈美國再平衡戰略與中國「一帶一路」〉，《中國社會科學網》，2016年6月20日，http://www.cssn.cn/zzx/gjzzx_zzx/201606/t20160620_3076796.shtml。

[34] 包道格，〈2017，習主席將進入全盛之年〉，《清華-卡內基全球政策中心》，2017年3月9日，http://carnegietsinghua.org/2017/03/09/zh-pub-68512。

[35] 有關中共如何運用經濟手段對他國進行強制、說服，或是吸引目標國，使其採取中共偏好的政策之行動，詳請參閱Robert D. Blackwill & Jennifer M. Harris, *War by Other Means: Geoeconomics and Statecraft* (Massachusetts: Harvard University, 2016). William J. Norris, *Chinese Economic Statecraft: Commercial Actors, Grand Strategy, and State Control* (London: Cornell University, 2016).

再者，歐盟（European Union）也開始擔憂接受「一帶一路」投資的歐洲國家受到中共的影響，在政治、人權等領域採取有利於中共的政策，因而造成內部分裂。前者如2016年7月，匈牙利與希臘對於歐盟針對南海仲裁案發表的聲明中批評中共的部分表示反對的意見；[36] 後者如2017年3月，匈牙利拒絕簽署聯合聲明，譴責中共對拘留律師採取侵犯人權的行為；同年6月，希臘阻止歐盟在聯合國人權理事會上發表批評中共人權問題的聲明。[37]

同時，如斯里蘭卡在2015年因無法償還債務，因而將港口與周邊土地租借給中共99年的例子，[38] 也引起西方與亞洲國家對「一帶一路」將造成國家主權的侵蝕之疑慮。新美國安全中心（Center for a New American Security）近期的報告便表示，「一帶一路」在斯里蘭卡、厄瓜多、阿根廷、辛巴威、以色列、巴基斯坦、緬甸等國家的投資項目中，存在著長期的不公平、長期租約或營運時間的現象，可能引發長期的政治影響力與依賴，或者在未來債務無法履行時，不得不讓渡部分的主權以抵銷積欠的債務。[39]

最後，「一帶一路」向西推進使得俄、中兩強在歐亞大陸的中亞地區—此區域被俄羅斯視為前蘇聯留下的傳統勢力範圍—互相碰撞。即使俄羅斯在衡量國內經濟持續衰退，以及與西方國家關係惡化下選擇加入「一帶一路」，在經濟領域上由平等關係逐步轉向扈從中共建立的體制，[40] 仍然積極運用其主導的歐亞經濟聯盟（Eurasian Economic Union）等區域組織對「一帶一路」倡議

---

[36] Robin Emmott, "EU's statement on South China Sea reflects divisions," Reuters, July 15, 2016, https://www.reuters.com/article/southchinasea-ruling-eu-idUSL8N1A130Y.

[37] Simon Denyer, "Europe divided, China gratified as Greece blocks E.U. statement over human rights," *The Washington Post*, June 19, 2017, https://www.washingtonpost.com/news/worldviews/wp/2017/06/19/europe-divided-china-gratified-as-greece-blocks-e-u-statement-over-human-rights/.

[38] Maria Abi-Habib, "How China Got Sri Lanka to Cough Up a Port?," *The New York Times*, June 26, 2018, https://www.nytimes.com/2018/06/25/world/asia/china-sri-lanka-port.html.

[39] Daniel Kliman et al, "Grading China's Belt and Road," Center for a New American Security, April, 2019, https://s3.amazonaws.com/files.cnas.org/CNAS+Report_China+Belt+and+Road_final.pdf. 對於「一帶一路」導致「債務陷阱」的論點，部分學者提出反對意見，主張「一帶一路」雖然在個別國家造成不同影響，不過，並未有確切證據顯示中共運用束到道國的債務問題獲取國家利益。有關討論詳請參閱Deborah Brautigam, "A critical look at Chinese 'debt-trap diplomacy': the rise of a meme," Area Development and Policy, Vol. 5 No. 1(December 2019), pp. 1-14. 薛健吾在〈中國「一帶一路」在第一個五年的進展與影響（2013-2018）〉，《遠景基金會季刊》，第21卷，第2期，2020年4月，頁1-54。

[40] 連弘宜，〈一帶一路框架下中俄之潛在競合關係〉，《歐亞研究》，第3期，2018年4月，頁34-39。

進行「軟制衡」（Soft Balancing），掌握政治話語權，以及爭取其他國家的支持。[41] 換言之，在現實的考量下，莫斯科與北京在「一帶一路」的合作乃權宜之計，雙方之間依然存在多重的矛盾，並不熱見中共的勢力深入該地區，威脅其國家利益。

二、經濟

「一帶一路」另一個飽受批評之處，在於運作與管理模式所帶來的經濟風險，以及造成的負面影響。一般而言，大型投資項目普遍具有成本超支、工期延宕等共同的風險。[42] 「一帶一路」所投資的東道國通常具有經濟基礎薄弱的特質，或是項目本身涉及政治因素，不具財政可持續性，使得風險進一步上升。此外，雖然強調「共商、共建、共享」的原則，但是，「一帶一路」從投資到營運卻存在著獨厚中共的供應商與承包商，[43] 中共的企業透明度普遍低於其他國家企業，[44] 以及可能導致前述的「債務陷阱」的現象。

就經濟風險而論，全球發展中心（Center for Global Development）的報告指出，吉布地、吉爾吉斯、寮國、馬爾地夫、蒙古、蒙特內哥羅、巴基斯坦與塔吉克八個國家從「一帶一路」的投資獲得大量建設基礎設施的資金；同時，八國積欠中共巨額的外債，債務危機的風險也顯著增加，如吉布地對中的負債額度已經高達國內生產毛額的九成以上。[45] 截至2020年中旬，中共在發展中國

[41] 有關俄羅斯如何看待「一帶一路」，以及進行「軟制衡」的過程，詳請參閱王家豪、羅金義著，《絲綢之路經濟帶，歐亞融合與俄羅斯復興》（台北：新銳文創出版社，2021年）。

[42] Bent Flyvbjerg, "What You Should Know about Megaprojects and Why: An Overview." *Project Management Journal*, Vol. 45, Issue 2, 2014, pp. 6-19. Bent Flyvbjerg, *The Oxford Handbook of Megaproject Managementz* (Oxford, UK: Oxford University, 2017), pp. 1-18.

[43] Richard Fontaine and Daniel M. Kliman, "On China's New Silk Road, Democracy Pays A Toll," *Foreign Policy*, May 16, 2018, https://foreignpolicy.com/2018/05/16/on-chinas-new-silk-road-democracy-pays-a-toll/. Christopher Balding, "Why Democracies Are Turning Against Belt and Road," Foreign Affairs, October 24, 2018, https://www.foreignaffairs.com/articles/china/2018-10-24/why-democracies-are-turning-against-belt-and-road.

[44] 有關內容，詳請參閱Barbara Kowalczyk-Hoyer et al, *Transparency in Corporate Reporting: Assessing Emerging Market Multinationals (2016)*, Transparency International, 2016.

[45] John Hurley, Scott Morris and Gailyn Portelance, *Examining the Debt Implications of the Belt and Road Initiative from a Policy Perspective*, Center for Global Development, 2018, https://www.cgdev.org/sites/default/files/examining-debt-implications-belt-and-road-initiative-policy-perspective.pdf.

家與新興國家所持有的債務總額已超越了「巴黎俱樂部」（Paris Club），其向海外輸出的資金大部分投資在「一帶一路」項目之上。這也使得積欠債務的國家提升了對中共的經濟與政治依賴。[46]

　　而造成此現象的原因，除了中共企業起步較晚，只能到投資風險高的國家發展，以及對於國際市場與東道國不熟悉，缺乏投資經驗及相關人才，導致風險評估能力不足，使得「一帶一路」投資屢發爭議之外，[47] 更重要的決定因素是「一帶一路」的投資往往蘊含濃厚的政治考量，而非基於經濟上的可行性。以在委內瑞拉的投資為例，中共為了擴展在拉丁美洲的政治影響力，儘管該國政治局勢混亂、財政紀錄不良，[48] 以及缺乏對項目完整的評估，依然在該國的基礎設施、能源等領域投入約650億美元的資金。同時，中共在當地的建設中，出現了迴避東道國法規、拒絕雇傭當地公司與勞工、環境破壞，以及涉及人權侵害等相關問題。[49]

　　次就「一帶一路」缺乏公平、開放以及透明度而論，2018年4月，27位歐盟（European Union）駐北京的大使即共同編寫了一份報告，批評「一帶一路」的運作機制阻礙自由貿易，中企在投資案中因此處於優勢地位。[50] 根據研究指出，相較於世界銀行（World Bank）等多邊開發銀行所主導的開發案，承包商比例相對平均的狀況（總部設在東道國當地公司占40.8%，中企占29%，以及不屬於前兩者的外國公司占30.2%），「一帶一路」的承包商則以中企為

[46] Matthias Diermeier et al, "Der chinesische Albtraum: Verschuldungsrisiken auf der Seidenstraße," *Institut der deutschen Wirtschaft*, Juli 26, 2020, https://www.iwkoeln.de/fileadmin/user_upload/Studien/Kurzberichte/PDF/2020/IW-Kurzbericht_2020_Der_chinesische_Albtraum.pdf.

[47] 王衛星，〈全球視野下的「一帶一路」：風險與挑戰〉，人民論壇編，《「一帶一路」：面相21世紀的偉大構想》（北京：人民出版社，2015年）頁63。

[48] Kinling Lo, "From oil to infrastructure, why China has plenty to lose from political turmoil in Venezuela," *South China Morning Post*, February 9, 2019, https://www.scmp.com/news/china/diplomacy/article/2185467/oil-infrastructure-why-china-has-plenty-lose-political-turmoil.

[49] María Antonieta Segovia, "The Chinese train derailed on Venezuela's plains," *Diálogo Chino,* March 3, 2021, https://dialogochino.net/en/infrastructure/40823-the-chinese-train-derailed-on-venezuelas-plains/.

[50] Dana Heide et al, "EU ambassadors band together against Silk Road," *Handelsblatt*, April 17, 2018, https://www.handelsblatt.com/english/politics/china-first-eu-ambassadors-band-together-against-silk-road/23581860.html?ticket=ST-1123772-cGeeLMba9w0BKKXFf2Ya-ap4.

主，占其中的89%，當地公司與外國公司僅各占7.6%和3.4%。[51]

「一帶一路」以中企承攬絕大部分的投資案的主要原因在於中共的「國家開發銀行」和「中國進出口銀行」為「一帶一路」的項目提供了大部分的資金，因此，投資案必須受國家發展與改革委員會、商務等等政府機構批准。在此機制上，北京得以在採購案中持續發揮影響力，或者在招標的過程中，將中企納入合約之中，亦或者規定貸款的國家必須選擇中企作為合約的承包商。另外，雖然在審查的內容中包含了如環境保護的項目，但是，國家開發銀行和中國進出口銀行卻並未公開相關的資料，以及提供明確的標準，[52] 這也使得其他國家的承包商在投標時處於資訊不對等的劣勢。

因此，由上述研究可見，由於「一帶一路」從投資到營運的過程中，基本上以中共的銀行、企業與勞工為主，不僅使得其他的國家與外企難以參與其中，同時在國際上形塑了一個由中共政府主導與支配的制度，增強了中共影響他國國內和國外政策的能力。其次，由於東道國藉著大規模建設帶動國家經濟發展的機會減少，以及阻礙了提升管理、技術等領域的效益。因此，東道國回收資本的速度減慢，或是根本無法從中獲利，陷入更大的債務之中。東道國在無法償還債務時，便面臨著被要求以石油、金屬礦產等自然資源，亦或者將建設項目轉讓給中共的可能局面，自然落入前述的「債務陷阱」中。

三、文化

面對外界對於「中國夢」與「一帶一路」的質疑，中共從上至下的回應出現了轉變，表現了強勢的一面。而這樣的改變也引發了各界的疑惑，究竟中共的崛起將會成長為傲慢的大國 還是自信的大國？重要的是，不同的認知結果，亦將導致其他行為者採取不同的行動，進而對「中國夢」的實現，產生不同的

---

[51] Jonathan E. Hillman, "China's Belt and Road Initiative: Five Years Later," CSIS, January 25, 2018, https://www.csis.org/analysis/chinas-belt-and-road-initiative-five-years-later-0. http://www.cscap.org/uploads/docs/CRSO/CSCAP2018WEB.pdf.

[52] Alison Hoare et al, "The Role of Investors in Promoting Sustainable Infrastructure Under the Belt and Road Initiative," *Chatham House*, May 11, 2018, https://www.chathamhouse.org/2018/05/role-investors-promoting-sustainable-infrastructure-under-belt-and-road-initiative-0/4.

效果。不幸的是，中共外交體系目前採取「戰狼外交」的作法，毫無疑問使其他國家對中共「包容、和平、合作與共贏」的論調失去信心，損害其長久經營的國際形象，反倒不利於「中國夢」在未來的實現。

　　隨著中共在2000年以來快速的經濟成長、綜合國力迅速提升，以及積極參與國際事務，「中國威脅論」的論調也隨之而來，並且從未消失。因此，中共為了降低外界的質疑，以及建立有利於中共整體發展的國際環境，第四代領導執政以來，提出「和平崛起」與「和諧世界」的理念作為指導對外政策的基礎。[53] 而習近平上台以後，雖然外交行動更為積極與強勢，但主要的原因還是在於應對國際環境改變，例如美國在全球領導力的下降，所作出的決策。最典型的例子即本文所探討的「中國夢」與「一帶一路」自提出以來，中共各界將其定調在包容、合作與共贏的論調之中。

　　不過，在2018年中美發生貿易戰以來，中共的外交作風漸趨強硬，並且在2020年新冠疫情爆發之後達到高峰。其中，最受關注的當屬以中共外交部發言人趙立堅為代表的外交官一系列對外攻擊性強烈的「戰狼外交」。吳心伯認為，這主要是中共對西方曲解與汙名化的反擊，以及回應過內民意的結果。[54] 而前美國駐北京大使芮效儉（J. Stapleton Roy）則認為，這些外交官不符外交慣例的作法，最主要的目的是為了討好北京，贏取中央的信任。[55]

　　在接受「一帶一路」投資的國家中，澳洲無疑是受到這股「戰狼外交」針對的代表性例子。[56] 自澳洲在2016年7月表示支持南海仲裁，以及之後擋下多起中共投資案、指控中共對澳洲滲透等事件以來，中澳關係便持續惡化至今。去年四月澳洲呼籲國際社會調查新冠疫情源頭，以及對於香港、南海與新

[53] 趙建民、許志嘉，〈中共第四代領導集體的「和諧世界觀」：理論與意涵〉，《遠景基金會季刊》，第10卷，第1期，2009年1月，頁1-44。

[54] 斯影，〈疫情之下大行其道的中國「戰狼」外交會不會成為新常態〉，《BBC中文網》，2020年5月13日，https://www.bbc.com/zhongwen/trad/chinese-news-52632979。

[55] 莉雅，〈中國駐外使節為何成為「戰狼」大使？(1)：上有所好下必甚焉〉，《美國之音》，2020年5月8日，https://www.voachinese.com/a/chinese-wolf-warriors-ambassadors-part-i-20200507/5410371.html。

[56] 斯影，〈中國「戰狼外交」再出擊，澳大利亞為何成為針對目標〉，《BBC中文網》，2020年12月3日，https://www.bbc.com/zhongwen/trad/world-55162589。

疆議題提出質疑後，中共便加強針對澳洲的外交動作，例如禁止澳洲向中共出口商品，或是對其課徵反傾銷和反補貼稅。[57] 今年4月，澳洲引用去年年底國會通過的「澳洲對外關係（州和領地協議）議案2020」（Australia's Foreign Relations〔State and Territory Arrangements〕Bill 2020），[58] 取消維多利亞州政府加入「一帶一路」的決定。[59] 因此，在中共駐澳使館表達了「強烈不滿和堅決反對」，[60] 以及駐澳公使表示「對協同合作秉持開放態度」，同時「會非常堅定地捍衛我們的國家利益」的言論之後，[61] 中共國家發展改革委員會在5月6日隨即發布無限期暫停中澳戰略經濟對話機制下一切活動的聲明。[62]

　　對於中澳關係持續惡化的過程中，中共外交官表現出強硬態度的現象，直接導致了澳洲反中的輿論。《德國之聲》的報導指出，這種一方面形塑自由貿易的正面形象，另一方面對挑戰其國家利益的國家實施懲罰的作法，實質上是在損害中共的國際形象。[63] 此外，由於近年來「一帶一路」引發的負面效應，以及對於中共外交動作的不滿，澳洲取消參與「一帶一路」的決策也可能將影響其他國家採取延緩，甚至是退出「一帶一路」的行動。[64]

---

[57] 毛遠揚〈澳大利亞與中國僵局：從「十四條」詔書到葡萄酒貿易戰〉，《BBC中文網》，2020年12月7日，https://www.bbc.com/zhongwen/trad/world-55187261。黃嬿，〈中澳貿易戰沒打到痛點，澳洲經濟衝擊有限〉，《科技新報》，2020年12月28日，https://technews.tw/2020/12/28/the-trade-war-of-china-and-australia/。

[58] 有關法案內容，詳請參閱 "Australia's Foreign Relations〔State and Territory Arrangements〕Bill 2020," *Federal Register of Legislation*, https://www.legislation.gov.au/Details/C2020B00125.

[59] 丘德真，〈澳中舌戰再起 澳洲堅持價值拒絕讓渡主權予中國〉，《中央社》，2021年4月28日，https://www.cna.com.tw/news/firstnews/202104230035.aspx。

[60] 〈駐澳大利亞使館發言人表態〉，《中華人民共和國駐澳大利亞聯邦大使館》，2021年4月21日，http://au.china-embassy.org/chn/sghdxwfb/t1870483.htm。

[61] 〈中國駐澳大利亞公使：中國不是奶牛，任何人都不該抱有「先擠奶後宰割」的幻想〉，《環球網》，2021年4月22日，https://www.sohu.com/a/462274943_162522。

[62] 〈國家發展改革委關於無限期暫停中澳戰略經濟對話機制下一切活動的聲明〉，《中華人民共和國國家發展改革委員會》，2021年5月6日，https://www.ndrc.gov.cn/xwdt/xwfb/202105/t20210506_1279171.html。

[63] Clifford Coonan，〈貿易領域的「戰狼」：中國露出獠牙〉，《德國之聲》，2020年11月28日，https://www.cna.com.tw/news/firstnews/202104230035.aspx。

[64] Thomas Kohlmann，〈中國的 "新絲綢之路" 走不通了？〉，《德國之聲》，2021年5月3日，https://www.dw.com/zh/%E4%B8%AD%E5%9B%BD%E7%9A%84%E6%96%B0%E4%B8%9D%E7%BB%B8%E4%B9%8B%E8%B7%AF%E8%B5%B0%E4%B8%8D%E9%80%9A%E4%BA%86/a-57416322。

四、地緣戰略

毫無疑問，受到地理與歷史的影響，中共為滿足外交政策的三大目標，恢復與維護領土的完整、在鄰國之間擴展影響力，防止外來力量對亞洲的控制，以及為經濟發展創造有利的國際環境，[65] 確實需要一個統籌全局的戰略計畫以完成上述的目標。而「一帶一路」的提出，正是達成這些目標的最佳工具。而從戰略的角度而言，中共推動「一帶一路」除了保障的國家發展、確保能源安全，以及強化周邊外交的成果以外，最重要的目的是為了回應美國在歐巴馬時期提出的「亞太再平衡」。[66]

事實上，「一帶一路」朝向西方與南方發展，不僅是避免與美國及其友好的盟邦，同時也是基於國家發展與安全考量下的不得不做出的選擇。隨著經濟的發展，中共的經濟將越發依賴海上交通線的安全，因此，必須向海洋進行發展。而中共藉由「一帶一路」的投資，將為中共逐漸增長的海上行動，以及海上武力的投射獲得必要的支持，例如獲得孟加拉的吉大港、緬甸的實兌港和可可島，斯里蘭卡的漢班托塔港，巴基斯坦的瓜達爾港，以及坦桑尼亞的巴加莫約港使得中共海軍在印度洋的行動獲得補給，以及修整。

此外，中共在海外的基礎建設，可能獲得了近似美國在冷戰期間於世界各地投資與建設基礎設施所帶來的情報優勢。中共透過「一帶一路」的投資，除了交通運輸之外，尚涉及了大量電信網路的合約，許多發展中國家都擁有由華為與中興通訊所建設的網路設施。因此，中共可能得以藉由這些基礎設施，進行情報的收集之用。不過，目前尚無確切的證據指出中共在「一帶一路」的投資項目中，確實運用相關的技術提供情報。

---

[65] 黎安友（Andrew J. Nathan）、施道安（Andrew Scobell）著，何大明譯，《沒有安全感的強國：從鎖國、開放到崛起，中國對外關係70年》（*China's search for Security*）（台北：左岸文化，2018年），頁74-75。

[66] Nadège Rolland, *China's Eurasian Century?: Political and Strategic Implications of the Belt and Road Initiative* (Washington, D.C.: National Bureau of Asian Research, 2017), pp.109-119.

肆、代結語：「中國夢」與「一帶一路」的可能性未來

從客觀形勢而論，「一帶一路」的推動並非暢通無阻；次就主觀認知而言，則是褒貶不一。誠然，「一帶一路」提供了部分國家—特別是不受國際投資青睞下的非民主與非自由市場經濟體制者—發展的機遇，但是，後續引發的衝擊卻往往抵銷了效益。

其次，本文透過政治、經濟、文化與地緣戰略四個面向的分析，認為「中國夢」與「一帶一路」在過去八年實踐的過程中，面臨重重困難，未能成為世界的夢，以及同世界相融。主要的原因在於中共的行動與主要大國產生了利益上的衝突和權力的競爭，以及中共參與國際事務時間尚短，整體條件如經驗及相關人才尚嫌不足。不過，「中國夢」與「一帶一路」會止步於此嗎？恐怕不然。

事實上，中共在面對外界對於「一帶一路」廣泛的批評，也開始進行調整。紐約時報〔The New York Times〕的專文便提到，中共對內開始要求承包商改變商業行為，對外則正在尋求其他國家，以及跨國金融金構的協助，改進在基礎設施項目上不足之處。由於「中國夢」與「一帶一路」的實踐正在變化之中，因此，目前也並非斷言兩者將會如何發展之良機。

從戰略思考的角度而言，究竟中共在推動「中國夢」與「一帶一路」的過程中，應朝何種方向邁進的問題，「道德性思考」或許將是最好的解答。所謂的「道德性思考」〔Ethical Thinking〕，意指在制定政策與戰略時，行為者評估其「正當性」〔Rightness〕的能力，使得該政策與戰略，以及所衍生的效應與結果，能被國內和國外的所有行為者所接受。世界銀行在2019年提出的報告《一帶一路經濟學：運輸走廊的機遇和風險》（*Belt and Road Economics: Opportunities and Risks of Transport Corridors*）便建議，在整體的規劃、私部門的參與、包容性、管理風險、治理風險、環境與社會風險等領域的改革，將有助於「一帶一路」推動的成功。而世界銀行提出的改革項目之中，有不少正是本文前述的「一帶一路」遭到質疑與攻擊的部分。

　　最後，「一帶一路」確實創立了一種新型的經濟合作模式，以及不同於西方國家與學者所提的全球治理。不過，其未來將建構何種樣貌的全球治理和國際關係的型態亦存在不確定性。因此，究竟中共的「中國夢」與「一帶一路」在未來的發展如何，尚待各界持續的觀察。

解析美中戰略互動下的兩岸關係

常漢青 *

壹、前言

　　2016年美國總統大選，川普在「讓美國再次偉大」（make American great again）的口號下贏得大選，成為第45屆美國總統。川普為落實美國優先的承諾，自2018年3月單方面對進口美國的所有鋼、鋁金屬採取提高關稅的政策[1]，進而轉變成為以美、中為主的貿易戰。然在中美貿易戰的過程中，亦逐漸演變為美、中的科技戰與軍事對抗。另自2018年3月中國上海國際能源交易中心掛牌，啟動中國石油期貨的人民幣計價，並可轉換成黃金的原油期貨交易。[2] 尤其2020年俄羅斯在美國的經濟制裁的壓力下，與中國的貿易直接採取本國貨幣交易，以避開國際美元結算的掣肘。[3] 此外，中國與伊朗於2021年3月24日簽署價值4,000億美元的25年合作協議，[4] 這項協議在伊朗仍持續遭受美國經濟制裁的狀況下，預料亦將採取本國貨幣方式交易。中國的人民幣在國際化的過程中，雖然在全球外匯存底貨幣的佔比僅為2.3%，遠低於美元59%、歐元的21.2%及日圓的6.03%。但隨著美元的佔比快速下降，更凸顯出人民幣對美元優勢的威脅。[5] 美國川普政府執政時期對世界各國所採取的積極性貿易戰，讓2020年的美國總統大選受到各國高度關注，由其是衝擊最大的中國。雖參拜登於2020年12月14日在美國國會選舉人團投票，以306票比232票確認當選第46屆

---

* 台灣戰略研究學會副秘書長

1　Swanson, Ana, "Trump to Impose Sweeping Steel and Aluminum Tariffs," The New York Times, March 1, 2018, <https://www.nytimes.com/2018/03/01/business/trump-tariffs.html>（檢索日期：2021年4月7日）

2　賀迎春、董菁，〈中國原油期貨正式掛牌交易〉，《人民網》，2018年3月26日，<http://energy.people.com.cn/n1/2018/0326/c71661-29889163.html>（檢索日期：2021年4月7日）

3　〈俄中金融聯盟？美國制裁領歐亞兩大國另闢蹊徑〉，《BBC NEWS/中文》，2020年10月30日，<https://www.bbc.com/zhongwen/trad/world-54756895>（檢索日期：2021年4月7日）

4　Fassihi, Farnaz, Lee Myers, Steven, "China, with \$400Billion Iran Deal, Could Deepen Influence in Mideast," The New York Times, March 27, 2021, <https://www.nytimes.com/2021/03/27/world/middleeast/china-iran-deal.html>（檢索日期：2021年4月7日）

5　林則宏，〈全球加速去美元化 人民幣積極挑戰美元霸權〉，《經濟日報》，2021年4月5日，< https://money.udn.com/money/story/5603/5367383?ref=tab20210405>（檢索日期：2021年4月7日）

美國總統。[6] 但是面對國內新冠病毒疫情的嚴峻挑戰，美、中貿易戰也似乎沒有立即改善的跡象。[7] 而且即使美國拜登政府上台後，對於美國海軍在東海及南海的軍事活動亦未減少，反而展現出對中國人民解放軍施加更大的壓力。[8] 因此，美、中關係在美國疫情結束之前，基本上會朝向緩和方向發展的機率不高。

台灣在國際層次研究兩岸關係上，通常會運用羅威爾・迪特莫（Lowell Dittmer）在1981所提出的美、中、蘇「戰略三角」（the strategic triangle）理論來探討美、中、台三邊關係。然可適用於戰略三角的條件，基本上比須具備下列兩個前提：一為要有三個主權獨立的行為者且彼此的戰略互動密切；二為三邊中任何一組的關係都會對其他兩組的關係產生影響。[9] 而對於成員國規範的界定，迪特莫認為有三個主、客觀要素：一是成員國均認為彼此各自獨立處於三角的各頂點，成員國在處理雙邊關係時均會考慮第三方的態度；二是某一成員國的國家實力足以獨立自主的選擇與其他成員國的關係；三是成員國在雙邊關係中不能成為長期的同盟關係。[10]

依據「戰略三角」理論前提與規範界定要素的定義觀點，若從美中台三邊關係互動的歷史過程來看，就事實面的而言，美中關係牽動著台美關係，而台美關係則影響著兩岸關係。然對中國來說兩岸關係是國家的內政問題，而美台關係則是在「一個中國」政策的非官方關係。台灣對於兩岸關係與台美關係的建構上，其實力尚不足以左右美、中關係。其反映出的意涵為兩岸關係與美台

---

[6] Linton, Caroline, "All 538 electors have voted, formalizing Biden's 306-232 win. Here's how the Electoral College works.," *CBS NEWS*, December 15, 2020, <https://www.cbsnews.com/news/electoral-college-votes-joe-biden-victory/ > (檢索日期：2021年4月4日)

[7] Davis, Bob and Hayashi, Yuka, "New Trade Representative Says U.S. Isn't Ready to Lift China Tariffs," *The Wall Street Journal*, arch 28, 2021, <https://www.wsj.com/articles/new-trade-representative-says-u-s-isnt-ready-to-lift-china-tariffs-11616929200> (檢索日期：2021年4月7日)

[8] 沈朋達，〈中國智庫指美軍艦現身長江口及南海 針對性明顯〉，《中央訓社》，2021年4月4日，< https://www.cna.com.tw/news/acn/202104040077.aspx>（檢索日期：2021年4月7日）

[9] Dittmer, Lowell, "The Strategic Triangle: The Elementary Game-Theoretical Analysis," World Politics, Vol.33, No.4, Jul. 1981, pp.490-491.

[10] Dittmer, Lowell, "The Strategic Triangle: A Critical Review," in Ilpyong J. Kim ed, *The Strategic Triangle: China, the United States and the Soviet Union* (New York: Paragon House Publisher, 1987), pp.29-33.

關係的操縱權，原則上掌握在美、中的手中而不是台灣。因此，美、中、台要形成所謂的戰略三角的情況是有其侷限性的，主要在於無法有效解釋美國將台灣當作與中國談判籌碼的過程中，台灣為何始終處於被動接受的狀況。

若從台灣的國家安全觀點分析，除非藉由台灣修改憲法跳脫中國主權的概念，將主權與治權合而為一，並且在與中國的獨立戰爭獲得勝利，以及獲得國際上主要強權國家的承認，進而成為聯合國的會員身分。否則台灣與美國及中國的關係，始終會受制於美、中對台灣所建構的「一個中國」框架所束縛。所以，美中關係建構在美台關係與兩岸關係之上是無法否認的事實。因此，本論文試著從兩岸關係的美國因素觀點，運用整合戰略與國際關係的研究途徑，從事實面、影響面、發展面、戰略面及執行面5個面向，解析美中戰略互動下的兩岸關係，以及探討台灣在兩岸關係中的戰略選擇。

貳、美、中關係現況與未來發展

自美國川普政府對中國採取強硬的對抗政策後，除了傳統的政治、軍事對抗外，經濟、科技及文化等也納入對抗的領域。而美國總統拜登的對中政策是否會有所改變，亦或是承襲川普政府的對中政策立場，除了檢視當前拜登政府的對中政策立場之外，也需回顧從他與川普在選舉過程中，所發表的對中政策立場的事實做解析。美國政府及社會在川普政府的主導下，強化了對於中國負面的觀感。[11] 對川普不斷批評拜登親中的立場，也迫使拜登不得不採取與川普一致批判中國的立場，並在《外交政策》（Foreign Policy）撰文強調美國必須以強硬的手段應對中國。[12] 然隨著拜登政府的正式運作，對於美、中關係的現況與未來發展之探討，本節將依據整合戰略與國際關係研究途徑之邏輯思考順序，從事實面的問題意識與問題性質的探討開始，以瞭解美中關係中的因果關係與未來發展。

---

[11] 辜樹仁，〈73％美國人討厭中國 不論誰當選，強硬對付中國已是美國新常態〉，《天下雜誌》，2020年11月2日，<https://www.cw.com.tw/article/5102544>（檢索日期：2021年4月8日）

[12] 鄧聿文，〈拜登的對華政策會如何變〉，《紐約時報中文網》，2020年11月23日，<https://cn.nytimes.com/opinion/20201123/biden-china-policy/zh-hant/>（檢索日期：2021年4月8日）

一、美中關係的問題意識

美國川普政府主政時期的對中政策，係採取單邊的現實主義權力平衡原則的對抗政策。但其所發動的貿易戰不僅僅直接挑戰中國，同時也引發傳統同盟國如歐盟、日本、韓國、加拿大與墨西哥等國的不滿。而新上任的拜登政府雖然目前未改變川普直接對抗的對中政策，但是更著重於強化與盟國的合作共同對抗中國。2021年2月4日美國總統拜登在國務院發表首次的重大外交政策演講時，其開宗明義即指出「美國回來了」（America is back），外交回到外交政策中心。美國將修復與盟國的關係，並再次與世界互動以迎接威權主義的新挑戰。尤其是美國競爭的對手中國，其日增的野心已對美國的繁榮、安全與民主價值帶來挑戰，以及俄羅斯企圖損害與分裂美國民主的決心。[13]

2021年2月19日拜登在參加德國慕尼黑安全會議時，再告訴世界「美國回來了」，強調美國必須重新獲得川普政府時代所失去的信任，並藉由建立更好的經濟基礎，以重新獲得美國在國際機構中的地位。除了樹立美國自己的價值觀，以及捍衛世界各地的價值觀。並且通過軍事能力的現代化與外交領導，強化美國的聯盟網絡和夥伴關係，為所有人創造更安全的世界。[14] 由此，美國面對中國持續在西太平洋的軍事擴張，將面臨東海中、日在釣魚台列嶼的主權爭端、台灣海峽緊張情勢及南海航行自由權的維護等3項重大挑戰，以及修補與日本、南韓的關係，並強化與印度的關係。2021年3月12日由美國主導的召開首次美、日、印、澳四方元首視訊會議，這次會議除了針對全球性的新冠病毒疫情、經濟合作、網路安全及氣候危機等議題提出討論。但四方會談所關注的核心議題仍是中國因素，尤其美國國家安全顧問約翰·沙利文（John Sullivan）於會後指出，美、日、印、澳四位元首確實討論中國所展現出的挑戰，並明確地表達他們對中國沒有幻想。[15]

---

[13] "Remarks by President Biden on America's Place in the World," *The White House*, February 4, 2021, <https://www.whitehouse.gov/briefing-room/speeches-remarks/2021/02/04/remarks-by-president-biden-on-americas-place-in-the-world/> （檢索日期：2021年4月9日）

[14] Garamone, Jim, "President Biden Tells World: 'America Is Back'," *US Dept. of Defense*, Feb. 19, 2021, <https://www.defense.gov/Explore/News/Article/Article/2509091/president-biden-tells-world-america-is-back/> （檢索日期：2021年4月9日）

[15] Jennett, Greg, "Australia joins US, India and Japan in 'unprecedented' deal for coronavirus vaccines after

2021年3月19日中國應美國邀請，在美國阿拉斯加安克拉治市舉行國安與外交2+2會談。在開放媒體採訪的對話中，美國國務卿布林肯在開場的談話中，明確表達對中國在香港、台灣及新疆所採取的行動作為，以及對美國網路攻擊與對盟國的經濟、軍事脅迫的關切。這些行動已威脅到維護全球穩定所需的規則秩序，而不是國家本身的內部問題。對於美中關係將會是該競爭時競爭，有合作之處就合作，需要對抗就對抗。美國國務卿沙利文則表達，美國方面的首要任務在維護處理世界事務與中國事務途徑的目的，旨在使美國人民受益，以及保護美國盟國與夥伴的利益。美國不尋求與中國的衝突，但歡迎嚴屬的競爭，這將是美國始終堅持的原則。[16]

而中共中央對外辦公室主任楊潔篪在回應美國的關切時，指出中國在2035年必將實現基本現代化，到2050年將實現全面現代化。強調中國的價值觀是和平、發展、公平、正義、自由與民主，這與人類價值觀是相同的。中國與國際社會所遵循的國際體系是以聯合國為核心的國際體系，是以國際法為基礎的國際秩序。而不是一小部份國家所鼓吹的規則為基礎的國際秩序。中美兩國都是大國負有對世界的和平與發展做出貢獻的責任，例如在抵抗疫情，經濟活動的恢復及因應氣候變化等議題上，中美兩國有很多的共同利益。應摒棄冷戰思維的零和遊戲，大國之間應該團結為人類的未來做出貢獻，以構建人類命運共同體，創造公平、合理和相互尊重新的國際關係。對於美國憑藉自身的武力及金融霸權，對其他地區的國家實施長臂管轄進行打壓，並濫用所謂國家安全的概念妨礙國家之間正常的貿易往來，以及煽動一些國家攻擊中國。

中國認為中美兩國在新的國際形勢下，應加強相互溝通、妥善管理分歧、努力推行合作，中美兩國不應該進行對抗。中國對中美關係的評價是「不衝突、不對抗」、相互尊重、合作共贏，而「不衝突、不對抗」也是兩國元首在電話通話中，美國總統拜登所表達的觀點，希望中美關係能夠重新回到健康、

historic Quad meeting" *ABC News*, March 13, 2021, <https://www.abc.net.au/news/2021-03-13/quad-australia-us-india-japan-in-massive-covid-vaccine-deal/13245198> （檢索日期：2021年4月9日）

[16] "How it happened: Transcript of the US-China opening remarks in Alaska," NIKKEI Asia, March 19, 2021, <https://asia.nikkei.com/Politics/International-relations/US-China-tensions/How-it-happened-Transcript-of-the-US-China-opening-remarks-in-Alaska> （檢索日期：2021年4月9日）

穩定、發展的軌道。另中國外交部長王毅再次強調中國過去、現在及未來絕對不會接受美國對中國的無端指責，並要求美國放棄動輒干涉中國內政的霸權行徑。[17]

在拜登政府執政以來首次與中國的正式對話中，雙方在開放式的對話所表達出強硬立場的爭鋒相對狀況，各至均有其對國內與同盟國的政治需求。此外，雙方對閉門會議的評論，美方認為氣氛平靜，對話的內容具體、嚴肅且直接。[18] 而中國認為雙方的交流儘管仍存在一些重要分歧，但是一個真誠、坦率和具有建設性的交流。[19]

美國學者專家認為中國不斷擴大的經濟影響力，以及在世界各地日益增長的戰略能力，其行動已直接影響到美國與世界各區域國家所建構的國際秩序。並認為中國學者專家的共同觀點是美國對中國的敵意根源是不會改變的，即使拜登政府也不會改變川普政府所制定的對中政策。但對中國而言，中國的戰略主要特點在於藉由緩和與美國的緊張關係、加強與鄰國的關係、深化與俄羅斯的關係，以及鼓勵歐盟展開戰略自主的行動，以打破美國對中國的戰略包圍行動。[20]

就國際關係理論的問題意識分析，可以明確的瞭解「中國崛起」已不是一個假設，而是一個現況的事實。依上所述的美中關係互動現況可明顯的可以看出，美國面對中國的崛起所採取的因應作為，仍會會採取強硬的對抗原則。儘管川普政府採取單邊主義對抗中國，拜登政府則著重採取聯盟的方式圍堵中

---

[17] 〈美中高層阿拉斯加首日會談 觸及台港和貿易問題？〉，《TVBS新聞》，2021年3月19日，YouTube自11分4秒至43分52秒，<https://www.youtube.com/watch?v=30VnUgZvPzU>（檢索日期：2021年4月9日）

[18] 馮兆音，〈中美阿拉斯加「艱難」會談落下帷幕 回顧雙方外交官如何唇槍舌戰〉，《BBC News/中文》，2021年3月19日，<https://www.bbc.com/zhongwen/trad/world-56464982>（檢索日期2021年4月9日）

[19] "China Talks About China-U.S. Hight-level Strategic Dialogue," *Ministry of Foreign Affair of the People's Republic of China*, 2021/3/21, <https://www.fmprc.gov.cn/mfa_eng/zxxx_662805/t1862857.shtml> (檢索日期：2021年4月9日)

[20] Hass, Ryan, "How China is responding to escalating strategic competition with the US," Brookings, March 1, 2021, <How China is responding to escalating strategic competition with the US (brookings.edu)> (檢索日期：2021年4月8日)

國。但就其美中關係互動的本質而言，現實主義的權力平衡政策仍將是美國對抗中國的戰略手段。相較中國而言，面對美國權力平衡的對抗政策，中國原則上處於被動因應美國權力平衡政策的情勢。但是中國似乎不願意陷入美國所建構的權力平衡對抗政策對抗中，而是期望透過自由貿易與全球合作的方式面對美國的壓力。換言之，中國避開美國權力平衡政策壓力的方式，所採取的作為傾向於新自由主義的國際機制、制度、規範與相互依存理論概念，以因應美國的挑釁。

二、美中關係的問題性質

依據美中關係發展的歷史過程分析，美中關係的變動基本上美國是處於主動的一方，相較於中國則原則上處於被動的一方。布林肯在阿拉斯加的美中對話時，表達美中關係將是該競爭時競爭，有合作之處就合作，需要對抗時就對抗的意義。就其意涵主要在釐清美國在權力平衡的對中政策上，如何處理美中關係。換言之，美國未來應是希望制定一個明確的對中政策行動指導方針。然就戰略研究的觀點，如何應美中關係對世界及對亞太區域，乃至於對兩岸關係的影響，都應從美中關係的問題性質來做解析。在美中關係的問題性質部分，若從範圍分析其包含全球性、區域性或針對性的個別國家。若從操作面向分析，其包含政治、經濟、科技、文化及軍事等領域。因此，未來美國將會針對不同的議題，在不同範圍採取不同的領域操作美中關係。分析如下：

（一）全球性議題

依據聯合國當前所關注的全球性議題，主要範圍為非洲、人口老化、愛滋病傳染、核能、大數據、孩童、氣候變化、非殖民化、民主、消除貧困、糧食、性別平等、健康、人權、國際法與司法、移民、海洋與海洋法、和平與安全、人口、難民、水資源及青年。[21] 但就國際政治的觀點，美、中在氣候暖化、環境汙染、生態改變、衛生與傳染病防治、毒品問題及恐怖主義、健康與

---

[21] "Global Issues," *United Nations-Peace, dignity and equality on a health planet*, <https://www.un.org/en/global-issues/> (檢索日期：2021年4月12日)

人類共同價值等政治爭議較少議題，在聯合國機制的運作下雙方合作的機率可能較高。如2021年4月14日美國氣候問題特使約翰‧凱瑞（John Kerry）前往中國訪問，並與中國達成依據〈巴黎協定〉共同合作減少碳排放量的合作共識。[22] 但對於生物醫學、金融、糧食、水資源、國際貿易、資訊科技與網路安全等議題，由於這些議題對於國家的權力、利益與安全具有一定的關聯性，故美中要達成合作共識的機率相對的也較低。

就以2021年的新疆民事件為例，美國以中國侵犯維吾爾族人權為由，宣布禁止進口新疆棉花和番茄產品。[23] 而對中國採取制裁行動的國家主要是傳統的西方強權（美國、歐盟、英國、加拿大、澳洲及紐西蘭）。[24] 而被中國指控幕後操控新疆棉花事件的「良好棉花發展協會」（BCI）上海辦事處則回應，在中國的團隊未在新疆發現強迫勞動的事實，企業行為與他們無關。[25] 然就新疆棉事件的問題本質分析，新疆人權的議題已西方強權國家長久以來批判中國的老議題。由於棉花是用途廣泛的世界主要經濟農作物，而近年來中國在新疆大量種植的棉花，其產量已佔全球棉花產量的20%。[26]

而國際棉花兩大交易所，一為1870年於美國紐約成立的棉花交易所，二為1990年於中國鄭州成立的商品交易所。美國棉花以出口為主，中國棉花以內需為主。[27] 基本上國際的棉花交易的定價權是掌握在美國。但自2004年6月1日，鄭州商品交易所推行「鄭棉期貨」交易試點後[28]，成交量逐漸擴大，漸漸具備

---

[22] Westcott, Ben and Hansler, Jennifer, "US and China agree to cooperatate on climate change after talks in Shanghai," CNN, April 18, 2021, < https://edition.cnn.com/2021/04/18/politics/us-china-climate-change-shanghai-intl-hnk/index.html> (檢索日期：2021年4月19日)

[23] Swanson, Ana, 〈美國宣布禁止從新疆進口棉花和番茄產品〉，《紐約時報中文網》，2021年1月14日，<https://cn.nytimes.com/usa/20210114/xinjiang-cotton-tomato-ban/zh-hant/>（檢索日期：2021年4月10日）

[24] 〈新疆維吾爾人權：美歐英家等多國聯合制裁中國〉，《BBC N/中文》，2021年3月23日，<https://www.bbc.com/zhongwen/trad/world-56493143>（檢索日期：2021年4月12日）

[25] 沈朋達，〈BCI捲新疆棉爭議 中國團隊指未發現強迫勞動〉，《中央通訊社》，2021年3月26日，< https://www.cna.com.tw/news/acn/202103250328.aspx>（檢索日期：2021年4月12日）

[26] 沙磊，〈新疆棉：新證據揭露時尚產業背後的強迫勞動〉，《BBC News/中文》，2020年12月17日，<https://www.bbc.com/zhongwen/trad/chinese-news-55344353>（檢索日期：2021年4月12日）

[27] 〈NYBOT棉花期貨價格〉，《財經M平方》，2021年4月12日，<https://www.macromicro.me/collections/3520/agri-cotton/27451/nybot-cotton-prices>（檢索日期：2021年4月12日）

[28] 〈關於棉花其貨運行情況的介紹〉，《中鵝證券監督管理委員會》，2005年6月28日，<http://www.

影響美國紐約棉花期貨交易的價格。因此，新疆棉事件看是美國等西方國家對中國人權的批評，但就其戰略研究的問題性質而言，此事件的問題核心還是關於美國等西方國家對主導國際經濟體系可能弱化的憂心。[29] 因此，在全球議題上，中美既有合作，也有競爭。而其競爭的面向，主要在美國如何因應中國對現有國際經濟體系的挑戰。

（二）區域性議題

就中國當前對國際事務所發揮的影響力而言，美中之間可能會有議題交會的區域，基本上主要是環繞在中國周邊的區域。分別為朝鮮半島和平問題、東海釣魚台列嶼主權爭端、台灣問題、南海自由航行權，以及中東伊朗核武協定問題。分析如下：

1. 朝鮮半島和平問題

北韓的核武與彈道飛彈發展是影響對朝鮮半島和平，以及東北亞區域穩定的關鍵。美國川普政府試圖排除中國朝鮮半島的影響力，藉由直接與北韓談判解決北韓發展核武的企圖，但在雙方再極度不信任的狀況下進展有限。對美國而言，北韓問題所牽動主要是日本的安全，其次才是南韓的安全。同樣的，也牽動著美國在東北亞的影響力。就美國與北韓的接觸與談判過程，不難看出中國對朝鮮半島的和平始終扮演關鍵性的角色，影響著北韓與南韓的政治取向。[30] 因此，在朝鮮半島的和平問題上，美中之間的關係原則上是處於政治競爭的態勢。

2. 東海釣魚台列嶼主權爭端

對台釣魚台列嶼的主權爭端，由於2013年4月29日台灣與日本已就釣漁台列嶼的漁業權簽訂〈台日漁業協議〉，雖不涉及主權爭議問題，而僅就漁業

csrc.gov.cn/pub/xinjiang/gzdt/200506/t20050627_88578.htm>（檢索日期：2021年4月12日）

[29] 孫恭正，〈觀點投書：「新疆綿」事件無關人權，關鍵在於「定價權」〉，《風傳媒》，2021年4月7日，<https://www.storm.mg/article/3582035>（檢索日期：2021年4月10日）

[30] 〈習近平訪韓、中美關係與半島無核化：王毅訪韓觸及的三大議題〉，《BBC News/中文》，2020年11月27日，<https://www.bbc.com/zhongwen/trad/chinese-news-55099791>（檢索日期：2021年4月13日）

合作部分達成協議。[31] 因此，台、日雙方對於釣魚台列嶼的主權爭議情勢相對較為緩和。而中、日之間對於釣魚台列嶼主權爭議，則成為長期性的對峙狀態。主要在於日本於2012年9月11日本單方面採取釣魚台列嶼「國有化」行動所致。[32] 為此，日本在與美國就〈美利堅合眾國與日本國之間互相合作與安全保障條約〉（Treaty of Mutual Cooperation and Security between the United States and Japan；簡稱：美日安保條約）內容討論時，經常發表美國同意將「釣魚台列嶼」納入其第5條的適用範圍。[33]

相對的，美國在此問題的表達上通常沒有明確的回應。2021年2月27日美國國防部發言人約翰·科比（John Kirby）為澄清美國對於釣魚台列嶼主權的政策立場沒有改變，同時亦明確指出美國將會依據「美日安保條約」第5條規定對日本進行防禦，包含釣魚台列嶼，反對任何改變現狀的單方面行動。[34] 因此，可以瞭解日本深知無法單獨在釣魚台列嶼主權爭端上與中國對抗，美國是支撐日本所認知有關領土主權安全的最大支柱。儘管在東海有主權爭議的國家是日本與中國，但對中國而言，其背後對抗的主要國家還是在美國。因此，美中關係在東海的問題性質主要是政治與軍事兩個面向。

3. 台灣問題

2016年台灣由民進黨執政，由於與中國的長期不信任感，且對於所謂「九二共識」認知有所不同的狀況下，兩岸政府之間的關係開始進入零互動的情勢。[35] 2017年1月20日美國總統川普上任，在其四年任期對中國採取強硬對

---

[31] 亞東太平洋司，〈亞東關係協會與公益財團法人交流協會漁業協議〉，2013年4月29日，<中華民國外交部全球資訊網－臺日漁業協議 (mofa.gov.tw)>（檢索日期：2021年4月13日）

[32] 童倩，〈日本政府完成釣魚台國有化〉，《BBC中文網》，2012年9月11日，<https://www.bbc.com/zhongwen/trad/world/2012/09/120911_japan_diaoyu_nationalized>（檢索日期：2021年4月14日）

[33] 〈【独自】「尖閣に安保適用」日米が共同文書明記へ…首脳会談で連携確認〉，《讀賣新聞》，2021年3月26日，<https://www.yomiuri.co.jp/politics/20210326-OYT1T50073/>（檢索日期：20212年4月14日）

[34] Moriyasu, Ken, "Pentagon correct statement on Senkakus: 'Np change in policy'," *Nikkei Asia*, February 27, 2021, <https://asia.nikkei.com/Politics/International-relations/Pentagon-corrects-statement-on-Senkakus-No-change-in-policy>（檢索日期：20212年4月14日）

[35] 陳俊華，〈兩岸溝通停擺 林全：不是完全零互動〉，《中央通訊社》，2016年6月28日，<https://www.cna.com.tw/news/aipl/201606280175.aspx>（檢索日期：2021年4月14日）

抗立場的過程中，藉由強化對台灣的關係作為對中國施壓的籌碼。[36] 台灣部分學者認為在蔡英文總統的第二任期間，兩岸仍將處於所謂冷對抗情勢。[37] 換言之，所謂「冷對抗」即為中國在政治上採取更加孤立台灣的政策；在軍事上改變往兩岸軍事對峙，但相互尊重的現況，尤其是中國已將台灣海峽、台灣防空識別區內的西南海域及台灣東部海域，納入中國海、空軍常態演訓區域；[38] 在經濟上，兩岸經貿往來確未因兩岸關係的緊張而降溫，尤其台灣對中國的出口值佔總出口值的43.8%。[39] 因此，美國在兩岸關係的冷對抗的情勢中，在政治與軍事上扮演平衡者的角色。

4. 南海自由航行權

自2012年中國與菲律賓發生黃岩島對峙事件後，2013年起中國即開始針對南海實際佔領的島礁實施擴建。目前所擴建的島礁設施，可以說已具備依據南海情勢需求，隨時轉換成海、空軍臨時基地的能力，有效提升中國在南海的制空及制海能力。[40] 美國在美中關係緊張對抗的川普政府時代，即以定期派遣海軍艦艇以「維護自由航行權」之名義巡弋南海。即使現在由民主黨執政的拜登政府，亦經常派遣海軍航母打擊支隊巡弋南海。並且運用海、空軍機、艦在東海、台灣海峽東部與西南部海域，對中國海、空軍機、艦實施近距離監控作為。[41] 因此，從美中在南海的海上軍事行動作為來看，美、中在南海的關係是處於直接軍事對抗的性質。

---

[36] 陳育晟，〈駐美外交官第一手觀察：美國國會為何如此挺台灣〉，《天下雜誌》，2021年3月1日，<https://www.cw.com.tw/article/5108572>（檢索日期：2021年4月14日）

[37] 趙春山，〈不要使兩岸「冷對抗」加溫〉，《財團法人國家政策研究基金會》，2020年2月20日，<

[38] 〈中國海軍遼寧艦航母編隊展開遠海訓練〉，《中華人民共和國國防部》，2021年4月5日，<http://www.mod.gov.cn/big5/topnews/2021-04-05/content_4882617.htm>（檢索日期：2021年4月14日）

[39] 〈我國對中國大陸（含香港）出進口統計〉，《經濟部國際貿易局經貿資訊網》，2021年3月18日，<file:///C:/Users/USER/Downloads/%E6%88%91%E5%9C%8B%E5%B0%8D%E4%B8%AD%E5%9C%8B%E5%A4%A7%E9%99%B8%E5%90%AB%E9%A6%99%E6%B8%AF%E5%87%BA%E9%80%B2%E5%8F%A3%E7%B5%B1%E8%A8%88.pdf>（檢索日期：2021年4月14日）

[40] 〈衛星圖片「顯示美濟礁新變化」中國被指推進「完全軍事基地化」〉，《BBC News/中文》，2021年2月25日，<https://www.bbc.com/zhongwen/trad/chinese-news-56198339>（檢索日期：2021年4月14日）

[41] Lendon, Brad, " US and China deploy aircraft carriers in south China Sea as tensions simmer," *CNN*, April 12, 2021, <https://edition.cnn.com/2021/04/12/china/south-china-sea-taiwan-military-tensions-intl-hnk/index.html>（檢索日期：2021年4月14日）

5. 伊朗核武協定問題

2018年5月8日美國川普總統正式宣布退出於2015年由美、俄、中、英、法、德六國與伊朗共同達成「聯合全面行動計畫」（Joint Comprehensive Plan of Action; JCPOA）的所謂伊朗核武協議，並恢復對伊朗實施經濟制裁。[42] 除引發中東的核武危機，以及美國與歐洲主要盟國的信任危機外，更促使伊朗與中國、俄羅斯達成更緊密的合作夥伴關係。2021年3月24日中國外長王毅訪問中東六國，提出中國願意為中東的和平發展做出「中國貢獻」，其主要的問題仍聚焦於伊朗核武問題，以及以、巴問題。[43] 但此次的訪問對美國來說，中國是有意挑戰以往由美國主導的中東事務。然就中國的軍事實力分析，中國海軍尚不具備遠洋兵力投射能力。因此，美中在中東事務上，基本上會處於政治競爭的性質。

依據上述對於美中關係的問題性質分析，可瞭解到在全球議題部分：美中在政治方面原則上處於合作關係，在經濟方面則處於競爭關係，尤其是資訊與精密工業等科技領域部分；在區域議題部分：美中在政治方面則處於競爭關係，在軍事方面始終會維持軍事對抗的狀態。

參、美、中戰略互動對兩岸關係的影響與未來發展

就台灣的立場在探討兩岸關係時，台灣學者如吳玉山及包宗和通常會將美國因素納入變項，並運用「戰略三角」理論，透過現實主義權力平衡與博弈理論做為解析兩岸關係的基礎。[44] 然從美中關係的事實面分析結果，可以發現傳統運用戰略三角分析美、中、台三邊關係時，無法明確說明「一中政策」對美台與兩岸關係的影響因素。台灣始終無法跳出在「一中政策」框架下，獨立決

---

[42] Maloney, suzanne, "Around the Hall: Brookings experts discuss the implications of President Trump's Iran nucleat deal announcement," Brookings, May 8, 2018, <https://www.brookings.edu/blog/order-from-chaos/2018/05/08/around-the-halls-brookings-experts-discuss-the-implications-of-president-trumps-iran-nuclear-deal-announcement/> （檢索日期：2021年4月14日）

[43] 〈王毅國務委員兼外長在結束訪問中東六國後接受媒體採訪〉，《中華人民共和國中央人民政府》，2021年3月31日，<http://www.gov.cn/guowuyuan/2021-03-31/content_5596904.htm>（檢索日期：2021年4月16日）

[44] 包宗和，〈戰略三解個體論檢視與總體論建構及其對現實主義的衝擊〉，包宗和、吳玉山主編，《重新檢視爭辯中的兩岸關係理論》，頁346-349。

定與另外兩邊的關係。而兩岸關係與美台關係的主導權又分別在中國與美國的手上，台灣則處於被動接受的情況。因此，「美台關係」與「兩岸關係」的定位與發展，主要的政策依據基本上仍是在「美中關係」的架構下考量。本節將依據整合戰略與國際關係研究途徑的影響面及發展面，探討美、中戰略互動對兩岸關係的影響與未來發展。

一、美、中戰略互動對兩岸關係的影響

台灣自2016年由民進黨執政以來，兩岸關係即進入所謂的「冷對抗」時期，雙方維持在可控制下不發生實際軍事衝突的狀態，而其對抗面向主要是政治性與軍事性的議題。美國在川普政府的領導下，美中關係的緊張情勢逐漸提高。美台關係也因美中緊張關係的加劇而強化，自2018年起美國國會制定〈台灣旅行法〉等一系列的所謂「友台法案」。相對的中國面對美國持續加強對台灣的外交與軍事關係，其所採取的因應行動除了在外交上對美提出抗議外，中國海、空軍機、艦亦增加與擴大在台灣周邊海域實施軍事演訓活動。[45] 更重要的是中國海、空軍機、艦的演訓活動，已改變兩岸軍事活動不越過所謂海峽中線的默契。[46]

對台灣來言，中國在軍事及政治上已形成逐漸加壓的態勢。雖然美國亦派遣軍艦及海上巡邏機在台灣海峽與台灣東部海域巡弋，以展現維護台海和平與牽制中國軍事行動的決心。[47] 2020年美軍艦艇通過台灣海峽已達13次之多。[48] 從上述事件可以檢證當中美關係發生緊張對抗的情勢時，美台關係相對的在美國的主導下而提升與強化，同時兩岸關係就陷入緊張的狀態。

---

[45] 涂鉅晏，〈今年共機擾台變本加厲/我機戰巡已4132架次 去年的2.3倍〉，《自由時報》，2020年10月6日，<https://news.ltn.com.tw/news/politics/paper/1404107>（檢索日期：2021年4月16日）

[46] Dowling, Hannah, "China flew Fighter jet into Taiwanese Territory amid US visit," *World of Aviation*, August 11, 2020, <https://worldofaviation.com/2020/08/china-flew-fighter-jets-into-taiwanese-territory-amid-us-visit/>（檢索日期：2021年4月16日）

[47] "7th Fleet Destroyer transits Taiwan Strait," *Commander, U.S. 7th Fleet*, April 7, 2021, <https://www.c7f.navy.mil/Media/News/Display/Article/2563410/7th-fleet-destroyer-transits-taiwan-strait/>（檢索日期：2021年4月16日）

[48] 〈美國2艘軍艦通過台灣海峽 今年第13度〉，《中央通訊社》，2020年12月31日，<https://www.cna.com.tw/news/firstnews/202012315002.aspx>（檢索日期：2021年4月16日）

2019年7月在美國川普政府的強硬對中政策，引發美國百名學者、專家在華盛頓郵報對川普發表公開信，表達中國不是敵人，但可以確認是一個競爭的挑戰者。[49] 其意涵即說明中國已然成為美國的戰略競爭者是美國兩黨的政治共識。2021年3月3日美國總統拜登公布〈臨時國家安全戰略指導〉（Interim National Security Strategic Guidance），認為中國是美國在世界上唯一可以將經濟、外交、軍事和技術力量相結合的競爭對手。其引發世界性的權力分配的改變，並對當前穩定和開放的國際體系構成挑戰，以及成為美國新的威脅。[50] 由此，可以瞭解美中的戰略互動即使有緩解的可能，但至少在美國新冠病毒（COVID-19）疫情緩解之前，美中之間的戰略互動仍將會延續川普政府先前所建構的對抗態勢。

2021年4月8日參議院外交關係委員會主席羅伯特・曼南德斯（Robert Menendez）宣布兩黨達成協議，共同制定針對中國戰略方針的「2021戰略競爭法」，以確保美國在未來十年與中國的競爭能力，其內容包含全方位的國家和國際力量。[51] 隔一天4月9日美國政府即發布「美台互動新準則」，強調開放與台灣的交流方式，但仍維持「非正式」關係與提供行政部門傳達一個有效維持「一個中國」政策的明確訊息。此一準則以「台灣關係法」、「三個聯合公報」及六項保證為指導，並經由「台灣保證法」之規定審查後所發布的新指南。其中重申美國對印太地區的盟友及伙伴的安全承諾。[52] 因此，依據對於美中關係的問題意識與問題性質的解析，顯現出美中關係在亞太地區的朝鮮半島和平、台灣問題及南海自由航行權問題，基本上處於政治與軍事對抗的戰略互動格局。

---

[49] Fravel, Taylor, ad., "Opinion: China is not an enemy," *The Washington Post,* July 3, 2019, < https://www.washingtonpost.com/opinions/making-china-a-us-enemy-is-counterproductive/2019/07/02/647d49d0-9bfa-11e9-b27f-ed2942f73d70_story.html> (檢索日期：2021年4月16日)

[50] The White House, "Interim National Security Strategic Guidance," March, 2021, pp.7-8.

[51] Pachon, Juan, "Chairman menendez announces bipartisan comprehensive China legislation," Foreign relations, April 8, 2021, < https://www.foreign.senate.gov/press/chair/release/chairman-menendez-announces-bipartisan-comprehensive-china-legislation> (檢索日期：2021年4月10日)

[52] Price, Ned, "New Guidelines for U.S. Government Interactions with Taiwan Counterparts," *U.S. Department of State*, April 9, 2021, <https://www.state.gov/new-guidelines-for-u-s-government-interactions-with-taiwan-counterparts/> (檢索日期：2021年4月16日)

　　然在台灣問題上，美中的戰略互動基本上是呈現軍事對抗的狀態。而中國遼寧號航空母艦於2012年9月加入中國海軍戰鬥序列之後，中國海軍遠洋作戰能力的評估即成為各國海軍關注的焦點。自2016年12月起以遼寧號為主體所編組的航母編隊，首次由宮古群島進入第二島鏈內海域實施遠海長航訓練[53]，而過去亦分別於2018年3月18-20日、2019年6月10-25日與2020年4月10-12日實施遼寧艦的航母編隊遠海長航訓練。[54] 2021年4月5日遼寧艦航母編隊，再次由宮古群島進入台灣以東的第二島鏈海域，實施遠海長航訓練任務。[55] 於此同時美國羅斯福號航母打擊支隊亦由麻六甲海峽進入南海，並與由馬金島號兩棲突擊艦所組成的兩棲準備支隊（Amphibious Ready Group: ARG）在南海實施演訓，以支持自由和開放的印太區域。[56] 因此，美中戰略對抗互動的影響面來看，對於兩岸關係的影響必須先解析美、中各自在亞太區域的主、客觀利益。

　　就美國在亞太地區的客觀利益而言，從美、中軍事對抗的角度，中國的軍事能力已具備在第一島鏈內有效抗衡美國海軍遂行制海作戰的能力。而且美國也認清這一事實，並在第二島鏈的關島及威克島強化軍事力量部署，以作為美國與中國軍事對抗的戰略前沿。[57] 然美國面對中國在西太平洋的不斷軍事擴張下，如果美國放棄在第一島鏈的軍事嚇阻優勢，若中國能夠在第二島鏈內海域與美國實施軍事抗衡，這將影響日本的海上交通線與台灣生存的安全，以及南

[53] "Hoe Does China's First Aircraft Carrier Stack Up ?," ChinaPower, <https://chinapower.csis.org/aircraft-carrier/> (檢索日期：2021年4月17日)

[54] 林柏州，〈中國「遼寧」艦編隊2020年遠海長訓析評〉，《財團法人國防安全研究院》，2020年4月14日，<https://indsr.org.tw/tw/News_detail/2173/%E4%B8%AD%E5%9C%8B%E3%80%8C%E9%81%BC%E5%AF%A7%E3%80%8D%E8%89%A6%E7%B7%A8%E9%9A%8A2020%E5%B9%B4%E9%81%A0%E6%B5%B7%E9%95%B7%E8%A8%93%E8%A9%95%E6%9E%90> (檢索日期：2021年4月17日)

[55] 〈中國航母遼寧艦入西太平洋 美國斯福號進南海 台灣需面對周邊軍事活動常態化〉，《BBC/中文》，2021年4月8日，< https://www.bbc.com/zhongwen/trad/chinese-news-56659815>（檢索日期：2021年4月17日）

[56] "Theodore Roosevelt Strike Group and Makin Island Amphibious Ready Group Conduct Integrated Operations in South China Sea," *Commander, U.S. 7th Fleet,* April 8, 2021, <https://www.c7f.navy.mil/Media/News/Display/Article/2566618/theodore-roosevelt-strike-group-and-makin-island-amphibious-ready-group-conduct/> (檢索日期：2021年4月17日)

[57] 許智翔，〈美國加強威克島機場建設意涵〉，《財團法人國家安全研究院》，2020年7月13日，<https://indsr.org.tw/tw/News_detail/2212/%E7%BE%8E%E5%9C%8B%E5%8A%A0%E5%BC%B7%E5%A8%81%E5%85%8B%E5%B3%B6%E6%A9%9F%E5%A0%B4%E5%BB%BA%E8%A8%AD%E6%84%8F%E6%B6%B5> (檢索日期：2021年4月17日)

海的航行自由權。為此美國印太司令部向國會提出「太平洋嚇阻倡議」案，希望於2022年至2027年的6年時間，在包含台灣、沖繩及菲律賓的第一島鏈上開發與部署精準打擊飛彈網，以面對中國企圖改變該地區的現狀，如台灣。[58]

2021年4月17日日本首相菅義偉訪問美國，在與美國總統拜登交換意見時，共同重申「台灣海峽和平與穩定的重要性」。[59] 從美、日元首對台灣問題的共同立場，可以看得出日本意識到台灣的現狀如果改變，對日本的安全是一項重大的挑戰。同樣的，對美國維持亞太安全的能力也會受到重大的挑戰。若從國家主觀利益的角度來看，通常大國之間在區域軍事力量的改變，將會影響地區國家的政治選擇。例如：面對北韓核武危機的南韓，在對美中關係的政治選擇上，從依賴美國轉變到親中的政策。因此，從美國的主觀利益來看，在「一中政策」的框架下，美國希望兩岸之間能夠維持分治的狀態，並在兩岸關係上扮演平衡者或仲裁者的角色。

對中國而言，由於與鄰國的陸上邊界爭議，除了印度之外，自1999年12月開始陸續與12個鄰國完成邊界線劃定。[60] 中國自2000年開始隨著經濟的快速發展，其安全構想已從陸權的國土防衛作戰構想，轉變為以「海權為主，陸權為輔」的國家安全戰略構想。[61] 換句之，中國21世紀的國家客觀利益，主要是防禦來自海上的威脅（主要是美國），並隨著中國海軍能力的提升，將海上防禦範圍向外延伸以增加防禦縱身，以確保中國為發展國內經濟發展時，所需一個外在和平穩定的國際環境。就中國的主觀利益而言，雖然中國將第二島鏈視為近海防禦的戰略目標，但影響的政治利益卻是改變西太平洋國家的政治傾向。當美國在第一島鏈的軍事優勢退出，並在第二島鏈形成與中國對治抗衡的狀

---

[58] Sharma, Aakriti, "US Military Draws Up ＄27 Billion Plan To Counter China In The Indo-Pacific Region," *The EurAsian Times*, April 17, 2021, <https://eurasiantimes.com/us-military-draws-up-27bn-plan-to-counter-china-in-the-indo-pacific-region/> (檢索日期：2021年4月17日)

[59] "Japan's Suga say he, Biden, reaffirmed importance of stability in Taiwan Strait," *Reuters*, April 17, 2021, <https://www.reuters.com/world/china/japans-suga-says-he-biden-reaffirmed-importance-stability-taiwan-strait-2021-04-16/> (檢索日期：2021年4月17日)

[60] 〈從「動盪之源」到亞洲安全之本〉，《中國文化研究院》，2018年9月27日，< https://chiculture.org.hk/tc/china-today/105>（檢索日期：2021年4月17日）

[61] 葛琳，〈對比中日海權思想 判未來發展走向〉，《中華人民共和國國防部》，2014年8月31日，<http://www.mod.gov.cn/big5/intl/2014-08/31/content_4533808_2.htm>（檢索日期：2021年4月17日）

況。這將贏種削弱美國在東北亞、台海及東南亞的政治影響力。

換言之，當美中陷入軍事危機期間，在美日於第一島鏈實施海上封鎖的狀況下，中國海軍艦隊要安全的進入第二島鏈海域以延伸防禦範圍，就必須先解決台灣問題。當台灣與中國達成和平協議，即使台灣不成為中國海、空軍的前進基地，而是在「一個中國」的概念下，保持中立的狀態。台灣不僅將成為美、日防禦中國海軍艦隊與空中武力進入第二島鏈的破口，亦將成為中國有效控制東海、台灣海峽與巴士海峽進入南海的遏制點。因此，中國對台灣的統一問題是中國必然的國家主觀利益。

依上所述，美、中在台灣問題上所呈現的是政治與軍事對抗的性質，美國採取加強與台灣的政治及軍事關係，以及強化美、日、澳聯盟對台灣和平與穩定現狀的支持，以確保台灣在兩岸關係上，持續維持「不獨、不統、不武」或所謂「主權獨立」的國家現狀。然美國等盟國持續在政治和軍事上介入兩岸事務，將會使中國採取更積極的抗衡行動。因此，當美、中在亞太區域處於政治與軍事對抗關係時，美國對中國所採取權力平衡對抗政策，使台灣成為美國對抗中國的籌碼是可預見的事實。中國也亦將採取相對應的行動，以抗衡美國來自海上的政治與軍事壓力。與此同時亦可預測中國將會壓迫台灣就「一個中國」表達政治立場。此時，兩岸關係是否會陷入緊張狀態，主要在於台灣是否接受中國所認知的「九二共識」。而台灣是否能夠抵擋或屈服於中國的政治和軍事壓力，仍在於美國是否會派遣軍隊介入兩岸可能的戰爭危機。

二、兩岸關係的未來發展

兩岸關係的未來發展，不僅僅取決於台灣政府與人民對於「一個中國」的認知與態度。也必須考量中國統一台灣在國家發展的優先順序，以及美國在兩岸關係中的國家利益。自2016年起至2021年上半年止，在美國的主導下，美中關係在亞太地區陷入政治與軍事對抗的狀態。相對的，美台關係在政治交往與軍事合作上則有明顯的提升與強化，而兩岸關係則始終陷入政治零和與軍事緊張對峙的狀態。本節將運用整合戰略與國際關係研究的發展面，藉由國內與國際兩個層次，解析兩岸關係的未來發展。

（一）在國內因素部分

2019年1月2日習近平在「告台灣同胞書40周年」談話中，強調中國是以「一個中國原則」為基礎，與台灣達成共同謀求國家統一的「九二共識」，推動兩岸關係從緊張對峙走向緩和改善，確立中國「和平統一，一國兩制」的兩岸關係基本方針。對於「一國兩制」的「兩制」認為是在一個中國原則下，為因應台灣的現況所提出制度性安排。在和平統一的進程上，強調兩岸長期存在的政治分歧，總不能一代一代傳下去，以及中國人不打中國人的承諾，並指出統一後的台灣不會損害任何國家在台灣的正當經濟利益。[62]

從中國的「一國兩制」的戰略觀點分析，「一國」所指的就是「一個中國原則」的中華人民共和國，[63] 也是中國推動兩岸關係的最終核心目標。而「兩制」則是兩岸在統一的中國下，中國大陸及台灣相互尊重各自實施其現有的社會制度。依此，「兩制」的安排是中國共產黨政府完整行使中國治理之前，對台政策的階段性戰略目標。中國對於兩岸統一的戰略途徑，雖然強烈表達以和平談判的方式推動兩岸統一。但並未排除最終會以武力的方式完成中國的統一，這可以說是因應台灣政府與人民可能推動法理台獨或獨台政策所設下紅線。中國的戰略目的是迫使台灣必須從「統一」的觀點思考兩岸關係，並對台灣及國際表達台灣如果選擇脫離中國而獨立，或美、日等強權國家以實際政治或軍事行動支持台灣脫離中國獨立，中國必將不惜一切代價採取軍事武力的手段統一台灣。

台灣面對中國強烈推動統一的決心，台灣政府及人民對於兩岸關係的選擇，必須從台灣人民身分的認同、台灣的政治及軍事實力及美國的國家利益三個面向思考。首先在台灣人民身分認同部分，自2000年民進黨首次執政以來，於2006年開始調整以台灣為主體的國、高中歷史教育課綱，將台灣史獨立於中

---

[62] 〈（現場實錄）習近平：在《告台灣同胞書》發表40周年紀念會上的講話〉，《新華網》，2019年1月2日，<http://www.xinhuanet.com/tw/2019-01/02/c_1210028622.htm>（檢索日期2021年4月19日）

[63] 〈一個中國原則〉，《中華人民共和國中央人民政府》，<http://big5.www.gov.cn/gate/big5/www.gov.cn/test/2005-07/29/content_18293.htm>（檢索日期2021年4月19日）

國史開始，[64] 到2019年歷史課綱的調整所引發的去中國化的疑慮。[65] 若依據自1992年6月到2020年12月止，政治大學選舉研究中心所做的有關台灣民眾台灣人/中國人認同的民調趨勢分析，2005年台灣民眾認為是台灣人的占45%，認為是台灣人也是中國人占43.4%，首次進入交叉點，此時是民進黨的陳水扁政府的第二任期。且自2008年起此差異不斷的擴大，除了國民黨執政時期對於歷史課綱採取放任的態度，以及2016年民進黨再度執政後的課綱修訂，2020年12月的民調顯示台灣民眾對於台灣人身分的認同已達64.3%，遠大於29.9%同時為台灣人與中國人的身分認同。[66]

　　另依據政治大學選舉研究中心對台灣民眾統獨立場的民意調查，2020年12月顯示維持現狀在決定佔28.8%，永遠維持現狀的佔25.8%，偏向獨立佔25.5%。[67] 由此可以看出大部分台灣民眾對兩岸關係的選擇上，統一不是主要的選項。換言之，2000年12-17歲的國、高中生到2020年時的年齡是32-37歲合計2,028,406人，[68] 雖然佔20歲以上總投票人口19,453,512人的約10.4%。[69] 再加上2000年以後出生的年輕人接受新修訂的歷史課綱教育，可以說明為何40歲以下的台灣年輕選民，會投向傾向訴求台灣獨立的民進黨的因素之一。尤其是台灣總統大選受到香港反送中運動的影響，在民進黨「亡國感」的操作下，台灣民眾除對台灣人的身分認同感更加提升，也更傾向於選擇台灣的獨立性。即使具有於未來不排除統一傾向的國民黨，為了選舉也表達反對中國所提「一國兩

---

[64] 高明士，〈歷史教育與教育目的〉，《歷史教育》，第14期，2009年6月，頁28。

[65] 許秩維，〈潘文忠：歷史新課綱為去中國化 強調分域關聯性〉，《中央通訊社》，2020年9月8日，<https://www.cna.com.tw/news/firstnews/202009080095.aspx>（檢索日期2021年4月19日）

[66] 〈臺灣民眾臺灣人/中國人認同趨勢分析（1992年6月~2020年12月）〉，《政治大學選舉研究中心》，2021年1月25日，<https://esc.nccu.edu.tw/PageDoc/Detail?fid=7804&id=6960>（檢索日期：2021年4月19日）

[67] 〈臺灣民眾統獨立場趨勢分析（1992年12月~2020年12月）〉，《政治大學選舉研究中心》，2021年1月25日，<https://esc.nccu.edu.tw/PageDoc/Detail?fid=7805&id=6962>（檢索日期：2021年4月19日）

[68] 〈重要性別統計資料庫109年度32-37歲人口〉，《性政院性別平等會》，<https://www.gender.ey.gov.tw/gecdb/Stat_Statistics_Query.aspx?sn=Jvwu1Ndiotx2AzKr6MD1kg%40%40&statsn=EcfUJy%24sRRPbnOe4TvO%24Jg%40%40&d=&n=93232>（檢索日期：2021年4月19日）

[69] 〈重要性別統計資料庫109年度20-100歲以上人口〉，《性政院性別平等會》，<https://www.gender.ey.gov.tw/gecdb/Stat_Statistics_Query.aspx?sn=Jvwu1Ndiotx2AzKr6MD1kg%40%40&statsn=EcfUJy%24sRRPbnOe4TvO%24Jg%40%40&d=&n=84280>（檢索日期：2021年4月19日）

制」和平統一方案。[70]

　　依上所述，台灣政府與人民對於兩岸關係的政治取向，傾向於被迫在國際「一個中國」政策下，維持與中國分治的政治現況。並朝向兩岸雙方降低軍事敵意及強化經貿往來，達到維護台灣人民安全與幸福的政治選擇。但2019年中國領導人習近平在「告台灣同胞書」40週年紀念會的講話內容，可以感受到中國對於兩岸的統一未來將會採取更加積極的行動。2021年3月21日中國國務院台灣事務辦公室於新聞記者會中，對於記者提問有關「九二共識」是定海神針底，以及「一中」的問題時，再次強調「九二共識」是兩岸平等協商的基礎，而「一中」就是大陸與台灣同屬一個中國，並在此基礎上通過對話協商尋求解決之道。[71] 因此，台灣內部面對兩岸關係的問題癥結點，若從獨立的觀點是沒有所謂「一中」的問題，只有是否採取實質「獨立」與否的問題。若從統一觀點，其關鍵因素在於如何界定「一中」，以及對於中國承諾的信任度。由於台灣的民進黨政府與中國政府在相互極度不信任，且國民黨重新獲得執政權的機會尚須極大的努力下，從兩岸當事者的觀點來看，兩岸關係未來在短、中期內能夠和平穩定的改善機率不大。

（二）在國際因素部分

　　中國及俄羅斯在美國不斷採取單邊主義經濟制裁與軍事對抗的壓力下，促成雙方領導人於2019年6月6日簽署同意建立「全面戰略協作夥伴關係」（strategic cooperative partnership），並強調維護以〈聯合國憲章〉宗旨和原則為核心的國際秩序和國際體系，秉持多邊主義原則處理國際事務。[72] 2021年4月18日大西洋理事會主席弗雷德・肯佩（Fred Kempe）認為中俄之間緊密合作

[70] 李宗憲、林祖偉、呂嘉鴻，〈中國威脅與亡國感〉，《BBC News/中文》，2020年1月6日，<https://www.bbc.com/zhongwen/trad/extra/sG2LG41QTX/taiwan-presidential-election-2020#group--lOgsjq3zMJ>（檢索日期：2021年4月19日）

[71] 〈國台辦新聞發布會輯錄(2021-03-31)〉，《中共中央台灣工作辦公室/國務院台灣事務辦公室》，2021年3月31日，<http://www.gwytb.gov.cn/xwdt/xwfb/xwfbh/202103/t20210331_12342228.htm>（檢索日期：2021年4月19日）

[72] 〈中俄發表關於發展新時代全面戰略協作夥伴關係聯合聲明〉，《俄羅斯衛星通訊社》，2019年6月6日，<http://big5.sputniknews.cn/politics/201906061028676300/>（檢索日期：2021年4月19日）

的關係，對現在和未來的美國總統拜登來說將面臨重大挑戰。當前美國同時面臨俄羅斯對烏克蘭，以及中國對台灣的軍事威脅。而這兩個區域的任何一個國家都沒有單獨抵抗中國與俄羅斯的能力，如果同時間發生軍事衝突，任何國家試圖透過與美國的結盟來對抗中國與俄羅斯，這將會發生災難性的後果。[73]

在亞洲與中國崛起有切身關係的國家，除了美國之外其次就是日本。以往日本在與美國討論西太平洋的安全議題時，通常會特別強調釣魚台列嶼納入「美日安保條約」第5條的適用範圍。然在探討美日對於台海安全問題是否納入「美日安台條約」的適用範圍時，需先瞭解1969年3月2日中、蘇爆發珍寶島衝突事件，促使美、中國期望改善關係的歷史事實，也是日後發生美國前國務卿季辛吉密訪中國事件的主因。[74]

然美國為安撫日本對美國在亞洲政策的疑慮，1969年11月21日日本首相佐藤榮作與美國總統尼克森發表共同聲明，表達維護台灣地區的和平與安全對日本的安全是極為重要的因素。而日本首相佐藤榮作針對此問題另再進一步表達，如果台灣受到外部的武力攻擊，美國若依據「中美共同防禦條約」履行防衛台灣的義務時，日本則會以國家利益觀點出發，採取應對措施。[75] 2021年4月17日新上任的美國總統拜登首度接待日本首相菅義偉的訪問時，日本應美國要求將「台灣海峽和平與穩定的重要性」納入共同聯合聲明中。依此可以看出美國對台海緊張情勢的憂慮，現今美國比起以往日本有更高的憂慮感。若從軍事的觀點，面對中國可能運用武力改變台海現狀的各種假設，美國也積極為可能的應對措施做準備。

依據2021年3月4日中國中央政府在「兩會」所提出的中國政府未來施政指導方針文件，〈中華人民共和國國民經濟和社會發展第十四個五年規劃和2035

---

[73] Kempe, Frederick, "Op-ed: China, Russia deepen cooperation in what could be Biden's defining challenge as president," *CNBC*, April 18, 2021, <https://www.cnbc.com/2021/04/18/op-ed-china-russia-cooperation-could-be-bidens-biggest-challenge.html> (檢索日期：2021年4月19日)

[74] Radchenko, Sergey, "The Island That Changed History," *The New York Times, March 4,* 2019, <https://www.nytimes.com/2019/03/02/opinion/soviet-russia-china-war.html?_ga=2.212976733.892836043.1618960607-1972870614.1570431049>(檢索日期：2021年4月21日)

[75] 段瑞聰，〈從日本角度看「中美共同防禦條約」〉，《國史館館刊》，第49期，2016年9月，頁123-124。

年遠景目標綱要〉報告。從報告中可以瞭解中國未來的發展仍是以國內的經濟、科技、環境、社會與軍隊現代化發展為主，對推進兩岸關係與統一的政策與以往並未有所改變。[76] 依此觀察，中國的國家安全與發展戰略目標，仍是以2035年達成國家基本現代化目標為主。若從中國的觀點，對於太平洋區域的主權爭議，在釣魚台列嶼主權問題上，主要是因應日本改變之前中、日擱置釣魚台列嶼主權爭議的共識所致；在提升台海軍事緊張情勢上，則是因應美國試圖將「一中政策」模糊化的應對作為；在南海主權問題上，主要在於因應美國挑戰中國在南海的主權聲張。因此，對於中國來說，對釣魚台列嶼、南海主權爭端及兩岸統一問題，應不會貿然採取積極性的行動，以解決領土爭議與國家統一的問題。

綜合上述從國內與國際兩個層次，對兩岸關係的未來發展分析。美國期望台灣處於在「一個中國」政策下，維持主權獨立的狀態，兩岸的關係在政治上有限度的交往，在軍事上保持對抗狀態。而台灣所期待的兩岸關係則是在維護國家的主權獨立的尊嚴下，在政治上與中國保持有限度的交流，在軍事上緩和軍事對峙的緊張情勢，在經濟上減低對中國的依賴。但中國對於兩岸關係的期待，將會隨著國家實力增強與軍事反擊能力的提升，或是被迫因應美國試圖改變對台灣的「一中政策」的狀況下，改變當前由美國與台灣所建構的台海情勢現狀。由於台灣當前政府否定中國所提的「九二共識」，所以在台灣沒有提出符合中國期待的有別於「九二共識」的兩岸關係主張之前，兩岸關係在政治與軍事對抗的型態不會改變，同時台灣經濟依賴中國的狀況也改變不了。

因此，未來試圖改變兩岸關係現狀的是中國，在台灣不接受中國所謂「一中原則」的情況下，軍事行動是改變兩岸關係的唯一手段。若從軍事的觀點，兩岸的軍事對抗，簡言之就是中美之間的軍事對抗。而未來核武大國之間對於是否發動戰爭的考量，「不在於是否會獲得戰爭的勝利，而在於是否能夠承擔戰爭後的損失。」因為當核武戰爭發生其結果是沒有贏家，所考量的是戰後

---

[76] 〈中華人民共和國國民經濟和社會發展十四個五年規劃和2035年遠景目標綱要〉，《中華人民共和國中央人民政府》，2021年3月13日，<http://www.gov.cn/xinwen/2021-03/13/content_5592681.htm>（檢索日期：2021年4月19日）

誰能最快復甦，重新領導新的世界秩序（先決條件是戰後的世界沒有被毀滅的話）。

肆、台灣的兩岸關係選擇

　　從兩岸關係的發展面分析結果，可得出如果民進黨政府在沒有提出符合中國期待，有別於「九二共識」的新兩岸關係主張之前，基本上兩岸關係沒有改善的機率，反而會在中美戰略互動的過程中，更朝向獨台的方向發展。因此，兩岸關係未來是走向統一或獨立，仍在於台灣的國家利益與安全戰略的選擇。而影響台灣對兩岸關係的戰略選擇的外在因素，一是中國會採取積極改變兩岸關係現況的時機；二是美國是否會採取實際的軍事行動介入兩岸的軍事衝突或戰爭。依此，將從台灣國家安全的戰略面思考與選擇，探討台灣的兩岸關係戰略選擇。

一、台灣的兩岸關係戰略思考

　　當思考國家安全戰略時，必須先確認國家主、客觀利益與安全。所謂「國家利益」，簡言之就是維護國家主權與人民生存與發展的權利。而「國家安全」，則是使國家與人民不感受到外的威脅。當前台灣討論兩岸關係的國家定位與走向時，可以發現執政的民進黨態度是「既不願意與中國統一，也不敢選擇脫離中國而獨立。」而在野的國民黨則是「既不願意脫離中國獨立，也不敢選擇與中國統一。」但當前兩岸現狀的維持仍受制於美中戰略互動的結果，而不是台灣自身的選擇。對於國家主權的定義，既無法改變當前所謂「一中」憲法，又不願承認憲法所呈現出主權與治權的分裂國家狀態。立場不同的政黨、團體及人民，對於國家主權、人民的身分與生存權的認知也就有不同的詮釋。尤其在國家的生存權的詮釋上，受限於台灣與國際法所定義的國家性質不同，相對的也影響台灣在闡述國家利益時所面臨的困境。

　　當台灣對國家利益的詮釋有所不同時，對於國家安全的感受也將不同。從國家利益的角度分析，若以統一的觀點，兩岸關係就有改善的空間。即使兩岸不會採取立即的統一行動，但對美國在亞太第一島鏈的安全布局將產生裂縫，

相對也會影響日本在東海對中國海空軍進入第二島鏈的圍堵戰略。由此，美、日對台灣的安全有可能構成挑戰，此時的台灣國家安全戰略思考，應是如何避免捲入中、美的軍事對抗。

　　若從台灣獨立的觀點，只要中國不放棄統一台灣的企圖，中國是敵人的身分就不會改變。而美國則是台灣唯一能制衡中國的大國。依此，如何促使美國改變「一中政策」是台灣國家安全戰略的主要思考方向，因為這是美國在考量是否採取實際軍事介入的關鍵因素。然而影響兩岸關係的主要因素是中美戰略互動的關係，所以台灣在實施國家安全戰略評估時，鑑於台灣在軍事上無法單獨與中國對抗，必會將下列兩個思考因素納入假設想定：一是為中國是否會付出巨大的代價武力統一台灣；二是當中國採取武力統一台灣的軍事行動時，美國是否會為確保對台灣安全的承諾派遣軍隊介入台海戰爭。

　　我們從兩岸關係發展面的解析，可以確定中國對於統一台灣不是決心的問題，而是落實統一行動的時間與方法問題。而美國是否願意介入台海戰爭，所考量的是戰爭的損失是否能夠承擔，以及承擔的程度。因此，台灣從國家安全戰略思考兩岸關係時，應建構在能成為美中戰略互動的權力平衡中，扮演砝碼的能動者角色，而不是美國在與中國的權力平衡對抗中的被動籌碼(台灣牌)。

二、台灣的兩岸關係戰略選擇

　　台灣從國家安全戰略的策略面思考兩岸關係的選項時，基本上就只有獨立、統一及維持現狀三種選項，即使有部分學者、專家提出各種所謂「邦聯」的論述，也許在論述上有所不同，但終究還是傾向於兩岸統一的概念。分析如下：

（一）獨立

　　依據政治大學選舉研究中心對台灣民眾對於台灣人的身分認同與統獨取向的民意調查結果顯示，與中國談「統一」不是大多數台灣民眾的選項，也是當前民進黨政府所希望極力排除的選項。若台灣在美國的支持下選擇獨立，兩岸無可避免的會發生戰爭。其結果不管台灣在戰爭中取得勝利與否，台灣人民的

生命、財產和現代化的基礎建設，必將遭受嚴重的破壞，可能需要花上數十年的時間重建，才能恢復現在的生活。這樣的結果對中國來說是可以承受之痛，而對美國來說只不過是國際格局的權力消長。

（二）統一

對於台灣是否與中國「統一」的問題，當前的台灣政府與社會氛圍已無法做理性的討論。尤其是此議題成為政黨選舉操作議題時，更無法讓台灣學者、專家能從國家安全的客觀角度，分析兩岸統一對台灣的利弊得失。造成即使不排斥兩岸走向統一選項的國民黨重新執政，短期內台灣人民主動選擇統一的可能性相當低。未來台灣會將統一納入選項的可能因素，主要來自於美國與中國的外在因素影響。而此外在因素，一為美國主動或被動放棄對台灣的軍事支持；二為中國採取有限度的軍事行動，造成台灣人民的生命及財產損失。然台灣政府會被迫同意中國所提出的兩岸統一條件的狀況，應是受到戰爭影響的台灣民眾，不願受到更大生命財產損失對政府所提出要求。因此，台灣會做統一的選項將是出於被迫，而不是自願的。相對的台灣在被迫統一談判的過程中，除將失去代表國家主權的國防與外交權力外，在自治的權利上亦將會有所限制。

（三）維持現狀

若台灣仍採取維持現狀的兩岸關係選項，則必須認清兩岸現狀維持的權力，掌控在美、中手上。換言之，美中的戰略互動決定兩岸關係的走向。自2016年美國總統川普開啟美、中對抗的局面之後，中國為回應美國對台關係的調整，試圖對台灣採取軍事壓力迫使台灣回到兩岸朝向統一的選項。如果台灣只能成為美國對抗中國的籌碼時，台灣終將無可避免的在統一與獨立之間做選項，沒有第三條道路的選項。因此，台灣要能在國際「一個中國」政策下，維持其獨立自主行使象徵國家的主權的權利與義務，必須能在美、中權力平衡上能動者的砝碼角色。而台灣能否成為美、中權力平衡上的能動者砝碼身分，除了必須依靠軍事實力來支持外，更重要的是具有對中國及美國關係的自主選擇權。

綜合上述分析，台灣在思考兩岸關係的選擇時，首先必須認清美國在西太平洋區域的絕對優勢已不若以往。面對中國統一台灣的決心，樂觀的說2050年是中國統一台灣的最後期限，以實現中國共產黨的中國夢與民族復興大業。但不可否認隨著中國海、空軍快速的現代化發展，2027年中國人民解放軍建軍100週年，對台灣來說是國家安全危機重大挑戰的時機。[77] 也正如美國印太司令部司令菲利普‧戴維森（Philip Davidson）上將，在美參議院軍事委員會答覆參議員詢問時表示，未來六年內有可能發生中國武力攻擊台灣的情勢。[78]

因此，台灣在兩岸關係的建構上，「統一」是無法排除的課題。雖然維持現狀是台灣當前最佳的選項，但如果無法跳脫唯美國為依靠的思維，兩岸關係對台灣始終是處於被動的狀態。若台灣政府必須在兩岸關係上選擇「統一」時，台灣政府應向台灣人民明確闡述與中國「統一」談判，不等於中華民國必須先「投降」，也不代表需要「放棄國防」。因為「國防」是台灣與中國談判的籌碼，目的是藉由展現國防的實力與決心，在兩岸和平談判中獲取台灣人民最大利益，也就是台灣人民生命安全、社會生活安定、經濟持續發展、保有現行政治制度與國防自主的權力。

伍、結論

本論文依據整合戰略與國際關係理論的研究途徑，從事實面、影響面、發展面、戰略面及政策面5個面向，分析美中戰略互動下的兩岸關係。從事實面的問題意識與問題性質分析部分：中國崛起已是無法改變事實，美中的戰略互動不會走向傳統現實主義的權力平衡狀態，而美國是在美中戰略互動中的主動者。美國將針對不同的區域與議題，對中國採取不同的應對方式，也就是在合作、競爭及對抗的三種型態中交互使用。

在影響面的國家利益分析部分：美、中在台灣問題上所呈現的是政治與軍

---

[77] 李海濤，〈為實現建軍百年奮鬥目標砥礪前行〉，《中華人民共和國國防部》，2020年12月14日，<http://www.mod.gov.cn/jmsd/2020-12/14/content_4875402.htm>（檢索日期：2021年4月21日）

[78] 〈美印太司令部上將：解放軍可能在六年內攻打台灣〉，《BBC News/中文》，2021年3月11日，<https://www.bbc.com/zhongwen/trad/chinese-news-56344323>（檢索日期：2021年4月21日）

事對抗的性質。兩岸關係是否會陷入緊張狀態，主要在於台灣是否接受中國所認知的「九二共識」。而台灣是否能夠抵擋或屈服於中國的政治和軍事壓力，仍在於美國是否會派遣軍隊介入兩岸可能的戰爭危機。

在發展面的國內與國際因素分析部分：未來試圖改變兩岸關係現狀的是中國，在台灣不接受中國所謂「一中原則」的情況下，軍事行動是改變兩岸關係的唯一手段。未來核武大國之間對於是否發動戰爭的考量，「不在於是否會獲得戰爭的勝利，而在於是否能夠承擔戰爭後的損失。」

在戰略面的因應對策與決策過程分析部分：台灣從國家安全戰略思考兩岸關係時，應建構在能成為美中戰略互動的權力平衡中，扮演砝碼的能動者角色，而不是美國在與中國的權力平衡對抗中的被動籌碼(台灣牌)。

在策略面的因應作為分析部分：台灣在兩岸關係的建構上，「統一」是無法排除的課題。雖然維持現狀是台灣當前最佳的選項，但如果無法跳脫唯美國為依靠的思維，兩岸關係對台灣始終是處於被動的狀態。若台灣政府必須在兩岸關係上選擇「統一」時，台灣政府應向台灣人民明確闡述與中國「統一」談判，不等於中華民國必須先「投降」，也不代表需要「放棄國防」。

因此，台灣在美中戰略互動的過程中，如何在兩岸關係中獲取主動。台灣政府必須認清以現今中國在國際的實力與影響力，國際的「一個中國」政策改變的可能相當的低。如何詮釋「九二共識」以外的「一中」是兩岸關係改善的關鍵鎖鑰，即使不是台灣政府與人民的意願，都不須做好準備以因應中國的軍事壓力，確保台灣人民的生命財產。

Taiwan and Sri Lanka: The Two Unsinkable Aircraft Carriers of the Indo-Pacific Strategy (IPS) of the United States and the Belt & Road Initiative (BRI) of China

Patrick Mendis、Joey Wang [*]

The Communist Party of China (CPC) places a great emphasis on the significance of anniversaries and political events, one of which is the May Fourth Movement against imperialism. Organized by students in 1919, the movement called for democracy, which led to the founding of CPC two years later. The 70th anniversary of the student movement was an inspiration for another group of Peking University students to once again lead pro-democracy protest. This protest was brutally suppressed and is now known in the annals of history as the 1989 Tiananmen Square Massacre.

More importantly for the CPC, however, is its national identity rooted in the defeat of Nationalist (or Kuomintang, KMT) forces by the Communists who then established the People's Republic of China (PRC) led by Chairman Mao Zedong in 1949. The KMT Generalissimo Chiang Kai-shek, and two million supporters who fled to Taiwan, claimed that he was the "President of all China" in the government of the Republic of China (ROC) in Taipei, supported by the United States.[1]

On January 1, 1979, when the United States formally recognized the People's

---

[*] Taiwan Fellow of the Ministry of Foreign Affairs and Distinguished Visiting Professor of Global Affairs, National Chengchi University
[1] Eric Pace, "Chiang Ching-Kuo Dies at 77, Ending a Dynasty on Taiwan," *New York Times*, January 14, 1988, https://www.nytimes.com/1988/01/14/obituaries/chiang-ching-kuo-dies-at-77-ending-a-dynasty-on-taiwan.html (accessed, March 7, 2020).

Republic of China (PRC) diplomatically from Taipei as the only lawful Chinese government, Beijing declared the ending of regular "artillery bombardment of Taiwan-controlled offshore islands close to the mainland and opened communications" with the ROC.[2] However, President Chiang Ching-kuo, the son of Chiang Kai-shek, refused Beijing's olive-branch offer with the "three-no's policy"— no contact, no compromise, and no negotiation with China—even as there is still no formal peace treaty with the two governments. [3]

The situation has now changed with successive governments as there is greater movements of people, ever-increasing commercial interactions, and growing cross-Strait investment in both directions.[4] Between 1988 and 2018, there were over 134 million cross-Strait visits. Two-way trade reached 2.6 trillion US dollars and the mainland has been Taiwan's largest market and top investment destination. This is complicated not only by the generations, who are being born with a unique Taiwanese identity, but also the claims of Taiwan's indigenous people that "Taiwan is a traditional area of the aboriginal peoples and is not part of China's territory," nor are they "ethnic minorities of the Chinese nation." [5]

II. THE NEW ERA OF PRESIDENT XI

With an iron hand in a velvet glove, Chinese President Xi Jinping conveyed in a

---

[2]CNBC Asia-Pacific News, "China to kick off year of sensitive anniversaries with a major speech on Taiwan," CNBC, December 30, 2018, https://www.cnbc.com/2018/12/31/china-to-kick-off-year-of-sensitive-anniversaries-with-major-speech-on-taiwan.html (accessed, March 7, 2020).

[3]Reuters World News, "China to kick off year of sensitive anniversaries with major speech on Taiwan," Reuters, December 30, 2018, https://www.reuters.com/article/us-china-taiwan/china-to-kick-off-year-of-sensitive-anniversaries-with-major-speech-on-taiwan-idUSKCN1OU049 (accessed, March 7, 2020).

[4]Xinhua, "Xinhua Headlines: Xi says "China must be, will be reunified" as key anniversary marked," Xinhua, January 2, 2019, http://www.xinhuanet.com/english/2019-01/02/c_137714898.htm (accessed, March 7, 2020).

[5]Anna Fifield, "Taiwan's 'born independent' millennials are becoming Xi Jinping's lost generation," Washington Post, December 26, 2019, https://www.washingtonpost.com/world/asia_pacific/taiwans-born-independent-millennials-are-becoming-xi-jinpings-lost-generation/2019/12/24/ce1da5c8-20d5-11ea-9c2b-060477c13959_story.html (accessed, March 7, 2020). See also: Keoni Everington, "Taiwan aborigines' message to Xi: 'Taiwan is not part of China,'" Taiwan News, January 8, 2019, https://www.taiwannews.com.tw/en/news/3612726 (accessed, March 7, 2020)

40th anniversary message to "compatriots in Taiwan" on January 1, 2019 that "China must be, will be reunified."[6] While Xi offered a five-point proposal for a "peaceful unification" with China, he also made no promise to renounce the use of force.[7] Invoking Deng's principles of "one country, two systems" as the best approach to realizing national reunification, Xi said, "it is a historical conclusion drawn over the 70 years of the development of cross-Strait relations, and a must for the great rejuvenation of the Chinese nation in the new era."

Nevertheless, with a growing Taiwanese national identity, a peaceful unification with a Chinese communist regime will be challenging.[8] Xi maintains that "Taiwan independence goes against the trend of history and will lead to a dead end." With the ongoing Hong Kong protests and American support for its democracy, the policy of ambiguity seems to have been all but abandoned.

As China approaches the centenary of the founding of the PRC in 2049, there is great concern in Taiwan that President Xi could become impatient and miscalculate— embroiling the US and destabilizing the entire region in the process.[9] The Chinese push for unification, the trade war, the emerging lessons from Hong Kong, and the fragility of the balance of power over the Taiwan Strait have now fully exposed the geopolitical rivalry between Beijing and Washington.

For Taiwan, the hedge against Beijing's aggressive campaign has no options but

---

[6]China.org.cn, "Message to Compatriots in Taiwan," *China.org,* January 1, 1979, http://www.china.org.cn/ english/7943.htm (accessed, March 7, 2020)

[7]Xinhua, "Xinhua Headlines: Xi says "China must be, will be reunified" as key anniversary marked," *Xinhua,* January 2, 2019, http://www.xinhuanet.com/english/2019-01/02/c_137714898.htm (accessed, March 7, 2020)

[8]Alice Su, "Must Reads: With each generation, the people of Taiwan feel more Taiwanese – and less Chinese," *Los Angeles Times,* February 15, 2019, https://www.latimes.com/world/asia/la-fg-taiwan-generation-gap-20190215-htmlstory.html (accessed, March 7, 2020). See also: Anna Fifield, "Taiwan's 'born independent' millennials are becoming Xi Jinping's lost generation," *Washington Post,* December 26, 2019, https://www. washingtonpost.com/world/asia_pacific/taiwans-born-independent-millennials-are-becoming-xi-jinpings-lost-generation/2019/12/24/ce1da5c8-20d5-11ea-9c2b-060477c13959_story.html (accessed, March 7, 2020).

[9]Nicholas Kristof, "This is How a War With China Could Begin: First, the lights in Taiwan go out," *New York Times,* September 6, 2019, https://www.nytimes.com/2019/09/04/opinion/china-taiwan-war.html (accessed, March 7, 2020).

the United States. The fate of Taiwan, because of history, is inextricably grounded in China's complex relations with America and its democratic alliance of the new Quadrilateral Security Dialogue (i.e., the Quad) with Australia, India, and Japan. The *South China Morning Post* has called the fate of Taiwan "the most disruptive factor in Beijing's complex relations with Washington." [10]

Against this backdrop, the Donald Trump administration has recently decided to sell Taiwan $8 billion worth of F-16 fighter jets. Last year, the US government also permitted Taiwan to import American submarine technologies to help develop the island's own submarine industry, initially building a $3.3 billion indigenous submarine in the southern port city of Kaohsiung. Taiwan is now also requesting US assistance in assessing its combat readiness in the face of increasing threats from the Chinese mainland. [11]

Since her famous phone call with then President-elect Trump in December 2016, Taiwan's President Tsai Ing-Wen has increasingly made efforts to develop Taiwan's own version of a military-industrial-complex, knowing that for China, all roads eventually lead to the unification of the island with the mainland. [12] Despite more than 80 percent of Taiwanese citizens opposed to unifying with China, Beijing has continued to claim sovereignty over Taiwan. [13]

---

[10] Shi Jiangtao, "Is Xi Jinping's Taiwan reunification push hastening a US-China clash?," *South China Morning Post*, January 13, 2019, https://www.scmp.com/news/china/diplomacy/article/2181600/xi-jinpings-taiwan-reunification-push-hastening-us-china-clash (accessed, March 7, 2020)

[11] Lawrence Chung, "Taiwan invites US to help gauge its military strength as analysts warn of growing threat from mainland China," *South China Morning Post*, October 30, 2019, https://www.scmp.com/news/china/military/article/3035625/taiwan-invites-us-help-gauge-its-military-strength-analysts (accessed, March 7, 2020).

[12] Ralph Jennings, "Meet the New Military-Industrial Complex in Taiwan," *VOA*, October 14, 2019, https://www.voanews.com/east-asia-pacific/meet-new-military-industrial-complex-taiwan, (accessed, March 7, 2020).

[13] RFA Radio Free Asia, "More Than 80 Percent of Taiwanese Reject China's 'Unification' Plan," *RFA*, January 9, 2019, https://www.rfa.org/english/news/china/more-than-80-percent-of-taiwanese-01092019115150.html (accessed, March 7, 2020).

III. THE GRAND PLAN AND THE SIX STABILITIES

It is important not to lose sight of the fact that the Taiwan issue is simply one step in China's greater strategic objective of pushing the United States as well as its influence out of Asia, similar to what Japan had tried to do after the turn of the 20th century. While securing Taiwan would be an important achievement, China's strategic objective cannot be achieved without other supporting objectives embedded in the Belt and Road Initiative (BRI), the Asian Infrastructure Investment Bank (AIIB), the control of the South and East China Seas as well as the Indian Ocean Region (IOR), and ensuring energy security for its citizens. Each of these objectives must be viewed through the regional diplomacy that Beijing is exercising throughout Africa, Latin America, throughout the Pacific, and the Caribbean basin. They must also be understood within the geopolitics of the regional balance of power through the Washington-led Quad.

Beijing's strategic objectives begins with the survival of the CPC and its very "legitimacy" to rule, according to President Xi's anti-corruption chief, Wang Qishan.[14] A report by the US-China Economic and Security Review Commission to the United States Congress in 2018 concluded that the primary goal of CPC is to "maintain its hold on power by ensuring domestic stability, protecting sovereignty claims, and defending China's territorial integrity."[15] In essence, these core policies are coded in the National Security Law of China.[16] Because China's strategic objective is predicated on this legitimacy, the CPC must provide economic and social

---

[14] The Economist, "A Crisis of faith: In their response to wobbly markets, China's leaders reveal their fears," January 16, 2016, https://www.economist.com/briefing/2016/01/16/a-crisis-of-faith (accessed, March 7, 2020).

[15] USCC.gov, "2018 Annual Report to Congress," *U.S.-China Economic and Security Review Commission*, November 2018, https://www.uscc.gov/sites/default/files/2019-09/2018%20Annual%20Report%20to%20 Congress.pdf (accessed, March 7, 2020).

[16] Eng.mod.gov.cn, "National Security Law of the People's Republic of China (2015) [Effective]," *Ministry of National Defense of the People's Republic of China,* March 3, 2017, http://eng.mod.gov.cn/ publications/2017-03/03/content_4774229.htm (accessed, March 7, 2020).

stability for its citizens. This cannot be achieved without an uninterrupted supply of energy resources to deliver not only economic growth but also sustain military operations if and when needed. This is key to the preservation of the CPC's power. It is no surprise that given China's current economic challenges, Beijing has called for the "six stabilities" of the economy—including employment, finance, trade, and investment—for the survival of the CPC. [17]

IV. DRIVEN BY NATIONAL IDENTITY

China's primary objectives of protecting sovereignty claims and defending its territorial integrity begin with Taiwan as part of its national identity. Even in 1951, the American Central Intelligence Agency (CIA) had recognized Taiwan in a now declassified report as "the last stronghold of the Nationalist regime" and that the Chinese were resolute in "capturing Taiwan in order to complete the conquest of Chinese territory."[18] The rest of the world has moved on, but for the Chinese, the civil war and the hostilities that followed have never formally ended.

For China, Taiwan is not merely a runaway province, it is a focal point in the first island chain that would provide China with a "permanent aircraft carrier" (as General Douglas McArthur first used the phrase) to project power out to the Western Pacific and the second island chain. This is generally referred to as Anti Access/Area Denial. [19]

In addition, Andrew Erickson of the US Naval War College and Joel Wuthnow

---

[17] Nicole Hao, "Top Chinese Leaders Call for focus on 'Six Stabilities' as Economy Lags," *The Epoch Times*, November 1, 2018, https://www.theepochtimes.com/as-chinas-economy-lags-top-leaders-call-for-six-stabilities_2706129.html (accessed, March 7, 2020).

[18] CIA.gov, "National Intelligence Estimate: Communist China," *Central Intelligence Agency*, January 17, 1951, https://www.cia.gov/library/readingroom/docs/CIA-RDP98-00979R000100150001-4.pdf, (accessed, March 7, 2020).

[19] Kevin Rudd, "How Xi Jinping Views the World: The Core Interests That Shape China's Behavior," *Foreign Affairs*, May 10, 2018, https://www.foreignaffairs.com/articles/china/2018-05-10/how-xi-jinping-views-world (accessed, March 7, 2020).

of the US National Defense University have quoted Chinese military sources, which conclude that without securing Taiwan, "a large area of water territory and rich reserves of ocean resources will fall into the hands of others" and "China will forever be locked to the west side of the first chain of islands in the West Pacific." [20] Another Chinese military publication has further concluded that "the biggest obstacle to the expansion of our national interests comes from the First and Second Island Chains set up by the United States." [21]

The United States, for its part, takes no official position with regard to the competing claims in the South China Sea.[22] However, the US and other Western nations do exercise their rights to conduct Freedom of Navigation Operations (FONOPS) under the United Nations Law of the Sea Treaty. Despite Chinese missiles and other emerging technologies that pose significant threats, France sent a warship through the Taiwan Strait in April, as did the US in May, Canada in September, and the UK in December, 2019.[23]

Thus, the Chinese establishment of its Air Defense Identification Zone (ADIZ) in the East China Sea and the island-reclamation followed by military buildup in the

---

[20] Andrew S. Erickson, Joel Wuthnow, "Barriers, Springboards and Benchmarks: China Conceptualizes the Pacific "Island Chains,", *The China Quarterly*, March 2016, https://www.cambridge.org/core/journals/china-quarterly/article/barriers-springboards-and-benchmarks-china-conceptualizes-the-pacific-island-chains/B46A 212145EB9D920616650669C697F0 (accessed, March 7, 2020).

[21] Ibid.

[22] Prashanth Parameswaran, "U.S. South China Sea policy after the ruling: Opportunities and challenges" *Brookings*, July 22, 2016, https://www.brookings.edu/opinions/u-s-south-china-sea-policy-after-the-ruling-opportunities-and-challenges/ (accessed, March 9, 2020).

[23] Idrees Ali, "U.S. Navy again sails through Taiwan Strait, angering China, *Reuters*, May 22, 2019, https://www.reuters.com/article/us-usa-china-navy/us-navy-again-sails-through-taiwan-strait-angering-china-idUSKCN1ST062 (accessed, March 7, 2020). See also: Idrees Ali, Phil Stewart, "Exclusive: In rare move, French warship passes through Taiwan Strait," *Reuters*, April 24, 2019, https://www.reuters.com/article/us-taiwan-france-warship-exclusive/exclusive-in-rare-move-french-warship-passes-through-taiwan-strait-idUSKCN1S027E (accessed, March 9, 2020), Ralph Jennings, "Canada Sends Warships Through Sensitive Taiwan Strait," *VOA*, September 13, 2019, https://www.voanews.com/americas/canada-sends-warships-through-sensitive-taiwan-strait (accessed, March 7, 2020), Teddy Ng, "British navy vessel passes through Taiwan Strait," South China Morning Post, December 7, 2019, https://www.scmp.com/news/china/diplomacy/article/3041076/british-navy-vessel-passes-through-taiwan-strait (accessed, March 7, 2020).

Spratly and Paracel Islands in the South China Sea—as well as the use of a maritime militia—are not only meant to claim *de facto* rights to the resources within the nine-dash-line, they are tactical steps toward employing coercive diplomacy with its neighbors, and establishing operational control over the region in its move toward unification with Taiwan.

A separate but related issue is that China likely believes it still has some unfinished business with Japan—both with respect to the Senkaku (Diaoyu in Chinese) islands in the East China Sea—as well as in the broader historical context of national humiliation.[24] Under President Hu Jintao during the years of 2003-07, for example, Japanese fighter jet scrambles against Chinese intrusions averaged about 37 per year.[25] However, under President Xi Jinping, the number of fighter scrambles have averaged 560 per year during 2013-17—a 15-fold increase. According to *Jane's Defense Weekly* in April 2019, the Japan Air Self-Defense Force scrambled 999 times in FY2018, and "523 times between April 1 and December 31, 2019"—"a 9% increase compared with the same period in 2018." [26] If there are any doubts about China's lingering and justifiable historical grievances, it is likely no accident that China's first indigenously built aircraft carrier has been commissioned the "Shandong"—a colonial province that was ceded to Japan after World War I.[27]

---

[24] David Lague, "China's hawks take the offensive," *Reuters Investigates*, January 17, 2013, https://www.reuters.com/investigates/china-military/ (accessed, March 7, 2020)

[25] Masataka Oguro, "Ensuring Japan's Future Air Security: Recommendations for Enhancing the JASDF's Readiness to Confront Emerging Threats," *Foreign Policy at Brookings*, September 25, 2018,

[26] Kosuke Takahashi, "Number of JASDF scrambles in FY 2018 second highest on record, says MoD," *Janes*, April 15, 2019, https://www.janes.com/article/87889/number-of-jasdf-scrambles-in-fy-2018-second-highest-on-record-says-mod (accessed, March 7, 2020). See also: Gabriel Dominguez, "JASDF sees rise in number of Chinese military aircraft approaching Japanese airspace,", *Janes*, January 29, 2020, https://www.janes.com/article/93981/jasdf-sees-rise-in-number-of-chinese-military-aircraft-approaching-japanese-airspace (accessed, March 7, 2020)

[27] Steven Lee Myers, "China Commissions 2nd Aircraft Carrier, Challenging U.S. Dominance," *New York Times*, December 17, 2019, https://www.nytimes.com/2019/12/17/world/asia/china-aircraft-carrier.html (accessed, March 7, 2020)

V. THREE ENABLERS OF NATIONAL REJUVENATION

The key enabler that has allowed Beijing to protect its sovereign claims and project its power has been China's explosive economic growth. As it cools, however, major programs such as the BRI will be critical to any future projection of power. As envisioned, the purpose of BRI is to promote the economic prosperity of the countries along the Belt and Road and regional economic cooperation, strengthen exchanges and mutual learning between different civilization, and promote world peace and development."[28]   Behind this heady mixture of material, economic, and cultural aspirations, however, there are other motivations not likely to be mentioned in official Chinese literature.

First, China also wants to decrease the dependence on its domestic infrastructure investment and begin moving investments overseas to address the industrial capacity overhang.[29] China's average growth rate over three decades up until 2010 had been 10 percent.[30] However, this was a growth rate that was largely recognized as unsustainable. And, in fact, has been trending downward near 6.5 percent over the past three years.[31] Chinese construction companies, equipment manufacturers, and other related businesses that had profited from the domestic building boom are now having to look elsewhere for growth opportunities. The key instrument of this investment transfer comes with the Chinese system of "state capitalism," which has further been solidified by President Xi.[32]

---

[28] The State Council of the People's Republic of China, "Visions and Actions on Jointly Building Silk Road Economic Belt & 21st-Century Maritime Silk Road,", March 28, 2015, https://www.beltandroad.news/action-plan/ (accessed, March 7, 2020)

[29] Moody's Analytics, "The Belt and Road Initiative – Six Years On," June 2019, https://www.moodysanalytics.com/-/media/article/2019/belt-and-road-initiative.pdf (accessed, March 7, 2020)

[30] BBC, "China's economy grows at slowest pace since 1990s," July 15, 2019, https://www.bbc.com/news/business-48985789 (accessed, March 7, 2020)

[31] The World Bank, Country Profile, https://data.worldbank.org/country/china, (accessed, March 7, 2020

[32] Robert J. Samuelson, "Why China clings to state capitalism,", Washington Post, January 9, 2019, https://www.washingtonpost.com/opinions/why-china-clings-to-state-capitalism/2019/01/09/5137c6d4-141e-11e9-b6ad-9cfd62dbb0a8_story.html (accessed, March 7, 2020)

When Deng Xiaoping began adopting market-oriented reforms in 1978, it allowed private enterprises to flourish. These private companies ultimately became the engine that has powered the Chinese economy as it has made significant contributions "to China's stellar output, employment and export growth." [33] Indeed, Xi himself recognized the role of market forces during the Third Plenum of the CPC's 18th Congress in 2013 that "the government must retreat from its current powerful role in allocating resources." [34]

Since then, however, President Xi appears to have had a change of heart. Rather than building on the conclusions of the Plenum, he has, instead placed emphasis on the role of state industrial policy and state-owned enterprises.[35] Among the BRI infrastructure development projects, Chinese companies accounted for 89 percent of the contractors, according to a five-year analysis of BRI projects by the Center for Strategic and International Studies in Washington.[36] A recent study by the *Washington Post* also suggests that BRI projects are generally more likely to be funded by state-owned enterprises (SOEs) than by the private sector. [37] The reason for this is that a private enterprise assesses a project in terms of its financial viability. If a project has no foreseeable return on its investment, then there is simply no business rationale for its undertaking. This is not the case for an SOE because SOEs exist to facilitate

---

[33] Nicholas Lardy, "Xi Jinping's turn away from the market puts Chinese growth at risk," *Financial Times*, January 15, 2019, https://www.ft.com/content/3e37af94-17f8-11e9-b191-175523b59d1d (accessed, March 7, 2020)

[34] Arthur R. Kroeber, "Xi Jinping's Ambitious Agenda for Economic Reform in China," *Brookings*, November 17, 2013, https://www.brookings.edu/opinions/xi-jinpings-ambitious-agenda-for-economic-reform-in-china/ (accessed, March 7, 2020)

[35] David J. Lynch, "Initial U.S.-China trade deal has major hole: Beijing's massive business subsidies," *Washington Post*, December 31, 2019, https://www.washingtonpost.com/business/economy/initial-us-china-trade-deal-has-major-hole-beijings-massive-business-subsidies/2019/12/30/f4de4d14-22a3-11ea-86f3-3b5019d451db_story.html (accessed, March 7, 2020)

[36] Jonathan E. Hillman, "China's Belt and Road Initiative: Five Years Later," Center for Strategic and International Studies, January 25, 2018, https://www.csis.org/analysis/chinas-belt-and-road-initiative-five-years-later-0 (accessed, March 7, 2020)

[37] Xiaojun Li, Ka Zeng, "Beijing is counting on its massive Belt and Road Initiative. But are Chinese firms on board?," *Washington Post*, May 14, 2019, https://www.washingtonpost.com/politics/2019/05/14/beijing-is-counting-its-massive-bridge-road-initiative-are-chinese-firms-board/ (accessed, March 7, 2020)

Beijing's broader political objectives. Therefore, the government has much greater control over companies.[38] As a tool of economic statecraft, profit maximization is not necessarily the critical deciding factor if, in fact, a deciding factor at all. It matters little that many of these companies are often called "zombie enterprises" as they continually incur losses and must rely on government or bank support to survive.[39]

Second, China wants to internationalize the use of its currency along BRI and with the new partners of Africa, Latin America, and the Caribbean basin. Making the Renminbi (RMB) a global currency in 2015 had been one of the highest economic priorities of Beijing's larger strategy.[40] China and some 65-BRI countries in Eurasia—which account collectively for over 30 percent of global GDP, 62 percent of population, and 75 percent of known energy reserves—are increasingly using the RMB to facilitate trade and infrastructure projects.[41] Pakistan, for one, has switched from the dollar to the RMB for bilateral trade with China after President Donald Trump publicly attacked Pakistan on Twitter for harboring terrorists.[42] And its foreign trade settlements in RMB surged in the first 11 months of 2019.[43] The China-Pakistan Economic Corridor (C-PEC) to Xinjiang—as one of the massive projects

---

[38] Ibid.

[39] W. Raphael Lam, Alfred Schipke, Yuyan Tan, Zhibo Tan, "Resolving China's Zombies: Tackling Debt and Raising Productivity," *International Monetary Fund*, November 27, 2017, https://www.imf.org/en/Publications/WP/Issues/2017/11/27/Resolving-China-Zombies-Tackling-Debt-and-Raising-Productivity-45432 (accessed, March 7, 2020)

[40] Keith Bradsher, "China's Renminbi Is Approved by I.M.F. as a Main World Currency," New York Times, November 30, 2015, https://www.nytimes.com/2015/12/01/business/international/china-renminbi-reserve-currency.html (accessed, March 7, 2020)

[41] The World Bank, "BRI at a Glance", March 29, 2018, https://www.worldbank.org/en/topic/regional-integration/brief/belt-and-road-initiative (accessed, March 7, 2020). See also: Karen Yeung, "China's yuan more popular as a reserve currency despite trade war, IMF data shows," *South China Morning Post*, October 2, 2018, https://www.scmp.com/business/money/wealth/article/2166652/chinas-yuan-more-popular-reserve-currency-despite-trade-war (accessed, March 7, 2020)

[42] CNBC World News, "Pakistan summons the US ambassador in protest after Trump's angry tweet," *CNBC*, January 2, 2018, https://www.cnbc.com/2018/01/02/pakistan-summons-the-us-ambassador-in-protest-after-trumps-angry-tweet.html (accessed, March 7, 2020)

[43] The News International, "'Foreign trade settlement in yuan climbs 2.5 time in 11 months'", December 21, 2019, https://www.thenews.com.pk/print/586199-foreign-trade-settlement-in-yuan-climbs-2-5-times-in-11-months (accessed, March 10, 2020)

under the BRI—can now depend upon a steady stream of Chinese capital. Pakistan can now also minimize the risk of Washington's threats such as cutting off economic assistance and military support.[44] Use of the RMB would also help authoritarian regimes like Iran, North Korea, and Sudan to undermine American-imposed "financial sanctions" on the violations of such norms as human rights, child labor, and human trafficking.[45] Furthermore, the success of BRI, if achieved, would establish Eurasia as the largest economic market in the world and the changing currency dynamics could initiate a shift in the world away from the dollar-based financial system by creating new geopolitical concerns for the United States. [46]

Third, China seeks to secure its energy resources through new pipelines in Central Asia, Russia, and South and Southeast Asia's deep-water ports. Beijing's leadership for some years has been concerned about its "Malacca Dilemma" as President Hu Jintao declared in 2003 that "certain major powers" may control the Strait of Malacca and China needed to adopt "new strategies to mitigate the perceived vulnerability."[47] The Strait of Malacca is not only the main conduit connecting the Indian Ocean and the Pacific Ocean to China via the South China Sea but also the shortest sea route between oil suppliers in the Persian Gulf and key markets in Asia.[48] In 2016, 16 million barrels of crude oil transited through the Malacca Strait

---

[44] Nyshka Chandran, "Pakistan is ditching the dollar for trade with China – 24 hours after Trump denounced the country," *CNBC Politics*, January 3, 2018, Updated January 4, 2018, https://www.cnbc.com/2018/01/03/pakistan-china-ties-strengthen-after-president-donald-trumps-rant.html (accessed, March 7, 2020)

[45] Keith Bradsher, "China's Renminbi Is Approved by I.M.F. as a Main World Currency," New York Times, November 30, 2015, https://www.nytimes.com/2015/12/01/business/international/china-renminbi-reserve-currency.html (accessed, March 7, 2020)

[46] Simeon Djankov and Sean Miner, Editors, "China's Belt and Road Initiative: Motives, Scope, and Challenges," *Peterson Institute for International Economics*, March 2016, https://www.piie.com/system/files/documents/piieb16-2_1.pdf (accessed, March 7, 2020)

[47] Ian Storey, "China's Malacca Dilemma," *The Jamestown foundation: China Brief*, April 12, 2006, https://jamestown.org/program/chinas-malacca-dilemma/ (accessed, March 7, 2020)

[48] U.S. Energy Information Administration, "The Strait of Malacca, a key oil trade chokepoint, links the Indian and Pacific Oceans," August 11, 2017, https://www.eia.gov/todayinenergy/detail.php?id=32452 (accessed, March 7, 2020)

each day, of which 6.3 million barrels were destined for China.[49] In 2017, China surpassed the United States as the world's largest crude oil importer, importing 8.4 million barrels/day. [50] Therefore, the sustainability and security of energy supplies is a key input not only to China's domestic stability and economic growth but also to its military operations and, concomitantly, the very legitimacy of the CPC. Initiatives under the BRI—such as the C-PEC to Xinjiang province, the Kyaukpyu pipeline in Myanmar that runs to Yunnan province, and the ongoing discussions for the proposed Kra Canal in Thailand—are of vital interest to China because they would provide alternative routes for energy resources from the Middle East directly to China that bypass the Malacca Strait.[51] The BRI will also support expansion of China's military bases across the Bay of Bengal, Indian Ocean, and the Arabian Sea.[52]

VI. BRI 2.0 AS SELF-CORRECTIONS

BRI has been a bumpy ride in its first six years. Since its inception in 2013, allegations of the lack of transparency, fraud, and corruption persist, as has capital flight, and "the frenzy of buying soccer clubs," all under the banner of BRI.[53] And

---

[49] U.S. Energy Information Administration, "World Oil Transit Chokepoints," July 15, 2017, https://www.eia.gov/international/analysis/special-topics/World_Oil_Transit_Chokepoints, (accessed, March 7, 2020). See also: U.S. Energy Information Administration, "More than 30% of global maritime crude oil trade moves through the South China Sea, August 27, 2018, https://www.eia.gov/todayinenergy/detail.php?id=36952 (accessed, March 7, 2020)

[50] U.S. Energy Information Administration, "China surpassed the United States as the world's largest crude oil importer in 2017," December 31, 2018, https://www.eia.gov/todayinenergy/detail.php?id=37821 (accessed, March 7, 2020).

[51] Michael J. Green, "China's Maritime Silk Road: Strategic and Economic Implications for the Indo-Pacific Region," *Center for Strategic and International Studies*, April 2, 2018, https://www.csis.org/analysis/chinas-maritime-silk-road (accessed, March 7, 2020). See also: Praveen Swami, "Thailand's move on Kra Canal alarms New Delhi as route will. Oost Chinese naval power in Indian Ocean," *First Post*, Nov 5, 2018, https://www.firstpost.com/world/thailands-move-on-kra-canal-alarms-new-delhi-as-route-will-boost-chinese-naval-power-in-indian-ocean-5507121.html (accessed, March 7, 2020)

[52] Andrew S. Erickson and Gabriel Collins, "Dragon Tracks: Emerging Chinese Access Points in the Indian Ocean Region," *Asia Maritime Transparency Initiative*, June 18, 2015, https://amti.csis.org/dragon-tracks-emerging-chinese-access-points-in-the-indian-ocean-region/ (accessed, March 7, 2020)

[53] Finbarr Bermingham, "China's lending tumbles as Belt and Road gets a refresh," *China Investment Research*, October 4, 2017, http://www.chinainvestmentresearch.org/media/4-october-2017-chinas-lending-tumbles-belt-road-gets-refresh/ (accessed, March 7, 2020). See also:

given state incentives for shipping and general trade imbalances, rail cars heading both outbound and inbound are often empty.[54] There have also been questions about the actual efficiencies of rail shipping.[55]

Nonetheless, BRI will not be going away anytime soon. If anything, China is expected to be taking the lessons of its first six years and applying them with greater discipline and rigor going forward. While BRI investments have significantly decreased, it should be viewed as one becoming more focused, rather than an altogether declining initiative.[56] Derek Scissors of the American Enterprise Institute has cautioned, in his testimony to the US Senate Committee on Finance that, in fact, the BRI is overhyped, and that the US should not overreact.[57] Scissors adds that while more countries are joining the BRI, the actual level of activity is decreasing, and that BRI will become more focused, with resources directed more toward a smaller group of priority countries of immediate interest to Beijing.[58] One priority is likely to be Greece's Piraeus port, where the Chinese shipping firm COSCO has purchased a majority stake in the port, and where China is seeking to make Piraeus not only a critical transshipment hub of trade between Europe and Asia, but also "the biggest harbor in Europe."[59] China is not only the EU's biggest importer, but also the

---

[54] Sidney Leng, "China's belt and road cargo to Europe under scrutiny as operator admits to moving empty containers," *South China Morning Post*, August 20, 2019, https://www.scmp.com/economy/china-economy/article/3023574/chinas-belt-and-road-cargo-europe-under-scrutiny-operator (accessed, March 7, 2020). See also: Reid Standish, "China's Path Forward is Getting Bumpy," *The Atlantic*, October 1, 2019, https://www.theatlantic.com/international/archive/2019/10/china-belt-road-initiative-problems-kazakhstan/597853/ (accessed, March 7, 2020)

[55] David Fickling, "Railways Put China on a Belt and Road to Nowhere," *Bloomberg Opinion*, November 5, 2018, https://www.bloomberg.com/opinion/articles/2018-11-05/trans-asian-rail-puts-china-on-a-belt-and-road-to-nowhere (accessed, March 7, 2020)

[56] Cissy Zhou, "China slimming down Belt and Road Initiative as new project value plunges in last 18 months, report shows," South China Morning Post, October 10, 2019, https://www.scmp.com/economy/global-economy/article/3032375/china-slimming-down-belt-and-road-initiative-new-project (accessed, March 7, 2020)

[57] Derek Scissors, "The Belt and Road is Overhyped, Commercially," *American Enterprise Institute*, June 12, 2019, https://www.aei.org/wp-content/uploads/2019/06/BRI-Senate-testimony-6.12.19.pdf (accessed, March 7, 2020)

[58] Ibid.

[59] Silvia Amaro, "China bought most of Greece's main port and now it wants to make it the biggest in Europe,"

EU's second largest export market. According to the European Commission, bilateral trade averages over a billion euros a day.[60]

Greece has desperately needed foreign investment for years, which China has been happy to provide. Now that Greece has formally joined the BRI, the depth of this relationship has become increasingly complex not only for NATO, of which Greece is a member, but also the EU and the larger UN. As a member of the EU, Greece stood behind China in 2016 at the tribunal at the Permanent Court of Arbitration on the disputes in the South China Sea by opposing any strong language in the EU statement. This not only prevented a unified EU statement, but also allowed China to drive a wedge between Greece and the EU. [61]

This occurred again when China's COSCO took majority ownership of the Piraeus Port Authority (PPA) in 2017, after which Greece "vetoed an EU condemnation of China's human-rights record at the United Nations."[62] Greece's government, under then Prime Minister Alexis Tsipras, then denied that the veto had been a *quid pro quo* for China's investment.

During the second Belt and Road Forum in April 2019, President Xi rolled out what is now called BRI 2.0. Recognizing the complaints that have been leveled against the BRI, as well as their significance for the BRI's long term success, China hopes to take the lessons learned since 2013, and applying them to BRI 2.0. Christine Lagarde, the then managing director of the International Monetary Fund (IMF), in her speech to the Belt and Road Forum on April 26, 2019, remarked that infrastructure

---

CNBC Economy, November 15, 2019, https://www.cnbc.com/2019/11/15/china-wants-to-turn-greece-piraeus-port-into-europe-biggest.html (accessed, March 7, 2020)

[60] Ibid.

[61] Theresa Fallon, "The EU, The South China Sea, and China's Successful Wedge Strategy," *Asia Maritime Transparency Initiative*, October 13, 2016, https://amti.csis.org/eu-south-china-sea-chinas-successful-wedge-strategy/ (accessed, March 7, 2020)

[62] Hellenic Shipping News, "China's Piraeus power play: In Greece, a port project offers Beijing leverage over Europe," July 9, 2019, https://www.hellenicshippingnews.com/chinas-piraeus-power-play-in-greece-a-port-project-offers-beijing-leverage-over-europe/ (accessed, March 7, 2020)

investments should be managed carefully to prevent a "problematic increase in debt." And that BRI 2.0 could benefit from "increased transparency, open procurement with competitive bidding, and better risk assessment in project selection." Adding that the BRI "should only go where it is needed," and only where it is sustainable "in all aspects."[63]

VII. SRI LANKA: A DEBT TRAP OR A BEIJING STRATEGY?

One issue that is consistently raised with the BRI is the suggestion that China is employing a "debt trap" strategy with host countries to get them hopelessly indebted to China—a new form of the tributary system of the Imperial China. While this prevailing view might have some justification, the truth is somewhat more nuanced.

For context, it is true that as a country with very deep pockets, a number of countries have found Chinese capital difficult to resist. China's state owned enterprise, China Harbor Engineering Company, had built a port for Sri Lanka in Hambantota in 2010, for which it was eventually unable to pay the $1.1 billion it owed China.[64] With few options, Sri Lanka was forced to grant a 99-year concession to China Merchants Port Holdings on the Hambantota Port, which includes approximately 15,000 acres of land nearby for an industrial zone.[65] In a joint venture with Sri Lanka Port Authority (SLPA), China Merchants Port Holdings would also hold a 70 percent stake in the operation of the port. While the Sri Lankan case of asset seizure is more of an outlier than a representative of the BRI projects surveyed, the Rhodium Group has warned that the Hambantota Port "serves as a cautionary tale

---

[63] Christine Lagarde, IMF Managing Director: Belt and Road Forum, Beijing, "BRI 2.0: Stronger Frameworks in the New Phase of Belt and Road," *International Monetary Fund*, April 26, 2019, https://www.imf.org/en/News/Articles/2019/04/25/sp042619-stronger-frameworks-in-the-new-phase-of-belt-and-road (accessed, March 7, 2020)

[64] Maria Abi-Habib, "How China Got Sri Lanka to Cough Up a Port," *New York Times*, June 25, 2018, https://www.nytimes.com/2018/06/25/world/asia/china-sri-lanka-port.html (accessed, March 7, 2020)

[65] Ranga Sirilal, Shihar Aneez, "Sri Lanka signs $1.1 billion China port deal amid local, foreign concerns," *Reuters*, July 29, 2017, https://www.reuters.com/article/us-sri-lanka-china-ports/sri-lanka-signs-1-1-billion-china-port-deal-amid-local-foreign-concerns-idUSKBN1AE0CN (accessed, March 7, 2020)

of the dangers attached to countries' overreliance upon Chinese financing."[66]

Some countries have now learned from the Sri Lanka experience and have recognized that the costs far outweigh the benefits. Bangladesh, for instance, has declined Chinese funding for the much needed "20km-long rail and road bridges over Padma river" and has instead opted for "self-generated funds."[67] Citing China's "unfair" infrastructure deals by his predecessor, Malaysia's former Prime Minister Mahathir Mohamad cancelled two projects; the East Coast Rail Link, which would have connected Port Klang on the Straits of Malacca with the city of Kota Bharu, and a natural gas pipeline in Sabah.[68] While the Sabah pipeline project remains cancelled, the East Coast Rail Link restarted again only after China agreed to cut the price by nearly $11 billion.[69] Thailand is also working to create a regional infrastructure fund via the Ayeyawady-Chao Phraya-Mekong Economic Cooperation Strategy (ACMECS) to reduce reliance on China. [70]

While the debt trap narrative persists, a number of recent studies, including those from the Rhodium Group, the Brookings Institution and Lowy Institute in Australia indicate that China has not, in fact, been deliberately engaged in "debt

---

[66] Agatha Kratz, Allen Feng, and Logan Wright, "New Data on the "Debt Trap" Question, *Rhodium Group*, April 29, 2019, https://rhg.com/research/new-data-on-the-debt-trap-question/ (accessed, March 7, 2020)

[67] Dipanjan Roy chaudhury, "Bangladesh avoids Chinese debt trap, building biggest bridge with own funds," *The Economic Times*, November 7, 2018, https://economictimes.indiatimes.com/news/international/world-news/bangladesh-avoids-chinese-debt-trap-building-biggest-bridge-with-own-funds/articleshow/66531817.cms (accessed, March 7, 2020)

[68] Amanda Erickson, "Malaysia cancels tow big Chinese projects, fearing they will bankrupt the country," Washington post, August 21, 2018, https://www.washingtonpost.com/world/asia_pacific/malaysia-cancels-two-massive-chinese-projects-fearing-they-will-bankrupt-the-country/2018/08/21/2bd150e0-a515-11e8-b76b-d513a40042f6_story.html (accessed, March 7, 2020)

[69] Malay Mail, "Daim leading negotiations with China to recoup RM8.3b payment for pipeline projects, says finance minister," *Malay Mail*, October 2, 2019, https://www.malaymail.com/news/malaysia/2019/10/02/daim-leading-negotiations-with-china-to-recoup-rm8.3b-payment-for-pipeline/1796360 (accessed, March 7, 2020)

[70] Yukako Ono, "Thailand plans regional infrastructure fund to reduce China dependence," *Nikkei Asian Review*, June 4th, 2018, https://asia.nikkei.com/Politics/International-relations/Thailand-plans-regional-infrastructure-fund-to-reduce-China-dependence (accessed, March 7, 2020)

trap" diplomacy.[71] However, they also caution that smaller economies with weak institutions, a lack of resilience to climate change, and disadvantaged by economic geography, make countries such as Tonga, Samoa, and Vanuatu particularly vulnerable to falling into a debt trap. According to the International Monetary Fund (IMF), while none of the Pacific countries are currently in debt distress, the risks are becoming higher over time. It is worthy of note that Beijing will often write off loans without a formal renegotiation process, even when there are few signs of debt distress.[72] A recent Working Paper from the US National Bureau of Economic Research (NBER) indicates that there have been 140 cases of debt write offs and restructurings with developing countries since 2000.[73] These cases of debt forgiveness likely include an effort, to some degree, to create good will with the recipient government and improve bilateral relations.[74] While Sri Lanka's debt-for-equity swap remains the only case of asset seizure, it remains a cautionary tale for debt sustainability.[75]

There are a number of common threads that run through these analyses. First, China's lending practices do not follow any specific pattern in the recipients' debt profiles or levels of governance. China has lent to both authoritarian governments as well democratic ones, and with varying degrees of debt. The BRI is not one

---

[71] Agatha Kratz, Allen Feng, and Logan Wright, "New Data on the "Debt Trap" Question, *Rhodium Group*, April 29, 2019, https://rhg.com/research/new-data-on-the-debt-trap-question/ (accessed, March 7, 2020). See also: David Dollar, "Understanding China's Belt and Road Infrastructure Projects in Africa," *Brookings*, September 2019, https://www.brookings.edu/wp-content/uploads/2019/09/FP_20190930_china_bri_dollar. pdf (accessed, March 7, 2020), Roland Rajah, alexander Dayant, Jonathan Pryke, "Ocean of Debt? Belt and Road and Debt Diplomacy in the Pacific,", *Lowy Institute*, October 21, 2019, https://www.lowyinstitute.org/publications/ocean-debt-belt-and-road-and-debt-diplomacy-pacific (accessed, March 7, 2020)

[72] Agatha Kratz, Allen Feng, and Logan Wright, "New Data on the "Debt Trap" Question, Rhodium Group, April 29, 2019, https://rhg.com/research/new-data-on-the-debt-trap-question/ (accessed, March 7, 2020)

[73] Sebatian Horn, Carmen M. Reinhart, Christoph Trebesch, "China's Overseas Lending," *The National Bureau of Economic Research*, Working Paper No. 26050, July 2019, https://www.nber.org/papers/w26050 (accessed, March 7, 2020)

[74] Agatha Kratz, Allen Feng, and Logan Wright, "New Data on the "Debt Trap" Question, *Rhodium Group*, April 29, 2019, https://rhg.com/research/new-data-on-the-debt-trap-question/ (accessed, March 7, 2020)

[75] David Dollar, "Understanding China's Belt and Road Infrastructure Projects in Africa," *Brookings*, September 2019, https://www.brookings.edu/wp-content/uploads/2019/09/FP_20190930_china_bri_dollar. pdf (accessed, March 7, 2020)

monolithic infrastructure agenda. Rather, it is "a blend of economic, political, and strategic agendas that play out differently in different countries, which is illustrated by China's approach to resolving debt, accepting payment in cash, commodities, or the lease of assets."[76] The strategic objectives are especially apparent where access to key ports and waterways are aligned with Beijing's investment.

Second, all have some, or a significant level of debt risk, both operational and economic. According to Moody's Investors Service, only 25% of the 130 countries that have signed BRI cooperation agreements have an investment grade rating "forty-three percent have junk bond status, while a further 32% are unrated." [77]

Finally, and likely the most important, it should be remembered that it is virtually impossible to achieve a high degree of precision on the level of China's loans overseas to the developing world. China generally funds BRI projects through its policy banks such as the China Development Bank and Export-Import Bank of China via China's SOEs.[78] And neither the banks nor the recipients themselves share the rates at which these loans are made. The Paris Club, which tracks sovereign bilateral borrowing, would normally track these transactions.[79] However, since China is not a member of the Paris Club, it "has not been subject to the standard disclosure requirements." Therefore, China does not provide its direct lending activities related to BRI. A National Bureau of Economic Research analysis suggests that nearly half of China's loans to the developing world are hidden.[80]

---

[76] Amaar Bhattacharya, David Dollar, Rush Doshi, Ryan Hass, Bruce Jones, Homi Kharas, Jennifer Mason, Mireya Solís, and Jonathan Stromseth, "China's Belt and Road: The new geopolitics of global infrastructure development," *Brookings*, April 2019, https://www.brookings.edu/research/chinas-belt-and-road-the-new-geopolitics-of-global-infrastructure-development/ (accessed, March 7, 2020)

[77] Moody's Analytics, "The Belt and Road Initiative – Six Years On," June 2019, https://www.moodysanalytics.com/-/media/article/2019/belt-and-road-initiative.pdf (accessed, March 7, 2020)

[78] Ibid.

[79] Sebatian Horn, Carmen M. Reinhart, Christoph Trebesch, "China's Overseas Lending," *The National Bureau of Economic Research*, Working Paper No. 26050, July 2019, https://www.nber.org/papers/w26050 (accessed, March 7, 2020)

[80] Ibid.

VIII. PERIPHERAL DIPLOMACY HEADS THE WESTERN OCEAN

What began as China's Peripheral Diplomacy is now a global initiative with participants in Asia, Africa, Europe, Latin America, and the Middle East.[81] The 21st Century Maritime Silk Road, as the second component of China's BRI agenda, now encompasses not only the sea lines of communication (SLOC) around the Arabian Gulf and the Bay of Bengal, but the greater Indian Ocean for which the famous Ming Admiral Zheng He called the "Western Ocean" in his seven voyages to the region, including Sri Lanka and East Africa. This is an increasing concern to Beijing because China is not only Africa's biggest trading partner currently, but also relies upon Africa for over a third of its oil imports.[82] And China is racing to get access to the growing populations and economies in Africa by investing in the continent now. The IMF reported that, of the twenty fastest growing economies in the world in 2017, seven were in Africa.[83]

The burgeoning economic relationship in Africa means that China must possess the ability to operate far from home in order to protect its SLOCs, which China recognizes as "the 'lifeline' of China's economic and social development."[84] As such, Chinese military literature opine that "once a maritime crisis or war occurs,

---

[81] Peter Cai, "Understanding China's Belt and Road Initiative," *Lowy Institute*, March 22, 2017, https://www.lowyinstitute.org/publications/understanding-belt-and-road-initiative (accessed, March 7, 2020)

[82] Elliot Smith, "The US-China trade rivalry is underway in Africa, and Washington is playing catch up," *CNBC Trade*, October 9, 2019, https://www.cnbc.com/2019/10/09/the-us-china-trade-rivalry-is-underway-in-africa.html (accessed, March 7, 2020). See also: Wade Shepard, "What China is Really up To In Africa," *Forbes*, October 3, 2019, https://www.forbes.com/sites/wadeshepard/2019/10/03/what-china-is-really-up-to-in-africa/#2b1923f45930 (accessed, March 7, 2020)

[83] Elliot Smith, "The US-China trade rivalry is underway in Africa, and Washington is playing catch up," *CNBC Trade*, October 9, 2019, https://www.cnbc.com/2019/10/09/the-us-china-trade-rivalry-is-underway-in-africa.html (accessed, March 7, 2020)

[84] Andrew S. Erickson, "Power vs. Distance: China's Global Maritime Interests and Investments in the Far Seas," *The National Bureau of Asian Research*, Strategic Asia 2019, http://www.andrewerickson.com/wp-content/uploads/2019/01/PLAN_China's-Global-Maritime-Interests-and-Investments-in-the-Far-Seas_NBR_Strategic-Asia-2019-China's-Expanding-Strategic-Ambitions.pdf (accessed, March 7, 2020)

China's sea transport lanes could be cut off," concluding that "the navy's future missions in protecting SLOCs and ensuring the safety of maritime transportation will be very arduous." [85]

Currently, the People's Liberation Army Navy (PLAN) has established ports in Djibouti in the horn of Africa and Sri Lanka.[86] And its port visits to Pakistan, Oman, and Yemen, as well as its escort and counter-piracy deployments around the Gulf of Aden and the Bay of Bengal not only help the PLAN develop proficiency in operating far from the mainland, but also identifies deficiencies in the PLAN's ability to project power at these distances. [87]

However, the PLAN will need to be able to operate even farther south and west to protect its interests. Naturally, this means it will require additional bases and facilities in forward deployed locations. The military dimensions of these issues, however, cannot be divorced from the political and economic dimensions as China needs to develop relationships with the archipelagic states in the region. That is clearly on the minds of the leadership in Beijing, given its investments in the Maldives and the Seychelles.[88]

---

[85] Ibid.

[86] Guest Blogger for Elizabeth Economy, "China's Strategy in Djibouti: Mixing commercial and Military Interests," *Council on Foreign Relations*, April 13, 2018, https://www.cfr.org/blog/chinas-strategy-djibouti-mixing-commercial-and-military-interests (accessed, March 7, 2020). See also: Maria Abi-Habib, "How China Got Sri Lanka to Cough Up a Port," *New York Times*, June 25, 2018, https://www.nytimes.com/2018/06/25/world/asia/china-sri-lanka-port.html (accessed, March 7, 2020)

[87] Andrew S. Erickson and Gabriel Collins, "Dragon Tracks: Emerging Chinese Access Points in the Indian Ocean Region," *Asia Maritime Transparency Initiative*, June 18, 2015, https://amti.csis.org/dragon-tracks-emerging-chinese-access-points-in-the-indian-ocean-region/ (accessed, March 7, 2020). See also: Associated Press, "Chinese navy ends Persian Gulf visits to Kuwait, Saudi Arabia, Qatar and united Arab Emirates," February 6, 2017, https://www.scmp.com/news/china/diplomacy-defence/article/2068445/chinese-navy-ends-persian-gulf-visits-kuwait-saudi (accessed, March 7, 2020)

[88] Lee Jeong-ho, "Why are China and India so interested in the Maldives?" *South China Morning Post,* September 25, 2018, https://www.scmp.com/news/china/diplomacy/article/2165597/why-are-china-and-india-so-interested-maldives (accessed, March 7, 2020). See also: Betymie Bonnelame, "China giving Seychelles $22 million in project support, including new technical-vocational school," *Seychelles News Agency*, June 19, 2019, http://www.seychellesnewsagency.com/articles/11167/China+giving+Seychelles++million+in+project+support%252C+including+new+technical-vocational+school (accessed, March 7, 2020)

IX. NEW REALITIES IN THE "WESTERN OCEAN" FOR NON-INTERFERENCE

The traditional Chinese policy of non-interference has now given way to current realities. Now codified in the new 2019 White Paper "China's National Defense in the New Era," the PLA must "address deficiencies in overseas operations and support, it builds far seas forces, develops overseas logistical facilities, and enhances capabilities in accomplishing diversified military tasks."[89] These deficiencies are likely rooted in Chinese military literature that increasingly refer to the strategy of "using the land to control the sea, and using the seas to control the oceans." [90]

Given that these events are taking place in India's backyard, India immediately responded after the White Paper was issued. Chief of the Naval Staff of India, Admiral Karambir Singh, announced that "We will have to watch China carefully," later adding that "we require long-term fiscal support to build a Navy, that is the only way we can plan. And, this has been my constant refrain." [91]

No doubt, India will be getting some help from other countries with similar concerns. Given China's diplomatic outreach, infrastructure investments, and military deployments in the Indian Ocean, nations have not only sat up and taken notice, they are also taking actions. It should come as no surprise that the UK has been looking at the possibility of establishing military bases in Asia and the Caribbean.[92] It

---

[89] Lu Hui, ed., "Full Text: China's National Defense in the New Era," *XinhuaNet*, July 24, 2019, http://www.xinhuanet.com/english/2019-07/24/c_138253389.htm (accessed, March 7, 2020)

[90] Andrew S. Erickson, "Power vs. Distance: China's Global Maritime Interests and Investments in the Far Seas," *The National Bureau of Asian Research, Strategic Asia 2019*, http://www.andrewerickson.com/wp-content/uploads/2019/01/PLAN_China's-Global-Maritime-Interests-and-Investments-in-the-Far-Seas_NBR_Strategic-Asia-2019-China's-Expanding-Strategic-Ambitions.pdf (accessed, March 7, 2020)

[91] TNN, "Need to respond to China's growing might in Indian Ocean: Navy Chief," *The Economic Times*, July 26, 2019, https://economictimes.indiatimes.com/news/defence/need-to-respond-to-chinas-growing-might-in-indian-ocean-navy-chief/printarticle/70389080.cms (accessed, March 7, 2020)

[92] J. Vitor Tossini, "A look at the considered locations for new British military bases overseas," *UK Defence Journal*, March 1, 2019, https://ukdefencejournal.org.uk/a-look-at-the-considered-locations-for-new-british-military-bases-overseas/ (accessed, March 7, 2020)

demonstrates that these are not concerns only for the United States and the countries in those regions.

Japan and the UK, in fact, had already conducted drills in the Indian Ocean in proximity to commercial sea lanes back in September 2018.[93] In April 2019, the US Navy's Seventh Fleet joined the Indian Navy to conduct Anti-Submarine Warfare (ASW) training and information sharing.[94] In May, France and India joined in its annual Varuna Exercise also focusing on ASW in the region.[95] In addition, they have established a strategic partnership to explore economic and development partnerships in the region.[96] French President Emmanuel Macron also announced a "3-pronged security partnership with India" in the southern Indian Ocean that includes maritime security, maritime surveillance, and the possible deployment of Indian Navy maritime patrols around Reunion Island.[97] India and Saudi Arabia will be planning to hold its first ever joint naval exercise in March 2020. [98] The HMS Defender made a port visit to Goa on November 12, 2019. And, strengthening UK-India Maritime cooperation, the "British High Commissioner to India, Sir Dominic Asquith announced in November that UK's aircraft carrier HMS Queen Elizabeth will operate in the Indian

---

[93] Teddy Ng, "Japanese, British warships carry out joint exercise in Indian Ocean in latest show of strength to China," *South China Morning Post*, September 27, 2018, https://www.scmp.com/news/china/military/article/2165968/japanese-british-warships-carry-out-joint-exercise-indian-ocean (accessed, March 7, 2020)

[94] U.S. 7th Fleet Public Affairs, "U.S., Indian Navies Practice Submarine Hunting in Indian Ocean," *Commander, U.S. 7th Fleet*, April 15, 2019, https://www.c7f.navy.mil/Media/News/Display/Article/1813999/us-indian-navies-practice-submarine-hunting-in-indian-ocean/ (accessed, March 7, 2020)

[95] Emanuele Scimia, "France, India bolster capabilities in Indian Ocean," *Asia Times*, June 3, 2019, https://asiatimes.com/2019/06/france-india-bolster-capabilities-in-indian-ocean/ (accessed, March 7, 2020)

[96] Dipanjan Roy Chaudhury, "India, France explore 3rd country projects in Western Indian Ocean region," *The Economic Times*, October 24, 2019, https://economictimes.indiatimes.com/news/defence/india-france-explore-3rd-country-projects-in-western-indian-ocean-region/printarticle/71743985.cms (accessed, March 7, 2020)

[97] Dipanjan Roy Chaudhury, "French President announces 3-pronged security partnership with India for Southern Indian Ocean," *The Economic Times*, October 26, 2019, https://economictimes.indiatimes.com/news/defence/french-president-announces-3-pronged-security-partnership-with-india-for-southern-indian-ocean/printarticle/71770145.cms (accessed, March 7, 2020)

[98] PTI, "India, Saudi Arabia to hold first joint naval drills in early March," *The Times of India*, October 31, 2019, https://timesofindia.indiatimes.com/india/india-saudi-arabia-to-hold-first-joint-naval-drills-in-early-march/articleshow/71836478.cms (accessed, March 7, 2020)

Ocean region on its maiden voyage" in 2021, and will "place a liaison officer in Indian Navy's information Fusion centre in Gurugram in Haryana." [99]

X. GLOBAL RESPONSE LED BY WASHINGTON

Geostrategically, as Beijing's intentions become clear, the continuing tensions have now revived with the US-led Quad. Each of these Quad members has its own economic and geostrategic concerns over balancing China's expanding power and influence with a host of counterstrategies. President Trump has, for example, signed into law the Asia Reassurance Initiative Act of 2018—a belated expression of America's commitment to the security and stability of the Indo-Pacific region. [100]

The United States Senate has also passed the Better Utilization of Investments Leading to Development (BUILD) Act of 2018 to reform and improve overseas private investment to help developing countries in ports and infrastructure. [101] It is also aimed at countering China's influence and assisting BRI countries with alternatives to China's "debt trap" diplomacy. [102] Most recently, the US International Development Finance Corporation (IDFC) launched the Blue Dot Network (BDN), which serves as a multi-stakeholder initiative that will harness governments, private sector, and civil society to "promote high-quality, trusted standards for global infrastructure development in an open and inclusive framework." [103]

---

[99] ANI, "HMS queen Elizabeth To Operate In Indian Ocean On 1st Voyage: UK Envoy," *NDTV,* November 15, 2019, https://www.ndtv.com/india-news/hms-queen-elizabeth-to-operate-in-indian-ocean-on-1st-voyage-uk-envoy-2133265 (accessed, March 7, 2020)

[100] Whitehouse.gov, "Statement by the President," December 31, 2018, https://www.whitehouse.gov/briefings-statements/statement-by-the-president-23/ (accessed, March 7, 2020)

[101] Whitehouse.gov, "Statement from the Press Secretary on H.R. 5105/S. 2463, the Better Utilization of Investments Leading to Development (BUILD) Act of 2018, July 17, 2018, https://www.whitehouse.gov/briefings-statements/statement-press-secretary-h-r-5105-s-2463-better-utilization-investments-leading-development-build-act-2018/ (accessed, March 7, 2020)

[102] Patricia Zengerle, "Congress, eying China, votes to overhaul development finance," *Reuters,* October 3, 2018, https://www.reuters.com/article/us-usa-congress-development/congress-eying-china-votes-to-overhaul-development-finance-idUSKCN1MD2HJ (accessed, March 7, 2020)

[103] U.S. International Development Finance Corporation, "The Launch of Multi-Stakeholder Blue Dot Network," November 4, 2019, https://www.dfc.gov/media/opic-press-releases/launch-multi-stakeholder-blue-dot-

Viewed in the context of history, China's rise has been nothing short of spectacular. In the 60-plus years since American Secretary of State John Foster Dulles declared the three principles that 1) the US would not recognize the People's Republic of China, 2) would not admit it to the UN, and 3) would not lift the trade embargo, China has grown from a veritable economic backwater to one that is now projecting its economic and military power around the world.[104] China now seeks to create a new set of global norms, while overturning the existing norms that Beijing claims it had no role in creating. That may be true, but China should remember that those existing international norms have also played a critical role in elevating China to where it is today.

XI. A NEW KIND OF WAR WITH OTHER MEANS

China's strategic objectives have remained much the same as they were in 1965 when the CIA concluded, *inter alia*, that the goal of CPC for the foreseeable future would be to "eject the West, especially the US, from Asia and to diminish US and Western influence throughout the world." [105] The CIA further reported that Beijing also aimed to "increase the influence of Communist China in Asia" as well as to "increase the influence of Communist China throughout the underdeveloped areas of the world."

This likely will not happen as long as the US continues to support Taiwan politically and militarily. Therefore, part of the Chinese strategy is to strip away recognition of Taiwan's sovereign legitimacy by using its wealth to isolate Taiwan

---

network (accessed, March 7, 2020)

[104] He Di, "The Most Respected Enemy: Mao Zedong's Perception of the United States," *The China Quarterly*, No. 137 (March., 1994), https://www.jstor.org/stable/655690?seq=1 (accessed, March 7, 2020). See also: World Bank, "GDP growth (annual %) – United States, China," https://data.worldbank.org/indicator/NY.GDP. MKTP.KD.ZG?locations=US-CN (accessed, March 7, 2020)

[105] CIA.gov, "National Intelligence Estimate: Communist China's Foreign Policy," *Central Intelligence Agency*, May 5, 1965, https://www.cia.gov/library/readingroom/docs/DOC_0001085117.pdf (accessed, March 7, 2020)

diplomatically. In Latin America, China peeled away El Salvador in August 2018 after peeling away the Dominican Republic in May.[106] And in September 2019, it peeled away the Pacific island nations of Kiribati and Solomon Islands in the same week.[107] The Pacific island of Tuvalu recently reaffirmed its support for Taiwan by declining a Chinese offer for development.[108] Currently, Taiwan is considered a "sovereign entity" by only the Holy See and 14 member-states of the United Nations.[109] No doubt, all eyes are now on Bougainville awaiting the final outcome of its referendum on independence from Papua New Guinea.[110] And with growing Chinese influence in Africa, the tiny landlocked country of Eswatini (formerly Swaziland) remains the only African country to recognize Taiwan after Burkina Faso established diplomatic relations with Beijing in May 2018.[111]

China is also expanding its presence and engagement in the Caribbean with capital investments and infrastructure financing, which have played significantly to China's advantage given the Caribbean's proximity to the hurricane belt in America's backyard. The Caribbean—so-called the "Third Border" of the United States, has been neglected even after Congress passed the US-Caribbean Strategic Engagement

---

[106] Carrie Kahn, "China Lures Taiwan's Latin American Allies," NPR, October 13 2018, https://www.npr.org/2018/10/13/654179099/china-lures-taiwans-latin-american-allies (accessed, March 7, 2020). See also: Chris Horton, "El Salvador Recognizes China in Blow to Taiwan," *New York Times*, August 21, 2018, https://www.nytimes.com/2018/08/21/world/asia/taiwan-el-salvador-diplomatic-ties.html (accessed, March 7, 2020), Austin Ramzy, "Taiwan's Diplomatic Isolation Increases as Dominican Republic Recognizes China," *New York Times*, May 1, 2018, https://www.nytimes.com/2018/05/01/world/asia/taiwan-dominican-republic-recognize.html (accessed, March 8, 2020)

[107] Lawrence Chung, "Taipei down to 15 allies as Kiribati announces switch of diplomatic ties to Beijing," *South China Morning Post*, September 20, 2019, https://www.scmp.com/news/china/diplomacy/article/3029626/taiwan-down-15-allies-kiribati-announces-switch-diplomatic (accessed, March 8, 2020)

[108] BBC, "Tuvalu: Pacific nations turns down Chinese islands and backs Taiwan," November 21, 2019, https://www.bbc.com/news/world-asia-50501747 (accessed, March 8, 2020)

[109] Taiwan Ministry of Foreign Affairs, "Diplomatic Allies," https://www.mofa.gov.tw/en/AlliesIndex.aspx?n=DF6F8F246049F8D6&sms=A76B7230ADF29736 (accessed, March 8, 2020)

[110] A. Odysseus Patrick, "Bloody past casts long shadow over Pacific islands poised to become world's newest nation," *New York Times*, December 13, 2019, (accessed, March 8, 2020)

[111] Salem Solomon, "Once Influential in Africa, Taiwan Loses All But One Ally," *VOA*, May 26, 2018, https://www.voanews.com/africa/once-influential-africa-taiwan-loses-all-one-ally (accessed, March 8, 2020)

Act of 2016.[112] The Trump White House has shown little interest in engaging the Caribbean basin.[113] Many observers think that the region is "too democratic and not poor enough" to get on the American foreign policy agenda even though Washington has long recognized the Chinese "inroads" in America's Third Border region.[114]

As China tries to expunge any reference to Taiwan around the globe, it is also wielding the power of its market to coerce its international partners into adopting Beijing's position. Consider the current production of the movie "Top Gun: Maverick."[115] The Chinese tech giant Tencent, which is a co-producer of the movie, is strongly suspected of having the Taiwanese and Japanese flags removed from actor Tom Cruise's flight jacket.[116] US airlines have also had to drop all references to Taiwan as a separate country in deference to China. And the National Basketball Association (NBA) recently groveled to China after China suspended licensing agreements with the NBA, and suspended streaming services of NBA games on state TV because a Houston Rockets executive's tweet defending protestors in Hong Kong had offended Beijing.[117]

---

[112] Hon. Ambassador Richard L. Bernal, "U.S. shouldn't ignore China's influence in the Caribbean," *Miami Herald*, August 15, 2018, https://www.miamiherald.com/opinion/op-ed/article216778165.html (accessed, March 8, 2020)

[113] Ben Tannenbaum, "Filling the Void: China's Expanding Caribbean Presence," Council on Hemispheric Affairs, April 3, 2018, http://www.coha.org/filling-the-void-chinas-expanding-caribbean-presence/ (accessed, March 8, 2020)

[114] Hon. Ambassador Richard L. Bernal, "U.S. shouldn't ignore China's influence in the Caribbean," *Miami Herald*, August 15, 2018, https://www.miamiherald.com/opinion/op-ed/article216778165.html (accessed, March 8, 2020). See also: Randal C. Archibold, "China Buys Inroads in the Caribbean, Catching U.S. Notice," *New York Times*, April 7, 2012, https://www.nytimes.com/2012/04/08/world/americas/us-alert-as-chinas-cash-buys-inroads-in-caribbean.html (accessed, March 8, 2020)

[115] Youtube, "Top Gun: Maverick (2020) – New Trailer – Paramount Pictures, https://www.youtube.com/watch?v=g4U4BQW9OEk (accessed, March 8, 2020)

[116] Sarah Whitten, "Tom Cruise's leather jacket in the 'Top Gun' sequel shows just how crucial China is as a movie market," *CNBC Entertainment*, July 19, 2019, https://www.cnbc.com/2019/07/19/tom-cruises-top-gun-jacket-shows-how-key-china-is-to-film-industry.html (accessed, March 8, 2020)

[117] Derek Thompson, "The NBA-China Disaster Is a Stress Test for Capitalism," *The Atlantic*, October 12, 2019, https://www.theatlantic.com/ideas/archive/2019/10/nba-china-disaster-stress-test-capitalism/599947/ (accessed, March 8, 2020). See also: Andrew Ross Sorkin, Michael J. de la Merced, Lindsey Underwood and Stephen Grocer, "The N.B.A.'s China Problem Is Getting Worse," *New York Times*, October 8, 2019, https://www.nytimes.com/2019/10/08/sports/basketball/nba-china.html (accessed, March 8, 2020).

XII. REPEATING HISTORY WITH NON-CLAUSEWITZIAN WAR

From the Chinese perspective, war is not Clausewitzian in nature. Instead, the Chinese approach will be to sow sufficient doubts in American leaders as to the likelihood of winning an armed conflict with China.[118] In the Chinese classic, *The Art of War* by Sun Tzu, it is focused on building dominant psychological and political positions such that the outcome of a conflict is all but a foregone conclusion.[119] By its own actions, the United States has helped China all along by sowing doubt among nations both in Europe as well as the Indo-Pacific as to Washington's own commitment to the security and stability of these regions.

China now sees history as its turn to be Asia's hegemon, much as Japan did soon after the turn of the 20th century. Looking back, there is little doubt that was one of the most devastating periods in the history of the world. And there is no interest in repeating that history, which will likely be far more devastating.

The United States must continue to engage allies and friends alike to maintain a consistent and persistent presence—irrespective of the administration in place—as an expression of its resolve, unity, and commitment to security, peace, and stability in the Indo-Pacific region. Vital archipelagic states should not be perpetually whipped around by the vicissitudes of geopolitics and by those who are not concerned about climate change, an issue upon which these states' very existence depends. Indeed, according to Stergios Pitsiorlas, Greece's deputy economics minister and one of the negotiators with China, one of the factors that made the Piraeus port partnership with China so attractive for Greece was that "they know what they want," and that unlike democratic nations that undergo leadership changes every few years, China has a "long

---

[118] Kevin Rudd, "How Xi Jinping Views the World: The Core Interests That Shape China's Behavior," *Foreign Affairs*, May 10, 2018, https://www.foreignaffairs.com/articles/china/2018-05-10/how-xi-jinping-views-world (accessed, March 8, 2020).

[119] Sun Tzu, "The Art of War," *MIT*, http://classics.mit.edu/Tzu/artwar.html (accessed, March 8, 2020).

and steady strategic view." [120]

In addition, the United States and its allies should firmly present a unified front in pressuring China and engaging Beijing to respect global norms in areas such as trade, technology transfer, and intellectual property theft. In this regard, it is quite clear that China's policies in pursuit of its Made in China 2025 goals is not only a concern for the United States but also for other advanced economies.[121] As China enters center stage, Beijing must recognize that in the long run it is in China's interest as well.

The current global coronavirus crisis no doubt has slowed Beijing's strategic objectives. Nonetheless, in light of his two goals to 1) build a "moderately prosperous society" by 2021, and 2) to become a "fully developed, rich, and powerful" country by its centennial in 2049, Xi Jinping is a man in a hurry.[122] Part and parcel of this will mean the unification of the "greater China." However, China should measure its historical grievances and ideological priorities against its costs. It cannot have the contradictory desires of peace on the one hand, and retribution on the other. If and when China and Taiwan unite, it will be based upon a mutual amity and belief that it is in the interest of all Chinese people to do so—not through coercion and aggression. Beijing cannot bend history to its will.

---

[120] Jason Horowitz, Liz Alderman, "Chastised by E.U., a Resentful Greece Embraces China's Cash and Interests," *New York Times*, August 26, 2017, https://www.nytimes.com/2017/08/26/world/europe/greece-china-piraeus-alexis-tsipras.html (accessed, March 8, 2020).

[121] Jost Wübbeke, Mirjam Meissner, Max J. Zenglein, Jaqueline Ives, Björn Conrad, "Made in China 2025," *Mercator Institute for China Studies*, December 2016, https://www.merics.org/sites/default/files/2017-09/MPOC_No.2_MadeinChina2025.pdf (accessed, March 8, 2020). See also: Ellen Nakashima, David J. Lynch, "U.S. charges Chinese hackers in alleged theft of vast trove of confidential data in 12 countries," *New York Times*, December 21, 2018, https://www.washingtonpost.com/world/national-security/us-and-more-than-a-dozen-allies-to-condemn-china-for-economic-espionage/2018/12/20/cdfd0338-0455-11e9-b5df-5d3874f1ac36_story.html (accessed, March 8, 2020).

[122] Graham Allison, "What Xi Jinping Wants," *The Atlantic*, May 31, 2017, https://www.theatlantic.com/international/archive/2017/05/what-china-wants/528561/ (accessed, March 8, 2020).

* **Patrick Mendis**, a former American diplomat and a military professor, is currently a Taiwan fellow of the Ministry of Foreign Affairs of the Republic of China and a distinguished visiting professor of global affairs at the National Chengchi University in Taipei. **Joey Wang** is a defense analyst. Both are alumni of the Kennedy School of Government at Harvard University. The perspectives expressed in this article, however, do not represent the organizations of which they are affiliated in the past or present.

從區域安全複合體理論角度
解析東協組織發展趨勢

施毓萱 *

壹、前言

本文主要探討的主題有三個：（一）區域安全複合體理論 （二）體系下的東協發展 （三）理論檢視下的東協。

首先，說明區域安全複合體理論（Regional Security Complex Theory）的內容，藉以分析東協的發展，因為該理論是以區域主義的角度，並將區域納入分析的一環中，以觀察體系由上、國家由下影響區域層次。該理論也被認為是發展較完整的，具有解釋力可分析區域的安全關係。[1]

第二，東協的發展往往會受到區域環境的影響，東協的成立也是在意識形態對立的情況下成立、東協區域論壇則有學者強調了中國崛起的因素，而東協共同體的實踐則是面對全球化、亞洲金融風暴後以及非傳統安全等問題，因此本文將了解東協發展過程中的全球體系、國內層次背景和東協發展結果。

第三，以理論檢視東協，分析東協作為區域安全複合體的特色，以及東協目前面臨何種問題，而該問題又與區域安全複合體理論有什麼關聯？

貳、區域安全複合體理論

區域安全複合體的定義為「一組單元之間的安全問題是密不可分，包含安全化與去安全化的過程都是彼此牽連」。[2] 該理論將體系區分為四個層次以作為分析架構，第一個層次為全球層次，代表全球大國所在的層次，以及如何在

---

* 淡江大學國際事務與戰略研究所碩士生
[1] 陳牧民、李賜賢著，〈國際安全研究中的區域主義：理論與發展簡介與評估〉，《全球政治評論》，第52期，2015年，頁81-82。
[2] Barry Buzan, Ole Waever, "Regions and Powers: The Structure of International Security" (United Kingdom：Cambridge University Press, 2003), P.44.

各個層次間做互動；第二個層次為區域間層次，表示區域安全複合體與鄰近區域的互動；第三個層次為區域層次，代表區域內國家間的關係建構模式；第四個層次為國內層次，關係著區域內國家的性質，如國內秩序穩定、民族等。[3]

　　區域安全複合體內擁有內核，而內核同時會建構起該區域安全複合體。其內核結構有四個部分，分別為「邊界」（boundary）、「無政府結構」（anarchic structure）、「極性」（polarity）、「社會性建構」（social construction）。總體來說，邊界可使各個安全複合體區隔開來；無政府結構是複合體的特性，且必須存在兩個國家以上；極性涉及單位間的權力分配與消長；社會性建構則涉及單位間的友好敵對模式。[4] 而本文主要針對社會性建構作為主軸，故僅針對此變量做討論。

　　社會性建構是區域安全複合體主要變量，是區域安全複合體的身份建構。社會性建構以國家間的友好敵對模式為參照。友好敵對可構成一光譜，其中構成身份關係的要素包含邊界領土爭議、種族爭端、意識形態、歷史因素等。

　　安全複合體社會結構的友敵模式類似建構主義的三種無政府文化，即霍布斯文化、洛克文化和康德文化，分別代表敵人、競爭對手、朋友的主要身份，在《國際安全概論》中則將社會結構分為衝突結構、安全機制結構和安全共同體結構，類似於上述三種無政府文化。[5]

　　在衝突結構中的區域安全複合體，國家間的利益、目標、意識形態都是對立的，在區域中也不存在穩定、以建構起來的安全機制，主要還是透過權力平衡來維持區域秩序。在安全機制結構中，國家間的利益、目標或價值觀並非完全對立或協調，而是透過安全機制來解決區域的安全困境問題，不僅僅是管理衝突或危機處理，也包含預防衝突，以避免螺旋式的上升。在安全共同體結構中，區域內的衝突在一定程度上解決了，國家間的利益一致且具有長期性，在這共同體中也難以發生暴力衝突，因而形成集體認同和共同利益。

---

[3] Barry Buzan, Ole Waever, "Regions and Powers: The Structure of International Security", P.51

[4] Barry Buzan, Ole Waever, "Regions and Powers: The Structure of International Security", P.51-53.

[5] 孫紅著，〈區域安全〉，王帆、盧靜主編，《國際安全概論》（北京：世界知識出版社，2010年），頁143。

這些模式，都和現有衝突解決程度和新出現的爭端性質相關，而友好敵對模式也和極性相關，極與極之間的關係，也就是國與國之間關係的改善或惡化都關係著區域安全複合體的身份結構，不僅僅是權力間的衡量。[6]

區域安全複合體的內核不僅僅是構成複合體的要素，也會決定複合體的未來發展。其未來發展方向可分為三種：「維持現狀」（maintenance of the status quo）、「內在變革」（internal transformation）、「外在變革」（external transformation）。

複合體發展方向由內核結構有無發生改變來決定之。維持現狀，代表該區域安全複合體的內核結構沒有發生重大改變；內在變革，代表內核結構的變革發生在該區域安全複合體的邊界之內；外在變革則與內在變革相反，變革的發生是因為其邊界的改變。

上述是區域安全複合體的基礎特性，但由於前述所提及的安全的主觀與客觀性，以及不同複合體間的四個變量以及安全關係之差異，所以安全複合體仍須有更詳細的分類，其種類劃分是以區域安全複合體內極性的差異來加以區分的，而Barry Buzan將其分為四種概要類型：「標準安全複合體」（standard）、「中心化安全複合體」（centred）、「大國安全複合體」（great power）和「超級安全複合體」（supercomplexes）。[7]

標準區域安全複合體，其內部極性由區域大國決定，且這類的安全複合體所囊括的行為體不包含全球性大國。其區域內國家間的安全政治主要是區域內的區域大國之間互動的關係，這樣的關係會影響較小的區域內國家，以及創造了全球性大國可以覆蓋或滲透該區域安全複合體的機會。

中心化區域安全複合體，Barry Buzan將其區分出三種主要會出現的種類。中心化區域安全複合體有兩種形式，一是極性為全球性大國主導，由超級大國或大國為結構中的單極，二是由制度將區域整合起來，所以三個種類分別為超

---

[6] 孫紅著，〈區域安全〉，王帆、盧靜主編，《國際安全概論》，頁143-145。

[7] Barry Buzan, Ole Waever, "Regions and Powers: The Structure of International Security", P.53-62.

級大國安全複合體、大國安全複合體、制度安全複合體。兩種形式截然不同，但如其名，前者中心化意旨內核結構為單極，後者則是一組國家建立一種集體化的制度，使其可以作為一行為體，甚至成為國際體系中的一極性，可以說兩者的共同點為這個區域的安全態勢是由誰來主導，即來自區域內部的中心。

　　大國區域安全複合體，內部極性由兩個以上的全球大國來決定。該類型的安全複合體並非單純處於區域層次，而是同時混雜著全球層次，所以大國區域安全複合體的安全態勢會同時混雜著區域和全球層次，而該區域安全複合體中的大國也可能打破區域的秩序，並以各種方式插手到其他鄰近區域，甚至可能形成擴溢的現象，使得邊界分明的各個區域安全複合體集合成一個「超級複合體」，並以一個以上的大國為中心，而這也是區域安全複合體的第四種型態。

參、東協統合發展

　　東協的形成是受到國際和區域環境所影響。在冷戰兩強相爭的背景下，東南亞區域也面臨共產與非共產國家之間的問題。首先，非共產國家對於蘇聯、中國和越南的共產革命擴散有著恐懼，而柬埔寨和寮國也有同樣的隱憂。[8] 於1955年的柬埔寨大選的施亞努（Shihanouk）在1958年承認了中共和北越政權，也接受了蘇聯的援助，更於1963年11月驅逐美國的軍事顧問、要求美國取消對其援助，最終在1965年斷交。[9] 寮國則是由於王室分裂引發國內政治不穩定，1957年到1967年期間二次組成的聯合政府都是勉強組成的，期間因為美國介入越戰，美國相繼進駐寮國、泰國等，使的寮國內部對立加劇，而1967年開始寮共開始進行武裝叛亂。[10]

　　1967年，東協成立，其成立之初，簽訂了《東協宣言》（The Asean Declaration, Bangkok Declaration），其中也包含了東協成立的目的和宗旨，該

---

[8] Ly Tuong Van, "The Vietnamese Revolution in the Cold War and its impact on Vietnam-ASEAN relations during the 1960s and 1970s", in Albert Lau ed., *Southeast Asia and the Cold War* (Oxon：Routledge, 2012), p.176.

[9] 顧長永著，《東南亞各國政府與政治：持續與變遷》（臺北：台灣商務，2013年），頁289-290。

[10] 顧長永著，《東南亞各國政府與政治：持續與變遷》，頁329-334。

宣言也初步規劃組織架構，此外宣言內容中還能抓到一關鍵，即文中表明「確保穩定與安全，不受任何形式的外來干涉」。[11] 不過東協在最初成立的時候，並無顯著的進展，處於在為不穩定的階段，其中有兩個原因，一是成員國之間的信賴程度需要經過長時間的磨合，二是因為成員國對於如何讓該區域保持沒有外來干涉這一目標一直無法達成共識。[12]

1971年11月27日東協簽署《東南亞中立化宣言》，即《吉隆坡宣言》，其目標為東南亞「和平、自由和中立區」之建立。[13] 其簽署背景在全球層次上牽涉到越戰，隨著美國涉入越戰的時間越來越長，美國逐步退出越南戰場，1968年5月美國與越南民主共和國和民族解放陣線進行巴黎和會談判，並於1969年7月美國宣布「尼克森主義」（Nixon Doctrine，又稱關島主義），意即美國會減少在東南亞的軍事活動，此意味著美國將退出越南戰事。另一方面，英國決定在1970年中撤出蘇伊士以東的軍事承諾，且蘇聯在1960年代末開始漸漸滲入東南亞區域。美國將撤出越南的舉動可能會在東南亞形成了權力真空，而面臨其他大國可能想填補該真空的可能性成為東南亞非共產主義國家面臨到的一迫切問題。[14]

除了上述的問題，在國內層次方面，也是1976年的巴厘會議背景另一個原因，即越南、柬埔寨和寮國共產主義的勝利。[15] 1973年美國完全撤離越南，1975年阮文紹總統辭職、南越被赤化，而1976年北越勞動黨和南越全國解放陣線合併為越南共產黨，成為越南的領導。[16]

在這些背景下，1970年代東協開始將「大國平衡」戰略納入東協對外關係

---

[11] The Association of Southeast Asian Nations, 〈The Asean Declaration〉, August 8, 1967.〈https://asean.org/the-asean-declaration-bangkok-declaration-bangkok-8-august-1967/〉.

[12] Frank Frost, "Introduction：ASEAN since 1967 – Origins, Evolution and Recent Developments", in Alison Broinowski, ed., *ASEAN into the 1990s*, pp.5-6.

[13] 陸建人著，《東盟的今天與明天：東盟的發展趨勢及其在亞太的地位》（北京：經濟管理出版社，1999年），頁9。

[14] Ly Tuong Van, "The Vietnamese Revolution in the Cold War and its impact on Vietnam-ASEAN relations during the 1960s and 1970s", in Albert Lau ed., *Southeast Asia and the Cold War*, pp.177-178.

[15] Shaun Narine, "Forty years of ASEAN: a historical review", *The Pacific Review*, Vol. 21, No. 4, 2008, pp.415.

[16] 顧長永著，《東南亞各國政府與政治：持續與變遷》，頁255-256。

中，而《東南亞中立化宣言》代表著大國平衡戰略。其表示著東協在權力真空的風險下，接納區域外國家並使它們在區域中保持均勢、相互牽制的狀態。[17]

不僅僅是越戰問題，越南與柬埔寨之間的衝突也成為東協的隱憂之一，也有國家曾經提出安全體系的概念，但最終都沒有被所有國家同意，而是在1976年簽署《東南亞友好合作條約》（Treaty of Amity and Cooperation in Southeast Asia, TAC）。[18] 該條約確立了東協原則，即互相尊重領土、主權完整性、不受外來干涉、不干涉內政、和平解決爭端、放棄武力威脅或使用武力等。[19] 而當時條約僅開放給東南亞各國，直到1987年東協對第18條進行修改，使東南亞以外的國家得以簽署。[20]

接下來東協面臨成員擴大的時期，時間落在1980年代到2000年前。汶萊、越南、寮國、緬甸和柬埔寨分別於1984、1995、1997、1999年加入東協。[21] 與此同時，東協也在1997到1998年面臨金融風暴以及911事件的影響，東協意識到後冷戰時代對於非傳統安全和全球化之需求，因此1990年代開始，東協逐漸改善其組織架構以及功能。除了成員國的擴大以外，東協加強其經濟整合、安全對話，以及區域間的連結。[22]

1992年，第四屆東協高峰會中，東協決定以「東協後部長會議」（ASEAN Post-Ministerial Conference, PMC）擴大延伸而成，這也是東協戰略與國際研究所於1991年提出的一項建議。[23] 東協區域論壇第一次會議確立

---

[17] 葛紅亮著，〈權力制衡到規範外溢：東協「大國平衡」戰略再評議〉，《全球政治評論》第68期，2019年，頁53-55。

[18] 陳鴻瑜著，《東南亞國家協會之發展》，頁62-63。

[19] The Association of Southeast Asian Nations，〈Treaty of Amity and Cooperation in Southeast Asia Indonesia〉，February 24, 1976，〈https://asean.org/treaty-amity-cooperation-southeast-asia-indonesia-24-february-1976/〉.

[20] The Association of Southeast Asian Nations，〈Protocol Amending the Treaty of Amity and Cooperation in Southeast Asia Philippines〉，December 15, 1987，〈https://asean.org/?static_post=protocol-amending-the-treaty-of-amity-and-cooperation-in-southeast-asia-philippines-15-december-1987〉.

[21] 林若雩著，《東協共同體的建構與成立 - 「4C安全文化」之理論與實踐》，頁65。

[22] 吳祖田著，〈「東南亞國家協會」組織之發展與回顧〉，《問題與研究》第37卷第8期，1998年，頁42。

[23] Mely Caballero-Anthony, "Regional Security in Southeast Asia: Beyond the ASEAN Way" (Singapore：ISEAS Publications, 2005), pp.126-127.

了議程,更重要的是《東南亞友好合作條約》被認為是管理國家間關係的準則,並藉此促進建立信心和預防性外交之原則。第二次的會議確立了兩件事情,一是採取雙軌制,第一軌道為官方性質,而第二軌道為非政府組織與學術界,二是採取漸進式的模式發展東協區域論壇,[24] 分為「促進建立信心措施」(Promotion of confidence-building measures)、「發展預防性外交機制」(Development of preventive diplomacy mechanisms)、「建立解決衝突機制」(Development of conflict-resolution mechanisms)三個階段。[25]

1990年代不僅是後冷戰時期區域環境的轉變,東協亦受到亞洲金融風暴的影響,東協開始思考如何因應全球化的挑戰。[26] 1997年的非正式領袖峰會產生了《東協2020願景》(ASEAN Vision 2020),緊接著1998年的東協峰會提出了《河內行動計畫》(The Hanoi Plan of Action, HPA),其中內容可見東協面對亞洲金融風暴做出的反饋,欲解決當前情勢以及促進經濟整合。[27]

2003年到2009年,分別完成了《峇里第二協約》(Bali Concord II)、《永珍行動計畫》(the Vientiane Action Programme, VAP)和《2009-2015東協共同體路徑圖》(Roadmap for an ASEAN Community 2009-2015)的簽署和制定,在這三個文件中包含了東協共同體的內涵,也代表這是東協區域整合的一里程碑。[28] 2006年,東協召開國防部長會議決定於2020年成立「東協安全共同體」(ASEAN Security Community, ASC),為了實現2003年《峇里第二協約》的目標。東協共同體包含三大主軸,分別是「東協政治－安全共同體」(ASEAN Political-Security Community, APSC)、「東協經濟共同體」

---

[24] 詳見Dominik Heller, "The Relevance of the ASEAN Regional Forum (ARF) for Regional Security in the Asia-Pacific", *Contemporary Southeast Asia*, Vol. 27, No. 1, April 2005, p.127-131.
Mely Caballero-Anthony, "Regional Security in Southeast Asia:Beyond the ASEAN Way", pp.127-128.
李昭賢著,〈東協區域論壇對北韓安全威脅因應之道:區域安全複合體理論觀點〉,《展望與探索》第16卷第10期,2018年10月,頁42-43。

[25] Mely Caballero-Anthony, "Regional Security in Southeast Asia: Beyond the ASEAN Way", p.128.

[26] 楊昊著,〈新憲章規範下的東協區域主義:回顧與展望〉,《台灣東南亞學刊》第5卷第1期,2008年,頁150。

[27] 楊昊著,〈新憲章規範下的東協區域主義:回顧與展望〉,頁152-153。

[28] 李瓊莉著,〈東南亞區域整合之基調、特色與挑戰〉,徐遵慈編,《東南亞區域整合－臺灣觀點》(臺北:財團法人中華經濟研究院、台灣東南亞國家協會研究中心,2012年),頁14。

（ASEAN Economic Community）以及「東協社會－文化共同體」（ASEAN Socio-Cultural Community, ASCC）。

肆、理論檢視下，東協問題

　　若從區域安全複合體的角度來解析東協，可從兩個面向分析，分別為東協機制和東協面臨的問題。

一、東協機制

　　東協機制可分為兩個部分，為組織內與組織外的機制，若從理論的層次分析途徑角度，組織內機制屬於區域層次，而組織外機制則屬於區域間層次和全球層次。組織內部的結構分為東協相關組織、區域內國家雙邊或多邊安全制度、區域內非國家行為體的參與；區域外的結構則劃分為官方機制和第二軌道機制。

　　東協組織內部的東協相關組織層次，包含東協高峰會、東協外長會議、東協秘書處等區域安全治理機構；區域內國家雙邊或多邊安全制度，其範圍大小各有所不同，從全區域性到兩國之間都有，例如《東南亞友好合作條約》、2002年馬菲東泰印的反恐協議、[29] 軍事安全合作機制等；非國家安全行為體參與的部分，則包含非政府組織和個人，例如東協戰略與國際問題研究所（ASEAN ISIS）。

　　東協組織外部的官方機制層次，是東協將區域外國家一同納入，建立對話以及制度，以求區域的穩定發展，包含東協區域論壇，以及區域外國家所簽署的《東南亞友好合作條約》等；第二軌道機制方面可以是由區域外非國家行為體參與。[30] 組織外部的官方機制可看出東協以大國平衡戰略，將區域外國家納入，並在東協方式下彼此建構關係。

---

[29] 2002年的協議包含航空旅客的掌握、著名罪犯名單掌握、強化邊境巡邏等。詳見參考宋興洲、林佩覽著，〈東南亞國協與區域安全〉，《全球政治評論》第25期，2009年，頁37。

[30] 金新、黃鳳志著，〈東協區域安全治理：模式、歷程與前景〉，《世界經濟與政治論壇》，第4期，2013年7月，頁2-4。

　　東協在漫長的發展過程中，雖然有許多受批評的部分，但其制度化的過程中仍然可以看到區域安全複合體在各個層次中互動的概念具體化。東協組織內部機制主要屬於區域層次，可分析國內層次和區域層次之間的互動，也能探討全球層次或區域間層次對於國內層次的影響如何反應在區域層次，即組織內部機制上。而東協組織外部機制，則可分析全球層次大國、外部區域大國和區域內國家間的互動關係，可表示為全球層次到國內層次的關係如何反應到區域層次上。最終的結果都會反饋到區域層次，再由區域層次對外或對內做出反應，形成一個循環在體系下運作。

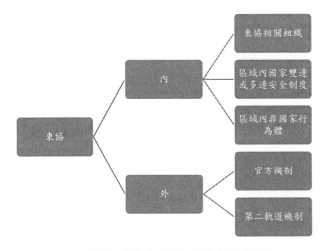

圖1　東協區域安全治理機制架構圖

資料來源：金新、黃鳳志著，〈東協區域安全治理：模式、歷程與前景〉，《世界經濟與政治論壇》，第4期，2013年7月，頁2-4。筆者自行繪製。

二、東協面臨的問題

　　其問題是東協各國的身分認同影響其利益和行為。東協身份意識不強，成員國在面臨國家利益考量時，將使東協共同目標和集體身份受到阻礙。[31] 儘管東協在1976年《東協和諧宣言》中意識到區域認同培養的重要，也公開其建

---

[31] 葛紅亮著，〈東協對南海政策的建構主義解析〉，《全球政治評論》第47期，2014年，頁118。

立一共同體之目標，[32] 並在2000年後逐漸將其制度化，有趨於成熟的模樣，但區域認同並非天生擁有的，那是不斷發展的、持續性的過程，必須由各個國家來共同努力，否則會成為一「想像的共同體」（imagined community），即Acharya所指「想像的並不會都和現實相符」。[33]

以南海爭端為例，2016年國際仲裁法院針對《菲律賓共和國和中華人民共和國之間的仲裁案》（Arbitration between the Republic of the Philippines and the People's Republic of China），做出裁決並命名為《南海仲裁案》（The South China Sea Arbitration）。[34]

馬凱頓指出，2012年金邊的東協外長會議，柬埔寨反對針對南海進行討論，而菲律賓和越南則是期望在聲明中提及。[35] 柬埔寨堅持共同聲明中不能提到關於《南海仲裁案》，以及要求刪除先前東協會議已有共識的部分，也認為此爭端應回歸雙邊機制。[36]

上述的情形，David Martin Jones和Nicole Jenne曾指出東協所面臨的問題根本，即東協成員國的組成和弱國結構。[37] 弱國缺乏調節國內的能力，所以它們主要關注在國內安全和區域的穩定性，如果無法改善調節國內的能力，則會在很大程度上依賴大國。另一方面，一個國家對於缺乏跨界威脅和解決區域內國家衝突的情況上，容易受到外部行為者的利用來維護利益。[38]

---

[32] ASEAN,〈The Declaration of ASEAN Concord〉, 1976.

[33] Amitav Acharya, "The Evolution and Limitations of ASEAN Identity", in Aileen Baviera and Larry Maramis ed., ASEAN @50 volume 4：Building ASEAN Community：Political-Security and Socio-cultural Reflections (Economic Research Institute for ASEAN and East Asia, 2017), pp29-31.

[34] 郁瑞麟著，〈南海仲裁案的評析與因應對策〉，《海洋事務與政策評論》第4卷第1期，2016年，頁36-37。

[35] Kishore Mahbubani, Jeffery Sug著，翟崑、王麗娜等譯，《解讀東協：前進東協，你不可不知道的經濟、政治、歷史背景，以及現況與未來》（The ASEAN Miracle: A Catalyst for Peace）（臺北：遠流出版事業股份有限公司），頁294。

[36] Joanne Chang,〈柬埔寨反對！東協外長會議聚焦南海爭端但無共識〉，《南洋誌》，2016年7月25日，〈https://aseanplusjournal.com/2016/07/25/20160725/〉。

[37] 弱國意指由於天然限制或基本經濟限制而使的國家有著固有弱點，或是因為國家內部不穩定、遭受外部攻擊而形成的衰弱，甚至也有兩者混合的情形。詳見Robert I. Rotberg, "Failed States, Collapsed States, Weak States: Causes and Indicators", in Robert I. Rotberg ed., State Failure and State Weakness in a Time of Terror (Washington, D.C.: Brookings Institution Press, 2003), pp.3-6 或 Robert E. Kelly, "Security Theory in the 'New Regionalism'", International Studies Review, Summer, Vol. 9, No. 2, 2007.

[38] David Martin Jones, Nicole Jenne, "Weak states' regionalism：ASEAN and the limits of security cooperation

　　根據The Fund of Peace在2019年所發行的"Fragile States Index"中，東南亞國家只有新加坡在sustainable一級，第二、三個狀況較好的國家馬來西亞和越南已經來到warning一級，其餘的東南亞國家則在這之後。[39] 也能從《東協憲章》中發現東協一直有意識到該問題，其對於縮小東協內部差異的目標代表著長期存在的難題。[40]

　　弱國結構會影響東協的共識與實際行為不一致，維護主權和無約束力的共識會成為東協整合之困難，而成員國政體、其他內部差異也會影響東協達成共識的效率。最後，David Martin Jones和Nicole Jenne認為東協規範亦會成為大國利用的管道來推廣其國家利益。[41]

　　另一方面，由於東協為共識決，中國只要突破一國家，就可以使東協成員國溝通和共識上遇到問題。[42] 如同前面所提及的柬埔寨反對東協共同聲明出現《南海仲裁案》，而東協共識決的性質使的當時會議僵持不下，最終菲律賓妥協同意撤掉決部分。[43]

　　該事件代表著中國與柬埔寨密切的關係，該國在施亞努國王或洪森執政時期間與中國有良好的往來，2004年洪森訪問北京時雙方達成16項協議，有八項是中國的無償援助。[44] 而中國也和柬埔寨於2019年簽訂「雲壤協定」，中國取得雲壤部分海軍基地港口的30年租借，也在七星海特區擁有99年租借期。[45]

　　上述的情形，區域內國家對於身分認同和對威脅來源抱持不同的態度，

in Pacific Asia", International Relations of the Asia-Pacific, Vol.16, No.2, 2015, pp.216-218.

[39] The Fund of Peace, "Fragile States Index Annual Report 2019", 〈https://fundforpeace.org/2019/04/10/fragile-states-index-2019/〉.

[40] ASEAN，《東協憲章中英對照本》(臺北：財團法人中華經濟研究院台灣東南亞國家協會研究中心，2011年)，頁13。

[41] David Martin Jones, Nicole Jenne, "Weak states' regionalism: ASEAN and the limits of security cooperation in Pacific Asia", pp.229.

[42] Daniel C. O'Neill, "Dividing ASEAN and Conquering the South China Sea: China's Financial Power Projection" (Hong Kong：Hong Kong University Press, 2018), p.16.

[43] Joanne Chang，〈柬埔寨反對！ 東協外長會議聚焦南海爭端但無共識〉，《南洋誌》，2016年7月25日，〈https://aseanplusjournal.com/2016/07/25/20160725/〉。

[44] 顧長永著，《東南亞各國政府與政治：持續與變遷》，頁318-320。

[45] 胡敏遠著，〈柬「中」雲壤協定 衝擊印太戰略〉，《青年日報》，2020年12月18日，〈https://www.ydn.com.tw/news/newsInsidePage?chapterID=1300402&type=forum〉。

可視為區域內國家對於外部行為體友好敵對關係之差異會造成東協所受到的阻礙。李昭賢曾針對國家面對區域安全威脅時的行為模式差異設計出一架構。雖然區域安全複合體的四個變量中包含社會性建構，但這只是關注於區域內國家間的關係建構，而新的架構包含了區域安全複合體國內層次與外部行為體的社會性建構，主要用以研究國家在無政府狀態下面臨安全威脅時，其認同偏好與行為動機之間的關聯與反應。不同區域內國家在因國家利益以及與外部威脅的關係有所差異，在行為動機上也會有積極和消極之分，因此各國出現的反應可能是支援、觀望、對抗或自保。[46]

若從區域內國家之社會性建構角度，若從歷史上來看，過去東協成員國因意識形態而區分成共產和非共產國家，也連帶因為如此明確的社會性建構關係造成區域內國家和其相應的意識形態大國建構關係，使的區域外大國得以滲透到東南亞區域。

而現今雖區域環境已有改變，東協也從衝突結構的一端，轉移到安全機制結構到安全共同體的光譜上，雖然其目標為建構安全共同體，但由於東協成員國之間還存在邊界、領土爭議，以及國家實力仍有著差距、政體的不同，也包含成員國對於主權、不干涉內政的重視會使的在議程內的議題有所停滯或成效不彰，所以距離建構安全共同體仍有著一段路程。

若要改善這樣的情況，東協的機制運作不能只有制度層面的表面實踐，而是需要實質透過制度運作，在面臨跨界衝突或國際爭端時也要保持對外一致。以邊界衝突為例，多數東協成員國在解決區域內衝突上，並無透過東協的爭端處理機制去處理，通常都是透過國際爭端處理方式解決。例如馬來西亞與新加坡的白礁島爭端、馬來西亞與印尼的斯巴丹島和利其丹島爭議、馬來西亞與新加坡間的供水問題，皆以國際機制解決。可以說東協內部尚未發展出對東協機制的一定程度信任或認同，所以才會面臨規範未有更深層、積極性的運作的問題。[47]

---

[46] 詳見李昭賢著，〈東協區域論壇對北韓安全威脅因應之道：區域安全複合體理論觀點〉，頁50-57。

[47] 宋興洲、林佩霓著，〈東南亞國協與區域安全〉，頁27-28。

綜上所述，東協以區域安全複合體的角度，其制度可應用在與各層次的互動上，並在這些機制上給予每個層次間的關係建構，但由於東協方式和東協基本原則的限制，制度層面的背後是國內層次與區域之間的連結不夠深厚的問題，即東協的弱國結構從國內不穩定以及區域內國家間的爭端等因素，建構出不同的國家利益和身分，以至於東協雖然想以東協中心主導區域事務，往往會因為東協身分建構之不完全而有所影響。

伍、結論

從歷史角度去看東協的發展過程，可發現東協一直受到區域外大國的影響。冷戰時期的美中蘇對抗，到現代中美的印太競逐，東協皆是區域外大國相互競爭之地。而從區域安全複合體的角度，當全球體系的大國之間的權力分配和區域安全複合體的區域安全有所關聯時，那將會給予大國滲透的可能性。[48]

根據本文透過理論的分析，東協主要的問題在於東協的身分建構並沒有與區域內所有國家相一致，沒有一定程度的認同也就無法建構東協所期望的身分，而身份的建構也會進一步影響友敵模式，不僅僅是區域層次，也包含跨界的區域間層次或全球層次。東協的弱國結構，也容易使的各國身分建構的差距拉大，不僅重視主權的維護，也需要區域外大國的援助。

而近期緬甸軍方叛變事件可能成為東協的考驗，測試東協是否有能力解決成員國的問題。即便東協在這次峰會後發出聲明，但東協的不干涉內政和尊重國家主權等基本原則是否會在這次事件上再度成為東協處理危機的阻礙，是東協當前要面對的問題。

---

[48] Barry Buzan, Ole Waever, "Regions and Powers: The Structure of International Security", P.46.

East Asian Economic Security Under US-China Competition

Pei-Shan Kao [*]

I. Macroeconomic Situation of East Asian Countries

On the definition of the International Monetary Fund (IMF),[1] East Asia is comprised of six countries and economies, that is, China, Japan, South Korea, Taiwan, Hong Kong and Mongolia. Since the GDP of Mongolia only occupies 0.06% that is relatively low in this region, this article will only focus on the economic performances of the other five countries and economies, including their gross domestic product (GDP), inflation, and foreign direct investment (FDI). According to the data and statistics of the IMF, East Asia is the second largest regional economy in the world in 2015 with nominal GDP of $17.5 trillion.[2] North America is the largest regional economy in the world with GDP of $20.8 trillion while West Europe ranked the third with GDP of $16.1 trillion. The IMF also forecasts that East Asia will become the largest economy in 2020 and that her GDP will be $26.5 trillion surpassing that of North America which is about $26.1 trillion in 2021. (See Figure 1) East Asia economy is characterized by a very high GDP per capita, a big trade surplus, low unemployment and inflation rates although it is an aging society with very low birth rate. Moreover, it is the main global manufacturing supply chain with a

---

[*] PhD in Government, University of Essex, UK; Associate Professor, Central Police University
[1] In the IMF regional economic outlook, it also use "Advanced Asia" refers to Australia, Hong Kong, Japan, Korea, New Zealand, Singapore, and Taiwan, and "Emerging Asia" refers to China, India, Indonesia, Malaysia, the Philippines, Thailand, and Vietnam. See "The Regional Economic Outlook: Asia and Pacific", *International Monetary Fund*, October 2019, p. viii, https://www.imf.org/en/Publications/REO/APAC/Issues/2019/10/03/areo1023#Introduction.
[2] On the annual world economic outlook can see the publications of the *IMF World Economic Outlook Reports*, http:// https://www.imf.org/en/Publications/WEO.

very high saving rate and foreign reserves, particularly holding many US government bonds. Although these countries and economies have some conflicts and discords with one another, for instance, the war happened between China and Japan during the World War II and the territorial disputes among China, Taiwan, and Japan on Diaoyu (Senkaku) Islands, and the Taiwan issue, etc., basically, they all maintain very good and strong cooperation and contacts on trade and economic issues.

The performance of these economies can see Table 1. For example, although in 2018 Taiwan's GDP Per Capita was $25,008 compared with that of China was $9,580, China's economy has grown quickly in the past few decades. Its real GDP average growth rate was 9.6% over the period of 2000 to 2015 compared with 2.0% of the United States at the same period. China's GDP is about 15% in world GDP in 2015 compared with only 2% in 1995 in terms of US dollars. But in terms of purchasing power parity (PPP), China's GDP share in the world was 17% more than that of 16% of the United States. After many years of high growing rates, China's leaders who took office in 2012 made plans to slow down economic growth rate due to concerns about excess reliance on low-end manufacturing and rising debt levels. (continued)

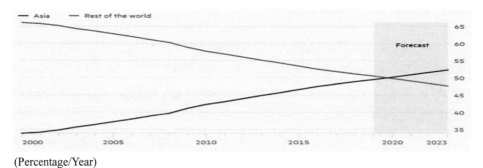

(Percentage/Year)

Figure 1.　Share of World GDP at PPP$ (2000-2023 Forcast)

Source:　Wang Huiyao, "In 2020, Asian economies will become larger than the rest of the world combined - here's how," *World Economic Forum*, July 25, 2019, https://www.weforum. org/agenda/2019/07/the-dawn-of-the-asian-century/

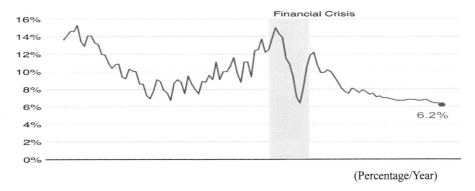

Figure 2. China's Economic Growth Rate (1992-2019)

Source: "China's GDP rate slumps to 6.2%, the lowest level of 27 Years," *Newsblare,* July16, 2019,
     https://www.newsblare.com/opinion/economic-finance/chinas-gdp-rate-slumps-to-6-2-the-
     lowest-level-of-27-years/.

They intend to reform their industrial structure, shifting the economy from heavily relying on investment to consumption, from export to local demand. The GDP growth rate has moderated from 7.9% in 2012 to 6.2% in the quarter of April-June of 2019 which is the lowest level of 27 years. (See Figure 2), and the trend still goes down as an "L-shape" as the Chinese government announced. In the past five years, that is to say, from 2015 to 2019, her GDP was 6.9%, 6.7% 6.9%, 6.5%, 6.1% according to China's National Bureau of Statistics.[3]

Figure 3 shows Japan's economic performance from 2000 to 2021. Although Japan's GDP Per Capita was $39,304 in 2018 and ranked the first in East Asia, Japan's economy was surpassed by China during the second quarter of 2010 although the United States is still the largest economy in the world. Japan has gradually lost its competitiveness with China. Japanese companies made many advanced products such as cars, ships, consumer electronics, textiles, steels, and petro chemical, etc., during the 1950s and 1980s. However, they lost out in the IT revolution due to a lack

---

[3]   About China's annual economic performance can see *National Bureau of Statistics,* http://www.stats.gov.cn/
    english/.

of innovation and a focus on copying older technologies. Therefore they failed to catch on to the PC wave in the 1980s, the internet in the 1990s, and mobile equipment in 2008. Another big problem was the huge appreciation of Japanese Yen over 86% from 1985 to 1988 arising from the Plaza Accord.[4] This negatively hurt Japan's exports and attracted speculation into its financial markets until the asset price bubble collapsed. The collapse made Japan fall into the "Lost 20 Years", during which it lost dominance in Asia, despite the government using quantitative easing, lower taxes, infrastructure priming, and other tools save its economy. When Premier Shinzo Abe took office in 2012, he announced a new economic policy; the so-called "three arrows" that included monetary, financial, and structural reforms. Although some economic data and figures have been improved and went up in a short-term time due to the depreciation of Japanese Yen, Japan is still struggling with deflation after the stimulus effect. So the GDP of Japan in 2008, 2009 and 2011 still saw negative growth, respectively. In 2018, Japan's GDP growth rate was 0.81% compared to the previous year [5] and it was estimated about 0.89% and 0.47, respectively, in 2019 and 2020.[6]

---

[4] The Plaza Accord resulted in a 50% depreciation of the US dollar relative to the Japanese yen and the deutschemark, see Yoichi Funabashi, "A U.S.-China 'Plaza Accord'?," *The Japan Times,* September 11, 2018, http://https://www.japantimes.co.jp/opinion/2018/09/11/commentary/japan-commentary/u-s-china-plaza-accord/#.Xm8vQagzaUk.

[5] See "Japan GDP," *Countryeconomy.com*, https://countryeconomy.com/gdp/japan?year=2018.

[6] See H. Plecher, "Gross domestic product (GDP) growth rate in Japan 2024," *Statista,* November 6, 2019, https://www.statista.com/statistics/263607/gross-domestic-product-gdp-growth-rate-in-japan/.

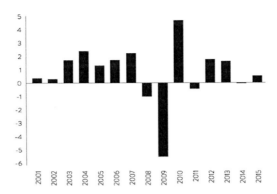

Figure 3. Japan's Economy (2000-2021)

Source: "Japan's Economy," *Asia Link Business*, https://asialinkbusiness.com.au/japan/getting-started-in-japan/japans-economy?doNothing=1

South Korea adopted different strategies from China and Japan during its economic growth period. While China made reformed industry structure from heavy to light industry, and Japan's invested heavily but lost in information technology, South Korean developed a step-by-step industry structure reform from labor-intensive light industry and capital-intensive heavy industry to information technology and communication industry. South Korea therefore became one of "the Four Asian Tigers" (the other three are Taiwan, Hong Kong and Singapore) in the 1980s by enhancing manufacture skills, utilizing foreign investment from Japan and the United States. Subsequently, South Korea was a victim of the 1997 Asian financial crisis due to its overleveraged economy. With the help of the International Monetary Fund (IMF) and other international organizations, South Korea merged many small companies to form bigger ones, leading to an economic structure dominated by big enterprises such as Samsung, LG, and Hyundai, etc. These big companies have improved South Korea's economy, especially on IC and IT industries, cars, ships and consumer electronic products, but this also worsened equity indices.

Table 1 Macroeconomic Statistics of East Asian Countries in 2018

Unit: $ and %

	China	Japan	South Korea	Taiwan	Hong Kong
GDP Per Capita ($)	9,580	39,304	33,320	25,008	48,451
Inflation Rate	2.1	1.0	1.5	1.5	2.4
Unemployment Rate	3.8	2.4	3.8	3.7	2.8
Current Account Balance/GDP	0.4	3.5	4.4	12.2	4.3
The Secondary Industry/ GDP	40.7	29.1 (2017)	38.7	35.2	7.5 (2017)
Gross National Savings/ GDP	45.2	28.0	35.8	33.1	26.0
Foreign Exchange Reserves ($billions)	3,073	1,210	393	462	-
Total fertility (2015-2020)	1.7	1.4	1.1	1.2	1.3

Note: Total fertility are estimated by the United Nations.

Source: IMF, IMD, UN, and the Central Banks of the listed countries.

Taiwan's story is, to an extent, quite similar like South Korea. Both adopted similar economic policies during the Cold War period when both faced a big adversary, respectively, China and North Korea. But there are still some differences between South Korea and Taiwan. The first difference is their business structure; South Korea is dominated by some big firms while Taiwan owns many small and medium enterprises (SMEs). In addition, Taiwan was not hurt by the 1997 Asian financial crisis, so it didn't need to merge SMEs to big firms. The second difference is that North Korea isn't as strong as China, so many countries could establish diplomatic relations with South Korea; it therefore could join many important international organizations such as the United Nations, the IMF, and the World Bank, etc. However, Taiwan cannot sign many agreements like South Korea neither could join these critical international organizations. The last difference is that South Korean companies prefer to create own brands and control the whole supply chain but

Taiwanese or Taiwan-based companies usually just join international companies' supply chain for contract manufacturing. For instance, Samsung produces own brand products like Note and Galaxy but Taiwan's Foxconn (Hon Hai) only makes iPhones for Apple Inc.

Hong Kong was always considered as the pearl on the Crown of the Queen before 1997. Its economy structure was closer to "City Economy" that owns more service sectors in economic structure than South Korea and Taiwan. Hong Kong had returned to China as a Special Administrative Region after 1997; hence, its economy was more heavily reliant on China. Many Hong Kong manufacturing businesses have moved to Shenzhen and Guangdong, as a result of which Hong Kong has become a shopping store, logistic and financial center. However, Hong Kong was number two in 2019 IMD world competitiveness ranking as Table 2 showed, the other East Asian countries (ranking) being Taiwan (16), China (14), Japan (30) and South Korea (28). Since Hong Kong takes advantage of a critical role entering into China, it attracts many foreign investments coming to set up financial, media, and other service centers that China blocks locally. However, though Hong Kong obtained many benefits from China's economy and grew quickly, the result of heavy dependency on China is that China can exert great influence to Hong Kong at the same time. Due to the decrease of global demand and US-China trade war, Hong Kong's GDP in Q4 2018 fell into 1.3% from 2.9% in Q3.[7] And then following with a half-year of protest on the Extradition Law, Hong Kong's economy fell and contracted by 1.2% in 2019.[8]

II. Macroeconomic Volatility within the Region

After explaining the development of macroeconomic situations of East Asian

---

[7] "Hong Kong's economy slumped 1.3% in Q4 2018: report," *Hong Kong Business,* March 20, 2019, https://hongkongbusiness.hk/economy/in-focus/hong-kongs-economy-slumped-13-in-q4-2018-report.

[8] "Hong Kong fell deeper into recession at end of 2019," Aljazeera, February 3, 2020, http://www.aljazeera.com/ajimpact/hong-kong-fell-deeper-recession-2019-200203084830989.html.

countries and economies separately, this section will discuss the economic changes in East Asian region in the past few years, including how they cooperate or compete with one another. There are two factors making volatility within the region from internal and external sides; the internal factor is that China's rapid economic growth brings many business opportunities and challenges to other economies in the region. The external factor is the global economic integration trend which attracts these economies to sign economic agreements with other countries or economies or join regional economic groups such as the Comprehensive and Progressive Agreement for Trans-Pacific Partnership (CTTPP) or the Regional Comprehensive Economic Partnership (RCEP).

Table 2   The IMD World Competitiveness Scoreboard in 2019

2019	Country	2018	Change		2019	Country	2018	Change	
1	Singapore	3	+2	⬆	33	Czech Republic	29	-4	⬇
2	Hong Kong SAR	2	-	-	34	Kazakhstan	38	+4	⬆
3	USA	1	-2	⬇	35	Estonia	31	-4	⬇
4	Switzerland	5	+1	⬆	36	Spain	36	-	-
5	UAE	7	+2	⬆	37	Slovenia	37	-	-
6	Netherlands	4	-2	⬇	38	Poland	34	-4	⬇
7	Ireland	12	+5	⬆	39	Portugal	33	-6	⬇
8	Denmark	6	-2	⬇	40	Latvia	40	-	-
9	Sweden	9	-	-	41	Cyprus	41	-	-
10	Qatar	14	+4	⬆	42	Chile	35	-7	⬇
11	Norway	8	-3	⬇	43	India	44	+1	⬆
12	Luxembourg	11	-1	⬇	44	Italy	42	-2	⬇
13	Canada	10	-3	⬇	45	Russia	45	-	-
14	China	13	-1	⬇	46	Philippines	50	+4	⬆
15	Finland	16	+1	⬆	47	Hungary	47	-	-
16	Taiwan, China	17	+1	⬆	48	Bulgaria	48	-	-
17	Germany	15	-2	⬇	49	Romania	49	-	-
18	Australia	19	+1	⬆	50	Mexico	51	+1	⬆
19	Austria	18	-1	⬇	51	Turkey	46	-5	⬇
20	Iceland	24	+4	⬆	52	Colombia	58	+6	⬆
21	New Zealand	23	+2	⬆	53	Slovak Republic	55	+2	⬆
22	Malaysia	22	-	-	54	Ukraine	59	+5	⬆
23	United Kingdom	20	-3	⬇	55	Peru	54	-1	⬇
24	Israel	21	-3	⬇	56	South Africa	53	-3	⬇
25	Thailand	30	+5	⬆	57	Jordan	52	-5	⬇
26	Saudi Arabia	39	+13	⬆	58	Greece	57	-1	⬇
27	Belgium	26	-1	⬇	59	Brazil	60	+1	⬆
28	Korea Rep.	27	-1	⬇	60	Croatia	61	+1	⬆
29	Lithuania	32	+3	⬆	61	Argentina	56	-5	⬇
30	Japan	25	-5	⬇	62	Mongolia	62	-	-
31	France	28	-3	⬇	63	Venezuela	63	-	-
32	Indonesia	43	+11	⬆					

Source: "IMD World Competitiveness Ranking 2019 Results," *IMD*, May 28, 2019, https://www. imd.org/contentassets/6b85960f0d1b42a0a07ba59c49e828fb/one-year-change-vertical. pdf.

The rise of China's economic power has made East Asian countries and economies face a dilemma in recent decades. On the one hand, the Chinese government offered many spaces and labors to the companies and firms from Japan, South Korea, Taiwan and Hong Kong. However, on the other hand, China also has become a competitor to these countries and economies due to the serious impact of the "red supply chain" which applied import substitutions in China. According to the IMF, China is the major recipient of global Foreign Direct Investment (FDI) and also a major provider of FDI outflows since the rise of its economy. In 2018, China received $139 billion, ranked the second largest recipient of FDI in the world after the United States. The largest foreign investor in China from 1997 to 2017 is Hong Kong, occupying 52.6% of total, Japan (6.1%), South Korea (2.7%), and Taiwan (1.2%) (see Table 3). The FDI from Japan is the third largest of the whole investment amount, South Korea is the 6th, and Taiwan is the 10th. It is worth noting that since 2005, China's FDI outflows also rose sharply and quickly and have surpassed FDI inflows in 2015. Now China is the world's second-largest provider of FDI outflows and Japan is the number 1.[9] It is worth noting that since 2005, China's FDI outflows also rose sharply and quickly and have surpassed FDI inflows in 2015. Now China is the world's second-largest provider of FDI outflows and Japan is the number 1.[10] In 2017, the major destination of Chinese investment goes to Hong Kong ($981 billion), the Cayman Islands ($251 billion), the British Virgin Islands ($122 billion), the United States ($67 billion), Singapore ($45 billion), Australia ($36 billion), and the United Kingdom ($20 billion).[11]

The same situation happens on trade as well. Table 4 shows the degree of foreign trade dependence (FTD), which refers to a country's import and export volume to its

---

[9] China's Economic Rise: History, Trends, Challenges, and Implications for the United States, *Congressional Research Service,* June 25, 2019, p. 21, https://fas.org/sgp/crs/row/RL33534.pdf.

[10] Ibid.

[11] China's Economic Rise: History, Trends, Challenges, and Implications for the United States, *Congressional Research Service*, June 25, 2019, p. 18, https://fas.org/sgp/crs/row/RL33534.pdf.

GDP ratio, between East Asian countries and economies with the United States and China, respectively, in 1996 and 2018. The FTD of Japan, South Korea and Taiwan with the United States are larger than their dependency on China in 1996. The largest is Taiwan's trade dependency with the United States that is 16.0% (see Table 4), following with South Korea's 9.2% and Japan's 4.1%. Their trade dependence with China was only 1.3%, 3.3% and 1.3% of their GDP, respectively. However, 20 year later, the FTD of each country increased quickly with China, the trade volume to GDP ratio of Japan was 6.4%, South Korea 15.6%, Taiwan 24% and Hong Kong 157.3%, larger than those with the United States by 4.5%, 7.63%, 12.2% and 20.7%, respectively, in 2018. This clearly indicates that East Asian countries are now more reliant on China since the rise of China's economy and due to the geopolitics. For China, the European Union is its largest trading partner, and the United States ranks the second in 2018, ASEAN, Japan, South Korea, Hong Kong, and Taiwan are also its major partners. The top three export markets of Chinese products are the United States, the European Union, and ASEAN countries. On the other hand, the EU, ASEAN and South Korea are the major sources for imports. China has the largest trade surpluses with the United States ($282 billion), Hong Kong ($274 billion) and the EU28 ($129 billion). [12]

Table 3 Top 10 Sources of China's FDI Inflows (1997-2017)

Unit: billions

Country	Amount	% of Total
Total	2,688	100
Hong Kong	1,241	46.2
British Virgin Islands	286	10.6
Japan	165	6.1
Singapore	108	4.0
Germany	87	3.2

---

[12] Ibid, p. 22.

Country	Amount	% of Total
S. Korea	73	2.7
U.S.	72	2.7
Cayman Islands	49	1.8
The Netherlands	37	1.4
Taiwan	33	1.2

Source: China's Economic Rise: History, Trends, Challenges, and Implications for the United States, *Congressional Research Service*, June 25, 2019, p. 17, https://fas.org/sgp/crs/row/RL33534.pdf.

Table 4   Trade Dependency of East Asian countries on the USA and China

Trade to	US		China	
Country　　Year	1996	**2018**	1996	**2018**
Japan	4.1	**4.5**	1.3	**6.4**
South Korea	9.2	**7.6**	3.3	**15.6**
Taiwan	16.0	**12.2**	1.3	**24.0**
Hong Kong	14.2	**20.7**	51.2	**157.3**

Note: Trade dependence is measured by the share of exports and imports to GDP

Source: The listed countries' Customs Statistics and the IMF, calculated by the authors.

China also offers a huge market for those foreign companies and firms who were searching for much more business opportunities which could face many risks if China's economy cools down. The IMF has estimated the impact of China rebalancing on Partner-Country for 13 countries and economies' growth transmitted through trade channel by measuring the impact of changes in China's consumption and investment.[13] Of the East Asian countries, Taiwan is the most vulnerable, followed by South Korea, Japan, and Hong Kong. Taiwan's greater vulnerability

---

[13] The partner countries stands for Japan, Korea, Taiwan, Hong Kong, Indonesia, Thailand, Philippines, Singapore, Malaysia, Vietnam, India, Australia and New Zealand.

is due to the lack of branded exports to China. As China's economic power has rapidly grown in recent years, the Chinese government has become dissatisfied with only performing subcontractor work and expressed their desire to become import substitutes to upstream countries like Japan, South Korea, and Taiwan. They also want to boost their capabilities to manufacture value-added products in different ways: through Chinese government's policy, via foreign direct investment or by hiring skilled professionals.

According to the statistics and data from the Chinese Customs, imports for subcontracting work decreased 14.0% from 40.6% in 2006 to 26.6% in 2015.[14] The import of panel products also decreased from $18.3 billion in 2012 to $13.0 billion in 2015. In the steel industry, the local manufacturing rates grew from 90.3% in 2001 to 98.7% in 2015, while imports from Japan, South Korea and Taiwan decreased 6.9%, 1.9% and 3.3% from 2008 to 2015. Similar patterns can be perceived and seen in petrochemical and panel industries. China's imports on ethylene products from Japan, South Korea and Taiwan went down from 13.5%, 18.5% and 14.7% in 2008 to 9.9%, 17.2% and 10.9% in 2015, respectively.[15] Facing the economic threats whether from China's economy landing, import substitutions or being their competitor, some economies try to diversify the dependency on China by different ways. For instance, Japan and South Korea joined the CPTPP and the RCEP to expand their economic cooperation, but since President Tsai In-Wen took office in 2006, Taiwan's government has proposed and suggested the "New Southbound Policy" to enhance their economic cooperation and trade relations with Asian countries including New Zealand and Australia. Although Hong Kong cannot expand its economic relations with other countries without China's approval, she wants to limit the number of tourists or immigrations from mainland China. Not only do other countries and economies want to diversify their economic ties with China, but also China tries

---

[14] See China Custom, http://english.customs.gov.cn/.

[15] Ibid.

to widely build connections with countries and economies outside the East Asian region. For example, China has already proposed the "One Belt One Road" (OBOR) initiative and strategy to expand its trade, investment, financial and other economic relations with South East Asia, South Asia, Central Asia, Europe and Africa. That will certainly help China to export its infrastructure products to these countries hence bring more pressure to Japan who leads the Asian Development Bank (ADB) to invest in many construction projects in many Asian countries. China also pushes the RCEP to compete with the CPTPP led by Japan.

III. The Role of the USA to East Asian Economy

On the role the United States played in the region of East Asia, it can be divided into two periods from 1945 to 2015. The United States led East Asian countries to build a supply chain union against the Communist group such as the Soviet Union (USSR), China, North Korea and Vietnam during the cold war period. At the same time, China made economic experiments for building a socialist country, but the experiments could not work due to occurrence of "the Cultural Revolution". The US-East Asian union collapsed after the practice of China's economic reform and open policy and the end of the cold war. These countries had many conflicts with China on currency, trade, and some open market issues. China therefore tried to coordinate and reconciled with the United States and learned some experiences from other East Asian countries.

In the first 35 years after the World War II, the United States helped their economic growth and transferred some industries to East Asia except China by the Official Development Assistance (ODA) projects, business investment, contract manufacture and trade opportunity during the cold war period. or political and economic reasons, the United States wanted to help these countries and economies to be rich and not influenced by the Communist group. The US businesses built low cost factories and made low price products in East Asia countries and exported them to American local market. East Asia countries except for China followed the "Flying Geese Model", Japan flew first, followed by South Korea, Taiwan, Hong Kong and Singapore to build manufacturing supply chain that produced consumer goods for exportation. These countries therefore received a lot of foreign investments, technology and product orders from the United States. This not only could develop their economy and maintain local political stability, but also could earn money for buying national defense equipment from the United States for protecting themselves.

It was a critical period to East Asian countries and economies to improve their economy over the period of 1978 to 1988. From the perspective of international politics, China practiced economic reform and proposed an open policy; the United States eventually suspended official diplomatic relations with Taiwan in 1978. China and the United Kingdom then reached agreement to return Hong Kong to China on December 19, 1984. These events pushed Taiwan and Hong Kong from the United States far away directly or indirectly, and the East Asian anti-Communist alliance seemed not to be so solid like before. On the economic issues, the "Plaza Accord" that was signed by the United States, the United Kingdom, France, West Germany and Japan in 1985 made the currency of Japan, South Korea and Taiwan to appreciate by 86%, 19% and 39%, respectively from 1985 to 1988.

Such rapid appreciation increased the cost of land and labor force in these countries and economies; many domestic manufacturers therefore had to leave their home countries to invest in the low-cost places. This made some countries' industrial sectors become empty. Taiwan is an example of this. One the other hand, currency appreciation and flows of hot money into these places also caused economic bubble in these countries and economies. When hot money went away, the bubble collapsed, and these countries therefore seriously suffered deep currency depreciation like Japan. During this period, the United States seemed to stand in the opposition side to Japan, South Korea, Taiwan and Hong Kong as asking East Asian countries to appreciate their currencies or put them into the "Special 301" watch list. At the same times, China's economic reforms were not as smooth as what people imagine now; many debates about the routes of capitalism or socialism often appeared and occurred. Problems like lack of protection of intellectual property rights, practice of modern company related laws, regulations and transparency of political system all made China's economic reform stumbled. In manufacturing sector, foreign companies hired the "three plus one" model to build their production capacity.[16] This non legal-person contracted manufacturing model developed China's industry in early period after 1978, and attracted some labor-intensive foreign investors until the happening of the 1989 Tiananmen Square incident.

China started the second economic reform since 1992. The former premier Zhu Ronji proposed a series of new laws of business and taxation to fit development of modern economic structure. South Korea built official diplomatic relations with China in 1991; on the other side, Taiwan gradually improved the relations with China by non-official contacts and exchanges since 1992. These internal and external factories made more investment to China from neighbor economies. To join the World Trade Organization (WTO) presented another chance for China to cooperate with East Asian countries and economies, and made her become the most popular investment place and potential market not only for East Asian countries but also for other countries such as the United States. The US information technology firms such as HP, Dell, and Compaq, etc., asked their OEM partner companies (most of them are from Taiwan) to invest in China. These firms expanded their production capacity in Yantzi River Delta and closed their home countries' factories hoping to attract other suppliers investing together. The supply chain hence moved to China from downstream to upstream and made China's high-tech manufacture ability to become stronger.

---

[16] The model contains custom manufacturing with materials, designs or samples supplied and compensation. Under this model, foreign companies and firms offered equipment, materials, samples, and were responsible for the final products export, Chinese enterprises offered lands, factories and labor forces. Both sides cooperated by contracts due to the government' s restraints to foreign legal person.

The economic power of the United States and China coordinated after global financial crisis in 2008. When the United States are involved in the financial crisis, China proposed a four-trillion Chinese Yuan investment project to improve her infrastructure, and restructured economy from relying on external trade, the State Owned Enterprises (SOEs) also invested in local consumption. Moreover, China also explored its outbound investment, proposed and joined the RCEP, asking more voting rights in the IMF, World Bank, and establishing te Asian Infrastructure Investment Bank (AIIB) based on the One-Belt-One-Road strategy. These efforts made China become more attractive to foreign investors and governments. The US influence in Asia-Pacific region seemed to decreased than before.

IV. Future Possible Scenarios under US-China Trade War

In order to understand future possible situation under US-China trade war since President Trump imposed tariffs on Chinese imports in March 2018, this research lists four possible scenarios in Table 5. In these cases, the Untied States could adopt two strategies such as "rebalance" as what President Obama did in recent years including the construction of the Trans-Pacific Partnership (TPP) that has been changed to Comprehensive and Progressive Agreement for Trans-Pacific Partnership (CPTPP). Or the United States could choose "isolation" to quit the CPTPP and tax Chinese export products for 45% tariff as President Trump claimed and suggested. Similarly, China could use the strategy of "maintaining the status quo" to propose what she did on the RCEP and "One Belt One Road" project, or she can use "expansion" to propose more strategies on economic integration and cooperation of East Asia countries such as the Free Trade Area of Asia Pacific (FTAAP).

Table 5   Possible Scenario for Future Macroeconomic Volatility in East Asia

		USA	
		Rebalance	Isolation
China	Maintaining the status quo	1	2
	Expansion	3	4

(This table is made by the author)

In Scenario 1, the United States and China adopt the strategy of "rebalance" and "maintaining the status quo", respectively. Under Obama's Administration, the United States attempted to rebalance American power in Asia, the TPP was proposed to Asia-Pacific countries and economies under the leadership of the United States. To compete with the American power and influence in this region, China also pushed the RCEP against and isolate US economic power. Some East Asian countries like Japan, South Korea want to join both economic and trade unions although the TPP is still their priority. For Taiwan, joining the TPP seems to be the unique choice due to its political opposition relations with the Chinese government.

In Scenario 2, after the victory of his election, President Trump has proposed his "America First" policy; he prefers to undertake bilateral negotiations and one-on-one tests of strength and guile. The United States therefore retreated from the TPP, these actions then left American's economic and trade partners in East Asia far away due to the lack of economic ties with the United States. The TPP turned out to be the Japan-led CPTPP that occupied 13.1% of global GDP and has already taken effect on December 30 in 2018.[17] The CPTPP always has considered as the competitor of

---

[17] The GDP of the eleven CPTPP members is close to $10.6 trillion, and its population is about 500 million people while the GDP of the RCEPT members is $23 trillion with a population of 3.5 billion. The RCEPT has occupied 60% of Taiwan's exportation. On the introduction of the CPTPP and the RCEP and all of their negotiation can see *the Bureau of Foreign Trade*, http://cptpp.trade.gov.tw and http://www.trade.gov.tw/Pages/List.aspx?nodeID=1557.

the RCEP that has finished all the negotiation and will be signed in 2020. Compared with the CPTPP, RCEP occupied 30% of global GDP. However, for Taiwan, due to current political situation with China, it is unlikely to join the RECP before accepting the 1992 Consensus proposed by the Chinese government. Taiwan can now only focus on IC and IT industries that occupied one third of her exportation under the benefits of the Information Technology Agreement (ITA) under the WTO; other industries will be in an inferior position by higher tariff. This will be very detrimental and harmful to the future development of Taiwan industries.

In Scenario 3, the United States and China can choose "rebalance" and "expansion". This means not only the United States but also China will still push its economic cooperation with Asian countries. China will release more policies and propose more incentives to attract her neighbor countries such as opening her market, investing in neighbor countries, or sending more tourists to travel. In this case, East Asian countries and economies will still try to join both trade agreements though will be closer economically to China. On the one hand, the economic power of the United States will be as strong as before, she has to take more actions to attract East Asian countries and economies to cooperate with her. For example, the United States will help Japan to compete with China to bid the infrastructure and construction projects in South East Asia and South Asia. However, East Asian supply chain will closely connect these countries and economies, the United States can only dominate few global branding companies like Nike, Apple and other non-branding products such as infrastructure, steel, petro chemical, textile etc. Chinese economic competition power will stand on a critical position in the supply chain.

Scenario 4 will be the most competitive situation for East Asia countries to push the United States and pull China. The United States will make many policies beneficial to herself hence will have more conflicts with East Asian countries on trade and currency issues just like what President Trump has done. However, China not only can offer economic integration with East Asia, but also can present more business opportunities for its neighbors. The result will definitely make East Asia be closer with China while Taiwan under the Democratic Progressive Party (DPP) rule may refuse this kind of gift from China to avoid falling into a dilemma on political and economic threats from China.

To evaluate the future volatility in East Asia, one can use the participation degree in GVCs to connect with the scenarios we listed above. In Table 6, the participation degree is measured by adding backward linkages (Foreign Value Added Share of Gross Export) and forward linkages (Domestic Value Added Embodied in Foreign Exports as Share of Gross Export). For example, the participation degree in GVCs of China is 34.8% in 2015, backward linkages is 17.3 which means its export contains 17.3% from import value added and forward linkages 17.5 means 17.5% of its export is for downstream country's intermediate goods. Compared with other East Asian countries and economies, China is much closer to the downstream of supply chain and Japan is in the upper side due to the forward and backward linkages in 2015.

Table 6 East Asia and US Participation Degree in Global Value Chains (2015)

Unit: %

	Backward Linkages (1)	Forward Linkages (2)	Participation Degree (1)+(2)
China	17.3	17.5	34.8
Japan	13.2	24.4	37.6
S. Korea	32.6	19.1	51.7
Taiwan	32.4	24.4	56.8
HK	26.6	15.7	42.3
USA	9.5	22.2	31.7

Source: OECD-WTO TiVA Data Base (Data extracted on 14 Nov 2019 from OECD.Stat)

Figure 4 shows the trend of East Asia and US participation degree in Global Value Chains from 1995 to 2015, the participation degrees of the countries and economies have risen except for China. South Korea, Taiwan and Japan sharply increased 12.3%, 10.3% and 8.2% between 1995 and 2011, compared with another three countries and economies, that is to say, the United States 0.8%, Hong Kong 4.8% and China -8.1%. The increase of forward linkages of Japan is 0.6% that is the lowest while that the increase of Taiwan is 8.6%, China 8%, the United States 2.8% and South Korea 2%. However, Hong Kong has negative growth of -0.1%. This means that Taiwan exported more materials and equipment for next stop manufacturing activities, and Hong Kong exported more final consumer goods than material or other intermediate goods.

On the backward linkages, the phenomena seem to be different; South Korea and Japan are the first highest economies with 10.3% and 7.6% that point difference between 1995 and 2015. Hong Kong and Taiwan increased 4.9% and 1.7%, respectively, while the United States and China both decreased with 2% and 16.1%. The differences show that South Korea, Japan, Hong Kong and Taiwan's exportation greatly rely on materials and semi-finished goods from other countries and economies more than before. The export of America and China not rely much on other countries and economies.

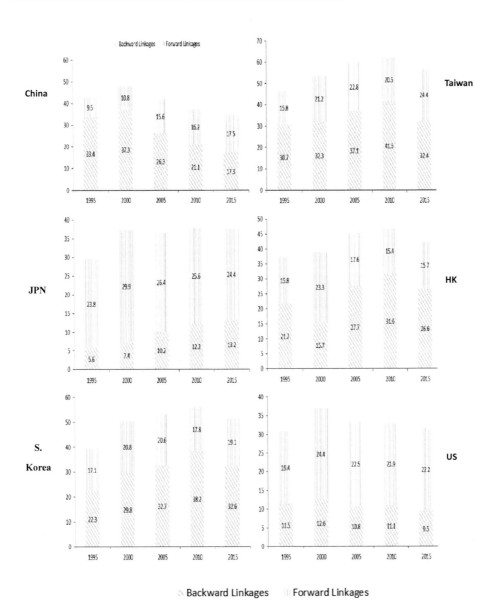

Figure 4   Trends of East Asia and US Participation Degree in Global Value Chains (1995-2015)

Note: Forward linkages show the domestic value added embodied in foreign exports as share of gross export; Backward linkages mean foreign value added share of gross export

Source: OECD-WTO TiVA Data Base (Data extracted on 14 Nov 2019 from OECD.Stat)

If one looks at the scenarios this research listed above, he will find that if the United States makes different economic policy to rebalance in Asia or isolate itself, she will not be affected so deeply because her participation degree (31.7%) is less than any East Asian countries and economies. It is also interesting to find that the forward linkages degree of America is 22.2% which means that the United States exports many products to other countries for next procedure. If the United States isolate herself and other country fight back to break the global supply chain, the United States will face more challenge hence damage her economy. In East Asia, anti-globalization will hurt Taiwan (56.8%) and South Korea (51.7%) deeply because of their participation degrees are the first two highest ranking in East Asia. That means any foreign trade change will bring benefit or hurt their economy deeply. Hong Kong is on the final product and assembly position while Japan is on the top place, China will deliver the impact from the United States-the final market to other countries in the supply chain. The CPTPP could be a tool to break the deliver channel from China if Japan, South Korea and Taiwan can have much more trade with the United States more directly once they all join the free trade agreement. However, since the United States already retreated from the free trade agreement, East Asian countries and economies may join the RCEP due to their high participation degree possibly. This certainly will be adverse to the United States.

V. Conclusions

To conclude, the macroeconomic volatility in East Asia could be divided into two periods from 1945 to 2015. During the first thirty-five years (1945-1980), the US connection with Japan, South Korea, Taiwan and Hong Kong were very deep against China. The United States provided them with funds, technology and management skills so that East Asian countries could form a supply chain to produce labor-intensive products to export. At that moment, China's economy had many troubles hence developed pretty slowly due to her domestic political problems. The second thirty-five years (1980-2015) happened when China opened to the outside world and practice economic reforms to attract foreign trade and investments. The United States just like other countries wanted to explore the huge potential market of China; she therefore strengthened economic ties with China and weakened her relations with her allies like Japan, South Korea and Taiwan. When China's economy grew up and became a competitor and consumer to other East Asian countries, Japan, South Korea and Taiwan faced new challenges. On the one hand, the relations between these East Asian countries and the United States were not as strong as before, sometimes many conflicts on trade or other economic issues happened. On the other hand, China is more powerful in political and economic areas, East Asia has to close to or rely on it to maintain their economic momentum.

Before the rise of China, the United States made her East Asian economic policies from political considerations since the World War II, like the examples for roping Japan, South Korea and Taiwan against China during the cold war period. After China's open to the outside world, the United States has developed deeply trade and economic relations with China and has obtained her support at the same time on some important world issues like what happened in recent years. That situation now has been reversed due to President Trump's China policy and the occurrence of US-China trade war. President Trump not only suggested that he would make diplomatic policy from economic or business consideration, especially on the trade issues with East Asian countries or economies. He also asked Japan and South Korea to pay for defending themselves. Under the impact of US-China trade war, East Asian countries will try to find other solutions hence will decrease their dependence with the United States. At the same time, the companies and enterprises from East Asian countries invested in China will be transferred to other countries so that their products would not be imposed taxes by the US government. Under this kind of situation, Taiwan is an exception, due to the downward of Cross-Strait relations since President Tsai In-Wen took office, the relationship between Taiwan and the United States is closer.

The US government and Congress have successfully passed a series of Acts and Regulations benign to Taiwan, Taiwanese invested companies in China have also transferred the supply chains, that is to say, the factories, to Taiwan. Taiwan therefore is facing a different situation with that of East Asian countries. From the geopolitics, China still is a giant trade partner who could not be ignored. How to balance their relations with China and the United States will be the most important topic for East Asian countries.

If the United States decides to maintain her layout in East Asia, she has to utilize market power to attract Asia countries to join the US-led trade alliance. For instance, to use innovation capability to lead Japan, South Korea, Taiwan and HK to build new supply chain or organize economic alliance to compete with China's all-in-one factory. Since President Trump and his Administration decided to retreat from the TPP and propose different economic and trade policies, this certainly will push East Asian countries to adopt different policies to response these changes. To sum up, political and economic interests cannot be separated and have to be considered carefully and comprehensively. The political and economic security of East Asia is part of American global interests; therefore, it will be unwise for the United States to retreat from Asia and only focus on her domestic politics. The US government perhaps can adjust its policies and design a new model to cooperate with East Asian countries; this will bring more benefits to the United States hence create a win-win situation to all the countries in this region.

解析新冠肺炎期間美中互動下的中國國家安全治理作為：「雙層賽局」與「安全化」融合研究途徑

翁明賢 *

壹、前言

2017年美國川普上台以來，一切以「美國優先」為原則，不僅改變傳統美國「世界警察」角色，要求其盟邦增加國防預算，並退出許多國際多邊條約與組織，更是以中國為對象，展開所謂「平等」與「互惠」的貿易談判，要求北京減少貿易逆差、智慧財產權保護、國有企業補貼與增加農產品採購問題。基本上，華盛頓「戰略」指導與具體「作為」就在於「抑制」一個崛起的中國，讓北京面臨複雜的與外在的國際安全格局影響。

雖然，在2020年1月15日，雙方簽署中美貿易談判第一階段協議，卻因2019年12月底，武漢地區爆發「新型冠狀肺炎」（COVID-19），使得雙方貿易爭議暫告一段落。面臨這場「新冠肺炎」風暴，2020年2月10日，習近平在北京調研指導新型冠狀病毒肺炎疫情防控工作時強調，「當前疫情形勢仍然十分嚴峻，各級黨委和政府要堅決貫徹黨中央關於疫情防控各項決策部署，進行打贏疫情防控的人民戰爭、總體戰、阻擊戰」[1] 2月14日，習近平在中央全面深化改革委員會會議上要求，把「生物安全」納入國家安全體系，因為「這次抗擊疫情是對『國家治理體系和治理能力的一次大考』，是以，必須盡快推動發布「生物安全法」，加快構建國家生物安全法律法規體系。[2] 事實上，針對

---

* 淡江大學國際事務與戰略研究所教授兼所長

[1] 習近平強調：「堅決貫徹堅定信心、同舟共濟、科學防治、精准施策的總要求，再接再厲、英勇斗爭，以更堅定的信心、更頑強的意志、更果斷的措施，緊緊依靠人民群眾，堅決把疫情擴散蔓延勢頭遏制住，堅決打贏疫情防控的人民戰爭、總體戰、阻擊戰。」，請參見：「習近平：堅決打贏疫情防控的人民戰爭總體戰阻擊戰」，中華人民共和國文化與旅遊部，https://www.mct.gov.cn/gtb/index.jsp?url=https%3A%2F%2Fwww.mct.gov.cn%2Fpreview%2Fspecial%2F8830%2F8837%2F202002%2Ft20200214_850771.htm。（檢索日期：2020/05/10）

[2] 習近平強調：「這次抗擊疫情是對『國家治理體系和治理能力的一次大考』。要研究和加強疫情防

「國家治理體系與治理能力」的挑戰，在於習近平2015年所強調的11種「綜合國家安全觀」，並沒有包括「生物安全」。習近平強調中國面臨的是一場「打贏疫情防控的人民戰爭、總體戰、阻擊戰」，顯示出重要性，牽涉到國內社會與經濟情勢的穩定，也攸關中國的「國際地位」與「國際聲譽」。

因此，北京透過雷厲風行的封城（1月23日武漢）、封省與邊境管控，而中國人民解放軍在湖北武漢地區興建火神山、雷神山醫院專門醫院，以及各類方艙醫院，集中管理、集中醫治。之後，在4月8日武漢解除封城之後，恢復部分地區與行業正常生產，逐步有效控制疫情至今。[3] 反觀美國的態度，初期川普輕忽疫情地區，雖率先進行武漢撤僑，以及邊境管制。但是，並沒有認真思考疫情的嚴重性。從3月份開始迅速傳染至全美各州，目前已經成為全球重災區。[4] 川普基於國內經濟、疫情管理、選情民意等考量，「究責中國」「隱瞞疫情」以及批判世界衛生組織「偏袒中國」，激發起美中計將貿易議題之後，另外一個爭議話題。加上，解放軍從2020年1月起，「常態化」海空機艦巡弋台海周邊，超越第一島鏈到南海地區，引發美國相對偵察機、戰機與轟炸機表達在亞太地區「軍事存在」與「前沿部署」的態勢。

事實上，中美兩國各有內部「國內層面」與不同「國際層面」的互動關係，從而影響兩國相互衝突與談判的因素。學者Robert Putnam提出了「雙層賽局」（two-level game）作為了解「國際政治」與「國內政治」如何相互影響的一個簡易分析架構。不管是「經濟議題」，還是純粹「軍事安全」，都會受到「國際關係」與「國內政治」的雙重約制問題。另外，國家決策者也會透過「安全化」議題操作，來轉移國際與國內安全議題的壓力。以中國為例，基於美中關係受到亞太戰略格局發展，特別是美國台海政策調整的衝擊，因此，在

控工作，從體制機制上創新和完善重大疫情防控舉措，健全國家公共衛生應急管理體系，提高應對突發重大公共衛生事件的能力水平。」，請參見：「習近平：把生物安全納入國家安全體系」，中央通訊社，https://www.cna.com.tw/news/acn/202002140343.aspx。（檢索日期：2020/05/10）

[3] 「新冠肺炎：武漢解除『封城』令『綠碼』居民可離開」，BBC新聞網，（檢索日期：2020/05/10）https://www.bbc.com/zhongwen/trad/chinese-news-52210928

[4] 基本上，全球各地確診感染病例400萬1437例，累計染疫不治27萬7127人，而美國境內總共130萬5544例確診，累計7萬8618人不治。請參見：「武漢肺炎全球最新情報 5/10」，中央通訊社，https://www.cna.com.tw/news/firstnews/202005100012.aspx。（檢索日期：2020/05/10）

取捨國際與國內事務的「優先性」時，主要相關「決策者」的客觀與主觀「利益」的判斷，可以從「安全化」角度，加以研判，從而解析中國安全治理政策推動的主導原則。

本文首先介紹Putnam提出的「雙層賽局」（two-level game）內涵與其核心概念：「底線集合」（win-set）要素，以及透過分析Ole Waever提出的「安全化」概念，加以融合，藉以理解不同「決策者」如何因應國家安全治理議題。繼而分析2012年習近平執政後的國家安全治理發展，包括：提出「總體國家安全觀」、設立「國家安全委員會」，推出「國家安全戰略綱要」與頒佈「國家安全法」等；第三、受到「新冠肺炎」影響下，中國面臨的國際與國內因素分析？第四、透過「雙層賽局」之「底線集合」與「安全化」檢證中國安全治理機制與作為。最後，本文提出2021年邁向第一個百年目標下，習近平的國家安全治理的後續走向。

貳、雙層賽局的意涵與安全化研究途徑

一、雙層賽局的概念與操作

Putnam認為一般「國際談判」可以從兩個層次加以分析，而「雙層賽局」的意涵是基於「國際層面」為「第一層次」（Level I, LI）：談判者之間的議價過程，引導出暫時初步性協議。與「國內層面」為「第二層次」（Level II, LII）經由國家內部特定程序（國會討論與投票）通過協議，就是一種「批准」（ratification）過程。其實，不僅藉由「國際談判」可以應用於「雙層賽局」分析途徑，一般「國際衝突」涉及兩個行為體「利益」的競合，一定程度也有參考的價值。

進一步討論，Putnam認為在「國內層次」方面，國家內部許多團體基於各自利益考量，形成政府決策時的壓力，決策者會跟相關團體結合以擴大其影響力，與滿足國內民意需求。在國際層次方面，一般國家與決策者擴大自身影響力為目標，並以決策者認定的「國家利益」，與他國進行談判。基本上，國家決策核心者面對兩個國際（他國談判者）與國內（個別政黨、利益團體、企業

代表等等）的談判桌，如果不滿意「第一層次」的談判協議還可以推倒重來，如果在「第二層次」無法滿足內部各種利益協調，直接會影響決策者聲望、權力或地位。因此，掌握「國內層面」最低妥協限度，就可以進一步掌握「國際層面」的「談判」或是「衝突」的可能結果。

二、雙層賽局下的底線集合

Putman指出「底線集合」（win-set）為「雙層賽局」的「核心理念」：係指「第二層面」對於「第一層面」所有初步暫時性協議中接受（批准）協議的範圍，「底線集合」越大，「第一層面」的協議就越有可能被批准，反之，國際談判的風險就越大；此外，各造「底線集合」的相對大小會影響到國際談判中共同利益的分配與談判策略，亦即：如果某一造已知的「底線集合」越大，他造就越容易加以利用。[5]

因此，「第二層面」的「底線集合」的範圍，基本上影響「第一層面」協議成功的關鍵。首先，更大的「底線集合」使得「第一層面」協議成功機會增高。第二、為何各方「底線集合」如此重要在於：第二層面中的各方的「底線集合」的大小，將會影響國際談判中，各造共同利益的分配與談判策略。[6] 其次，Putman認為存在三個因素影響「底線集合」的變化，首先，「第二層面」選民間的權力分佈（不同利益團體或路線間的權力關係），政治偏好與可能的結盟。其次，「第二層面」的政治制度（political institutions），以及第三、「第一層面」談判者採取的策略。[7] 事實上，「第一層面」的「談判者」必須考量於「第二層面」的利益導向，尤其是主要「決策者」的「意向」會影響「第二層面」：「國內層面」的走向。

[5] Robert D. Putnam, "Diplomacy and Domestic Politics: The Logic of Two-Level Games," International Organization, Vol. 42, No. 3. (Summer, 1988), pp. 427-460，此處pp.437-440. http://www.guillaumenicaise.com/wp-content/uploads/2013/10/Putnam-The-Logic-of-Two-Level-Games.pdf (2020/05/06)

[6] Robert D. Putnam, "Diplomacy and Domestic Politics: The Logic of Two-Level Games," International Organization, Vol. 42, No. 3. (Summer, 1988), pp. 427-460，此處pp.437-440. http://www.guillaumenicaise.com/wp-content/uploads/2013/10/Putnam-The-Logic-of-Two-Level-Games.pdf (2020/05/06)

[7] Robert D. Putnam, "Diplomacy and Domestic Politics: The Logic of Two-Level Games," International Organization, Vol. 42, No. 3. (Summer, 1988), pp. 427-460，此處pp.442-450. http://www.guillaumenicaise.com/wp-content/uploads/2013/10/Putnam-The-Logic-of-Two-Level-Games.pdf (2020/05/06)

三、底線集合與安全化概念

學者Ole Waever反對傳統安全研究上的兩種前提，首先，「安全」（security）是一種先於「語言」存在的「事實」，是一種外在的東西，而且主張安全越多越好。第二、Waever也反對Johan Galtung等學者主張，將生存、發展、自由與認同都被認為是安全的範疇，以至於安全成為一個無所不包、但卻空洞無物的概念。[8] 是以，Waever 提出一個研究命題：「究竟是什麼使一件事物變成了安全問題？」（What really makes something a security problem）[9] Waeve發現在國家政策實踐中，將某種發展變化成為「安全問題」，國家當局就可以要求一種特殊的權力，通常都是由國家及其菁英所確認。[10]

因此，如果我們透過「語言理論」可以視「安全」（security）是一個「語言行為」（speech act），就是表達一種事實。那些掌握國家權力的決策者通過「安全」這一個「語言」給某事物貼上一個「標籤」，主張使用一種特殊的權力，去動用各種資源來抵擋那些對於「安全」的威脅。[11] 一言之，Waever認為某些事物之所以被定義為一種安全事務，因為一旦如此就會使得這個問題比其它問題都更為重要，國家決策者就可以採取一種非常方式進行處理，從而打破常規的政治規則。[12]

其次，影響「底線集合」大小的因素在於「決策者」的主觀利益思考，透過「安全化」加以進行「安全治理」：「一系列行政部門基於提供戰略方向、確保目標達成，發掘風險，並確認企業的資源正確運用的一種責任與實務操作的過程。」[13] 是以，「安全化」的研究命題與「雙層賽局」的「底線集合」

---

[8] Ole Waever, "Securitization and De-securitization," in Ronnie D. Lipschutz eds., On Security (New York: Columbia University Press, 1995), p. 46, p.48.

[9] Ole Waever, "Securitization and Desecuritization," in Ronnie D. Lipschutz eds., On Security (New York: Columbia University Press, 1995), p. 54.

[10] Ole Waever, "Securitization and Desecuritization," in Ronnie D. Lipschutz eds., On Security (New York: Columbia University Press, 1995), p. 54.

[11] Ole Waever, "Securitization and Desecuritization," in Ronnie D. Lipschutz eds., On Security (New York: Columbia University Press, 1995), p. 55.

[12] Ole Waever, "Securitization and Desecuritization," in Ronnie D. Lipschutz eds., On Security (New York: Columbia University Press, 1995), p. 55.

[13] "Security Governance", Science Direct, accessed at: https://www.sciencedirect.com/topics/computer-science/security-governance (2020/03/12)

有關，成為本為研究架構的主軸概念，透過後續中美各自相互政策變化，以及主要受到近期「新冠肺炎」影響下的不同「國際」與「國內」互動反映的分析結果，再予以檢證說明。

在中美各自「底線集合」的部分，「國內層面」選民間的權力分佈（不同利益團體或路線間的權力關係），在中國方面僅僅是中國共產黨主導國家整體利益與權力分配關係；在美國方面，政治偏好與可能的結盟，包括共和黨、民主黨、各州，及其他利益團體態度。其次，「國內層面」的政治制度（political institutions）：中國方面則是一黨權威體制，雖然存在各級「政治協商會議」，基本上僅僅發揮建言諮詢功能角色。美國屬於兩黨自由民主制度，透過行政、立法與司法三權相互制衡；第三、「國際層面」談判者採取的策略，在美國方面主要以川普總統為首，在中國方面雖然以習近平，都具有絕對對外權威。

是以，透過上述分析「雙層賽局」的「國際層面」與「國內層面」要素、「底線集合」與「安全化」關係，本文整理出下列表一：「表一：「雙層賽局」的「底線集合」與「安全化」相互影響關係表」，呈現A與B兩方，面對國際、國內與底線集合相互作用之下，出現九種相互影響的「象限」關係：[14]，以為本文後續分析架構。

表1：「雙層賽局」的「底線集合」與「安全化」相互影響關係表

A方/B方	國際層面	國內層面	底線集合（安全化）
國際層面	談判者之間的議價過程，引導出暫時初步性協議。 A/B兩方國際層面的影響	國家內部程序（國會討論投票）通過協議：「批准」（ratification）。 A方國內層面對B方國際層面影響	「第二層面」對於「第一層面」所有初步暫時性協議中接受（批准）協議的範圍。 A方「底線集合」對B方國際層面影響

[14] 此九種關係如下：1.A/B兩方國際層面的影響；2.A方國內層面對B方國際層面影響；3.A方底線集合對B方國際層面影響；4.B方國內層面對A方國際層面影響；5.A/B兩方國內層面相互影響；6.A方底線集合對B方國內層面影響；7.B方底線集合對A方國際層面影響；8.B方底線集合對於A方國內層面影響；9.A/B兩方底線集合相互影響。

A方/B方	國際層面	國內層面	底線集合（安全化）
國內層面	國家內部程序（國會討論投票）通過協議：「批准」（ratification）。B方國內層面對A方國際層面影響	A/B兩方國內層面相互影響	A方「底線集合」對B方國內層面影響
底線集合（安全化）	「第二層面」對於「第一層面」所有初步暫時性協議中接受（批准）協議的範圍	B方「底線集合」對於A方國內層面影響	A/B兩方「底線集合」相互影響

資料來源：整理Robert D. Putnam, "Diplomacy and Domestic Politics: The Logic of Two-Level Games," International Organization, Vol. 42, No. 3. (Summer, 1988), pp. 427-460，以及Ole Waever, "Securitization and De-securitization," in Ronnie D. Lipschutz eds., On Security (New York: Columbia University Press, 1995), pp.46-86作者整理自製。

亦即：透過「雙層賽局」與「安全化」概念融合的解析途徑，本文提出下列三個問題意識：

（一）中國是否已經建立類似西方國家處理國家安全議題的決策機制與執行機構？面對新型冠狀肺炎疫情，北京是否有序啟動上述「安全治理機制」與檢討實際作為？還是北京另起「爐灶」，彈性調整加以因應？

（二）中國在新型冠狀肺炎疫情期間，不僅要內防疫情擴散，外防境外輸入，還要顧慮國家經濟發展受到延宕帶來的衝擊，還有美國因為「疫情」持續嚴重，為了轉移國內「民意壓力」，轉向「甩鍋」從「國際層面」問責中國、抵制中國，北京如何化解此種國際壓力，並兼顧國內發展需求：國際抵制整體形象、公共衛生居民健康、經濟成長百業復工，以及小康社會扶貧計劃目標？

（三）藉由雙層賽局：「國際層面」與「國內層面」關聯下的「底線集合」，以及「安全化」操作概念，是否能有效的「理解」中美在新冠肺炎疫情互動下，中國如何操作其「底線思維」，配合「國內層面」，以強化與美國在「國際層面」的抗衡？

參、習近平主政時期國家安全治理發展

一、習近平治國理念：中華民族復興夢

　　2012年中共召開18大，習近平接任總書記、中央軍委主席，次年接任國家主席，為回應國際社會關切，增進國際社會對中國發展理念、發展道路、內外政策的認識和理解，中國國務院新聞辦公室會同中共中央文獻研究室、中國外文出版發行事業局編輯了《習近平談治國理政》一書。[15] 另外，習近平在中國浙江省委書記任職期間（2002—2007年），[16] 也發表一些時政文章，被集結出了一本書「之江新語」。[17] 2012年11月，中共18屆1中全會之後，面對記者會，習近平指出：「人民對美好生活的嚮往，就是我們的奮鬥目標。我們的責任，就是要團結帶領全黨全國各族人民，接過歷史的接力棒，繼續為實現中華民族偉大復興而努力奮鬥。」[18]

　　2017年10月27日，經過五年「中國夢」的國家發展主軸之下，中共召開19屆全國代表大會，習近平首度提出「政治報告」強調：「不忘初心，牢記使命，高舉中國特色社會主義偉大旗幟，決勝全面建成小康社會，奪取新時代中國特色社會主義偉大勝利，為實現中華民族偉大復興的中國夢不懈奮鬥。」[19] 在國家安全方面，習近平強調：「堅持總體國家安全觀。統籌發展和安全，增強憂患意識，做到居安思危，是我們黨治國理政的一個重大原則。」[20]

---

[15] 《習近平談治國理政》一書收入習近平在2012 年11 月15 日至2014 年6 月13 日這段時間內的重要著作，共有講話、談話、演講、答問、批示、賀信等79 篇。請參見：「習近平談治國理政」，共產黨員網，http://syss.12371.cn/2015/05/15/ARTI1431670482542560.shtml。（檢索日期：2020/05/09）

[16] 事實上，從2003年2月至2007年3月，習近平在《浙江日報》「之江新語」專欄發表短論232篇，多數文章二三百字，最長不過500餘字，語言簡潔明快，觀點敏銳清晰，形式生動活潑，講道理淺顯易懂，文風樸實，涉及政治、經濟、文化、社會、生態和黨建等各個方面。請參見：「4年間發表232篇『之江新語』」，大公報，http://www.takungpao.com.hk/news/232108/2019/1110/371812.html。（檢索日期：2020/05/09）

[17] 「習大大10年前寫的《之江新語》 你知道多少」，sina全球新聞，http://dailynews.sina.com/bg/chn/chnpolitics/sinacn/20150527/04026688293.html。（檢索日期：2020/05/09）

[18] 「黨的十八大以來大事記」，中國共產黨新聞網，http://cpc.people.com.cn/BIG5/n1/2017/1016/c414305-29588223.html。（檢索日期：2020/05/09）

[19] 「習近平：決勝全面建成小康社會奪取新時代中國特色社會主義偉大勝利──在中國共產黨第十九次全國代表大會上的報告」，中華人民共和國中央人民政府，http://www.gov.cn/zhuanti/2017-10/27/content_5234876.htm。（檢索日期：2020/05/09）

[20] 總體安全觀的內涵為：「必須堅持國家利益至上，以人民安全為宗旨，以政治安全為根本，統籌外

二、總體國家安全觀的意義與類別

2014年1月24日，中共中央政治局會議研究決定設置「中央國家安全委員會」，由習近平任主席，李克強、張德江任副主席，將其定位為中共中央關於國家安全工作的決策和議事協調機構，向中央政治局、中央政治局常務委員會負責，統籌協調涉及國家安全的重大事項和工作。

2014年4月15日，習近平在中央國家安全委員會第一次會議上提出「必須堅持總體國家安全觀，以人民安全為宗旨，以政治安全為根本，以經濟安全為基礎，以軍事、文化、社會安全為保障，以促進國際安全為依托，走出一條中國特色國家安全道路。」[21] 習近平強調，要堅持「總體國家安全觀」，[22] 並將其確立為新時期國家安全工作的指導思想，走出一條中國特色國家安全道路。[23]

2015年1月23日，中共中央政治局會議審議通過《國家安全戰略綱要》，以及出台《中共中央關於加強國家安全工作的意見》，明確了國家安全的總體部署。中國學者李偉強調：整體國家安全觀闡述中提到的很多新觀點和新辦法，是國家安全頂層設計的重要步驟。此外，中共政治局會議還指出，要打造一支高素質的國家安全專業隊伍。[24] 同時，把國家安全保障能力建設納入

部安全和內部安全、國土安全和國民安全、傳統安全和非傳統安全、自身安全和共同安全，完善國家安全制度體系，加強國家安全能力建設，堅決維護國家主權、安全、發展利益。」請參見：「習近平：決勝全面建成小康社會奪取新時代中國特色社會主義偉大勝利——在中國共產黨第十九次全國代表大會上的報告」，中華人民共和國中央人民政府，http://www.gov.cn/zhuanti/2017-10/27/content_5234876.htm。（檢索日期：2020/05/09）

[21] 「習近平：堅持總體國家安全觀 走中國特色國家安全道路 李克強張德江出席 2014年04月16日 07:22來源：人民網－人民日報」，中國共產黨新聞網http://cpc.people.com.cn/n/2014/0416/c64094-24900492.html (2020/03/12)

[22] 「總體國家安全觀」強調「以人民安全為宗旨，以政治安全為根本，以經濟安全為基礎，以軍事、文化、社會安全為保障，以促進國際安全為依托，統籌發展和安全、外部安全和內部安全、國土安全和國民安全、傳統安全和非傳統安全、自身安全和共同安全，維護各領域國家安全，構建國家安全體系，走中國特色國家安全道路。」，請參見：陳文清（中央國安辦分管日常工作的副主任、國家安全部部長）、「牢固樹立總體國家安全觀在新時代國家安全工作中的指導地位」，中國共產黨新聞網，http://theory.people.com.cn/BIG5/n1/2019/0417/c40531-31033877.html。（檢索日期：2020/05/09）

[23] 「黨的十八大以來大事記」，中國共產黨新聞網，http://cpc.people.com.cn/BIG5/n1/2017/1016/c414305-29588223.html。（檢索日期：2020/05/09）

[24] 「政治局會議通過《國家安全戰略綱要》」，人民網，http://politics.people.com.cn/n/2015/0125/c1001-

「十三五」發展規劃綱要，更好適應了國家安全工作需要。建立重點領域國家安全工作協調機制，凝聚維護國家安全合力。

2015年7月1日　中國十二屆全國人大常委會第十五次會議通過《中華人民共和國國家安全法》，以法律的形式確立「總體國家安全觀」的指導地位，訂定每年4月15日為全民國家安全教育日，組織開展形式多樣的國家安全宣傳教育活動。[25] 2017 年2月17日，習近平以國安委主席名義召開「國家安全工作座談會」強調：「黨的十八大以來，黨中央高度重視國家安全工作，成立中央國家安全委員會，提出總體國家安全觀，明確國家安全戰略方針和總體部署，推動國家安全工作取得顯著成效。」[26]

此外，習近平更強調「堅持黨對國家安全工作的領導，是做好國家安全工作的根本原則。各地區要建立健全黨委統一領導的國家安全工作責任制。」[27] 是以，2020年4月16日，北京市委國家安全委員會會議強調，「要加強黨對國家安全工作的集中統一領導，強化統籌協調、督促指導，推進首都安全體系和能力建設。嚴格落實國家安全責任制，加強安全風險源頭防范和應對處置，形成維護首都安全整體合力。」[28]

事實上，由習近平主政以來，親任組長/主席的小組共有6個，這類小組受中共中央領導，小組辦公室大多是政治局或軍委會直屬機構，處理包括深化改革、國安、網絡安全、軍事等議題。[29] 2018年3月起，中國推動「黨和國家

26445047.html。（檢索日期：2020/05/09）

[25] 陳文清（中央國安辦分管日常工作的副主任、國家安全部部長），「牢固樹立總體國家安全觀在新時代國家安全工作中的指導地位」，中國共產黨新聞網，http://theory.people.com.cn/BIG5/n1/2019/0417/c40531-31033877.html。（檢索日期：2020/05/09）

[26] 習近平強調：「國家安全工作歸根結底是保障人民利益，要堅持國家安全一切為了人民、一切依靠人民，為群眾安居樂業提供堅強保障。」，請參見：「習近平主持召開國家安全工作座談會」，中華人民共和國國防部，http://www.mod.gov.cn/big5/leaders/2017-02/17/content_4772897.htm。（檢索日期：2020/05/09）

[27] 「習近平主持召開國家安全工作座談會」，中華人民共和國國防部，http://www.mod.gov.cn/big5/leaders/2017-02/17/content_4772897.htm。（檢索日期：2020/05/09）

[28] 「北京市委國家安全委員會召開全體會議 蔡奇講話」，中國共產黨新聞網，http://cpc.people.com.cn/BIG5/n1/2020/0416/c64094-31675888.html。（檢索日期：2020/05/09）

[29] 蔡文軒（中研院政治所助研究員），「一、中共新設高層領導小組及委員會觀察」，大陸與兩岸情勢簡報，2017.2，http://www.mac.gov.tw/public/Attachment/73101737208.pdf（檢檢索日期：2020/05/09）

的機構改革」，更進一步將一些「領導小組」改制為「委員會」，使其成為決策產生單位，更具備將『會議決議』透過文件下發至各相關部委，形成行政指令之權。」[30] 目前，習近平自任中共「全面深化改革委員會」、「財經委員會」、「國家安全委員會」、「外事工作委員會」以及「網絡安全和資訊化委員會」等組長，透過「小組政治」應處相關內外情勢，各委員會近期多次全體會議及工作會議內容，也因此備受外界關注。[31]

肆、新冠肺炎期間的國際與國內因素

一、中國所處國內層面防疫議題

2019年12月31日，中國武漢地區爆發新型冠狀肺炎疫情，2020年1月20日，中國國務院副總理孫春蘭主持召開「國務院應對新型冠狀病毒感染的肺炎疫情聯防聯控工作機制」簡稱：「國務院聯防聯控機制」的首次會議，這是中國為應對2020年初突發的新冠肺炎疫情，而啟動的中央人民政府層面的多部委協調工作機制平台，由中國國務院「國家衛生健康委員會」領頭建立，屬於一個應對新型冠狀病毒感染的肺炎疫情聯防聯控工作機制，成員單位共32個部門。[32] 聯防聯控工作機制下設疫情防控、醫療救治、科研攻關、宣傳、外事、後勤保障、前方工作等工作組，分別由相關部委負責同誌任組長，明確職責，分工協作，形成防控疫情的有效合力。[33] 換言之，早於1月20日，中國國務院

---

[30] 「《No. 107006》中共「黨和國家機構改革」之觀察（Observing on the Reform of the Party and State Institutions）of CPC」，亞太和平研究基金會，https://www.faps.org.tw/files/5858/90692B37-61B3-41C8-AB77-F8956BA5560A。（檢檢索日期：2020/05/09）

[31] 「《No. 107006》中共「黨和國家機構改革」之觀察（Observing on the Reform of the Party and State Institutions）of CPC」，亞太和平研究基金會，https://www.faps.org.tw/files/5858/90692B37-61B3-41C8-AB77-F8956BA5560A。（檢檢索日期：2020/05/09）

[32] 相關部門包括：「國家衛健委、中共中央宣傳部、外交部、國家發展改革委、教育部、科技部、公安部、財政部、民政部、人力資源和社會保障部、交通運輸部、中國鐵路總公司、工業和信息化部、農業農村部、商務部、市場監管總局、生態環境部、中國民航局、國家林業和草原局、國家藥品監督管理局、文化和旅遊部、中國紅十字總會、中央軍委後勤保障部衛生局等」，請參見：請參見：「國務院應對新型冠狀病毒肺炎疫情聯防聯控機制」，維基百科，https://zh.wikipedia.org/wiki/%E5%9B%BD%E5%8A%A1%E9%99%A2%E5%BA%94%E5%AF%B9%E6%96%B0%E5%9E%8B%E5%86%A0%E7%8A%B6%E7%97%85%E6%AF%92%E8%82%BA%E7%82%8E%E7%96%AB%E6%83%85%E8%81%94%E9%98%B2%E8%81%94%E6%8E%A7%E6%9C%BA%E5%88%B6。（檢索日期：2020/05/13）

[33] 「國務院聯防聯控機制」，百度百科，https://baike.baidu.com/item/%E5%9B%BD%E5%8A%A1%E9%99%

已經啟動「國務院聯防聯控機制」，[34] 其後1月26日，中共中央才成立「中央應對新型冠狀病毒感染肺炎疫情工作領導小組」，提升防疫指揮層級，超越「國務院」指揮與治理範疇，由李克強擔任小組長主持會議。

此一「國務院聯防聯控機制」在「中央應對新型冠狀病毒感染肺炎疫情工作領導小組」指導下，設立各省、自治區、各省、自治區、直轄市人民政府、國務院各部委、各直屬機構下達類似「國務院應對新型冠狀病毒感染肺炎疫情聯防聯控機制關於做好新冠肺炎疫情常態化防控工作的指導意見國發明電〔2020〕14號」，表達疫情「防控工作已從應急狀態轉為常態化。按照黨中央關於抓緊抓實抓細常態化疫情防控工作的決策部署，為全面落實『外防輸入、內防反彈』的總體防控策略，堅持及時發現、快速處置、精準管控、有效救治，有力保障人民群眾生命安全和身體健康，有力保障經濟社會秩序全面恢復。」[35] 2020年4月8日，「國務院聯防聯控機制」又印發《關於進一步做好重點場所重點單位重點人群新冠肺炎疫情防控相關工作的通知》，強調結合當前疫情防控形勢，落實分區分級防控要求，推進生產生活秩序逐步恢復。[36]

其實，中國自2020年1月23日，為了應對「新型冠狀肺炎」，採取「武漢封城」措施以來，根據1月25日習近平主持中央政治局常委會會議並作重要講話，對疫情防控工作進行再研究、再部署、再動員。1月26日，決定成立「中央應對新型冠狀病毒感染肺炎疫情工作領導小組」，在中央政治局常委會領導

99%A2%E8%81%94%E9%98%B2%E8%81%94%E6%8E%A7%E6%9C%BA%E5%88%B6/24316908。（檢索日期：2020/05/13）

[34] 2011年，中國國家衛生部制定發布《國家流感大流行應急預案》，規定「國務院建立國家應對流感大流行聯防聯控工作機制（以下簡稱國家聯防聯控工作機制），負責指導和協調我國內地流感大流行應急響應階段和恢復評估階段的防控工作。」，請參見：「國務院應對新型冠狀病毒感染肺炎疫情聯防聯控機制」，維基百科，https://zh.wikipedia.org/wiki/%E5%9B%BD%E5%8A%A1%E9%99%A2%E5%BA%94%E5%AF%B9%E6%96%B0%E5%9E%8B%E5%86%A0%E7%8A%B6%E7%97%85%E6%AF%92%E8%82%BA%E7%82%8E%E7%96%AB%E6%83%85%E8%81%94%E9%98%B2%E8%81%94%E6%8E%A7%E6%9C%BA%E5%88%B6。（檢索日期：2020/05/13）

[35] 「國務院應對新型冠狀病毒感染肺炎疫情聯防聯控機制關於做好新冠肺炎疫情常態化防控工作的指導意見國發明電〔2020〕14號」，中華人民共和國中央人民政府，http://www.gov.cn/zhengce/content/2020-05/08/content_5509896.htm。（檢索日期：2020/05/13）

[36] 「國務院聯防聯控機製印發通知 進一步做好重點場所重點單位重點人群新冠肺炎疫情防控相關工作」，中國共產黨新聞網，http://cpc.people.com.cn/n1/2020/0409/c419242-31666551.html。（檢索日期：2020/05/13）

下開展工作，加強對全國疫情防控的統一領導、統一指揮。會議強調，要突出重點，進一步加強湖北省和武漢市疫情防控。因此，中國中央向湖北派出「指導組」，推動加強防控一線工作。[37]

1月29日，中共應對新型冠狀病毒感染肺炎疫情工作領導小組由李克強主持召開領導小組會議，進一步研究疫情防控形勢，部署有針對性加強防控工作。[38] 2月3日，習近平在中央政治局常務會議針對疫情作出「阻擊戰」的判斷，把疫情防控工作作為當前最重要的工作來抓，真抓實幹，把落實工作抓實抓細，堅決遏制疫情蔓延勢頭，堅決打贏疫情防控阻擊戰。[39]

2月10日，習近平在北京調研指導新型冠狀病毒疫情強調，現在疫情防控正處於膠著對壘狀態，要堅決把擴散蔓延勢頭遏制住，堅決打贏疫情防控的人民戰爭、總體戰、阻擊戰。[40] 2月12日，習近平主持中共政治局常委會議，聽取中央防疫領導小組匯報，聲稱疫情防控取得成效，已到「最吃勁關鍵階段」，要「確保打贏疫情防控的人民戰爭」，努力實現今年經濟社會發展目標任務。[41] 習近平最後強調，統籌做好疫情防控和經濟社會發展，既是「大戰」，也是「大考」；在大戰中踐行初心使命，在大考中交出合格答卷，確保打贏疫情防控的人民戰爭、總體戰、阻擊戰，努力實現全年經濟社會發展目標的任務。[42]

---

[37] 「中央應對新型冠狀病毒感染肺炎疫情工作領導小組會議召開」，中華人民共和國國防部，http://www.mod.gov.cn/big5/topnews/2020-01/27/content_4859098.htm。（檢索日期：2020/05/13）

[38] 事實上，「會議要求，國務院聯防聯控機制要在中央應對疫情工作領導小組的領導下，完善每日例會制度，會商研究疫情發展趨勢，指導督促地方加強防控工作，建立疫情防控物資全國統一調度制度，及時協調解決防控工作中遇到的緊迫問題，及時協調調度防控所需醫護人員、醫用物資和居民生活必需品等，統籌增加相關物資進口，確保防控工作有力有序、科學周密推進。」請參見：「李克強主持召開中央應對疫情工作領導小組會議」，中華人民共和國國防部，http://www.mod.gov.cn/big5/topnews/2020-01/29/content_4859227.htm。（檢索日期：2020/05/13）

[39] 「李克強主持召開中央應對新型冠狀病毒感染肺炎疫情工作領導小組會議」，中國共產黨新聞網，http://cpc.people.com.cn/BIG5/n1/2020/0204/c64094-31571042.html。（檢索日期：2020/05/13）

[40] 「習近平：堅決打贏疫情防控人民戰爭總體戰阻擊戰」，聯合新聞網，https://udn.com/news/story/120936/4334998。（檢索日期：2020/03/12）

[41] 「習近平：打贏疫情人民戰爭 實現全年發展目標」，自由電子報，https://news.ltn.com.tw/news/world/paper/1351711。（檢索日期：2020/03/12）

[42] 「習近平：打贏疫情人民戰爭 實現全年發展目標」，自由電子報，https://news.ltn.com.tw/news/world/paper/1351711。（檢索日期：2020/03/12）

　　2月14日，在中共「中央全面深化改革委員會」第十二次會議上，習近平指出，「這次抗擊新冠肺炎疫情，是對國家治理體系和治理能力的一次大考。」[43] 在此處習近平提出兩個關鍵性概念：國家「治理體系」與「治理能力」，習近平強調：「要從保護人民健康、保障國家安全、維護國家長治久安的高度，把生物安全納入國家安全體系，系統規劃國家生物安全風險防控和治理體系建設，全面提高國家生物安全治理能力。要盡快推動出台生物安全法，加快構建國家生物安全法律法規體系、制度保障體系。」[44]

　　3月2日，李克強主持召開領導小組會議指出，要加強疫情形勢研判，做好分區分級精准防控，採取務實措施關心關愛防控一線城鄉社區工作者，統籌疫情防控和春耕生產，進一步拓展防控形勢積極向好態勢。[45] 3月10日，習近平在疫情爆發後，第一次前往武漢視察，[46] 他表示「疫情防控已取得階段性成果」，意味著中國已控制住至今仍在全球肆虐的「新冠肺炎」（COVID-19）疫情。[47]

　　4月9日李克強召開小組會議，強調「有針對性動態加強境外疫情輸入防範，進一步完善防範境內疫情反彈的措施，鞏固我國疫情防控階段性成效，全面推進復工復　，加快推進生 活秩序全面恢復。」[48] 4月13日，根據防疫小組會議指出：「繼續抓緊抓實抓細各項防控工作。動態完善防範疫情跨境輸入舉

---

[43]「習近平主持召開中央全面深化改革委員會第十二次會議」，中華人民共和國國防部，http://www.mod.gov.cn/big5/shouye/2020-02/14/content_4860469.htm。（檢索日期：2020/03/12）

[44]「習近平主持召開中央全面深化改革委員會第十二次會議」，中華人民共和國國防部，http://www.mod.gov.cn/big5/shouye/2020-02/14/content_4860469.htm。（檢索日期：2020/03/12）

[45]「李克強主持召開中央應對新冠肺炎疫情工作領導小組會議」，中國共產黨新聞網，http://cpc.people.com.cn/BIG5/n1/2020/0302/c64094-31613390.html。（檢索日期：2020/05/13）

[46] 根據新華社報導，習近平此行將「看望慰問奮戰在一線的廣大醫務工作者、解放軍指戰員、社區工作者、公安幹警、基層幹部、下沉幹部、志願者和患者群眾、社區居民」。考察湖北及武漢的疫情防控工作。請參見：「中國疫情警報解除？習近平終於親赴武漢視察！」，風傳媒，https://www.storm.mg/article/2385042（檢索日期：2020/03/12）

[47]「疫情爆發後 習近平首次視察武漢」，中時電子報，https://www.chinatimes.com/newspapers/20200311000225-260202?chdtv。（檢索日期：2020/03/12）

[48] 相關李克強談話都必須加上「要認真貫徹習近平總書記主持召開的中央政治局常委會會議精神，按照中央應對疫情工作領導小組部署」等用語，請參見：「李克強主持召開中央應對新冠肺炎疫情工作領導小組會議」，中華人民共和國中央人民政府，http://big5.www.gov.cn/gate/big5/www.gov.cn/premier/2020-04/10/content_5500948.htm。（檢索日期：2020/05/13）

措，提高針對性有效性。加強精准防控和規範處置，排查消除疫情反彈風險隱患，為推進全面復工復產創造有利條件。」[49]

2020年5月9日，中共中央在中南海召開黨外人士座談會，就新冠肺炎疫情防控工作聽取各民主黨派中央、全國工商聯和無黨派人士代表的意見和建議，習近平在會中表示：境外疫情暴發成長態勢仍持續，外防輸入壓力持續加大，疫情反彈風險始終存在，要落實和完善「常態化」疫情防控舉措，嚴密防範疫情反彈。[50]

總之，中國自武漢暴發疫情以來，中共中央連續召開了10次常委會議，為中國的防疫以及疫後工作指明方向。最近一次中央政治局常委會會議針對國際嚴峻形勢對中國的衝擊，強調要嚴守「底線思維」，亦即習近平所說：「面對嚴峻複雜的國際疫情和世界經濟形勢，我們要堅持底線思維，做好較長時間應對外部環境變化的思想準備和工作準備。」[51] 易言之，「底線思維」就是「堅持和發展中國特色社會主義，全面建成社會主義現代化強國，實現中華民族偉大復興的中國夢，是我們堅持底線思維的總體價值目標。」，同時，「任何影響中國特色社會主義發展、影響中華民族偉大復興中國夢實現的風險都要堅決防範化解。在這個問題上絕不能有絲毫動搖、猶豫和彷徨。」[52]

二、新冠肺炎期間的國際環境

從2020年1月23日，中國開始強力主導「武漢封城」，避免疫情持續擴大，並從2月開始，由解放軍出動人力、物力支援，興建火神山、雷神山醫院，以及為數眾多的方艙醫院，並由北部、東部、西部、南部五大戰區調派

---

[49] 「李克強主持召開中央應對新冠肺炎疫情工作領導小組會議 部署調集專家和防疫物資增強邊境地區疫情防控能力 進一步擴大檢測范圍做好精准防控和推動全面復工復產」，中國共產黨新聞網，http://cpc.people.com.cn/BIG5/n1/2020/0414/c64094-31672107.html。（檢索日期：2020/05/13）

[50] 「習近平：境外疫情仍暴發成長 要完善常態化防控」，中央通訊社，https://www.cna.com.tw/news/acn/202005080364.aspx。（檢索日期：2020/05/09）

[51] 紀碩鳴，「**以底線思維爭取抗疫全勝**」，超訊，http://www.supermedia.hk/43579/%E4%BB%A5%E5%BA%95%E7%B7%9A%E6%80%9D%E7%B6%AD%E7%88%AD%E5%8F%96%E6%8A%97%E7%96%AB%E5%85%A8%E5%8B%9D/（檢索日期：2020/05/04）

[52] 「深入學習貫徹習近平新時代中國特色社會主義思想：**深刻認識堅持底線思維**」，中國共產黨新聞網，http://theory.people.com.cn/BIG5/n1/2019/0625/c40531-31177911.html。（檢索日期：2020/05/04）

4500各醫療人員進駐武漢，及自3月中旬以來，大體控制疫情發展。反倒是中國以外地區，疫情持續增長，擴及全球五大洲。根據美國約翰霍普金斯大學（Johns Hopkins University）統計，美國的新冠肺炎確診數3月27日為止，達到85991例，超過中國的81782例、義大利的80589例，已經躍居全世界確診數最多的國家。[53] 是以，3月28日零時開始，中國禁止外籍人士進入中國，實行「境外管制」，4月初湖北省解除交通管制，4月8日起，武漢也解除近兩個月以來的「封城」行動。

在美國方面，起初川普輕忽疫情，強調美國不擔心，2月26日，他和美國「疾病控制與預防中心」（CDC）官員出席新聞發布會，闡述美國政府對新冠疫情的應對措施。之後，川普任命副總統彭斯全權負責新冠病毒工作。川普表示，美國新型冠狀病毒蔓延的風險「非常低」；疫苗研發「進展迅速」；美國已經做好「一切應對措施」。[54] 川普還肯定中國在疫情方面表現，並指責美國民主黨人「誇大疫情」、「恐慌市場」，因為每年美國死於流感的人數在2.5萬到6.9萬之間，相比之下，只有2700人死於新冠肺炎。[55]

不過，3月以來新冠肺炎的發展，卻出乎川普的樂觀想像，將近一個月時間，川普一直在淡化疫情在美國的嚴重性。[56] 根據3月28日，《紐約時報》以「美國錯失的一個月」為標題分析美國延誤因應時機，[57] 其結論是美國本來可能可以擋住病毒，但是因為技術失誤、法規阻礙和政府領導無方，讓新冠病毒

---

[53] "Coronavirus COVID-19 Global Cases by the Center for Systems Science and Engineering (CSSE) at Jo", Johns Hopkins University, Johns Hopkins University & Medicine, Coronavirus Resource Center, accessed at: https://coronavirus.jhu.edu/map.html(2020/03/27)，「『武漢肺炎』美國確診數破8.3萬 超越中國成全球最多」，自由電子報，https://news.ltn.com.tw/news/world/breakingnews/3114231。（檢索日期：2020/03/27）

[54] 「川普：美國流感每年致死數萬人，致死率比新冠肺炎高，我都驚了」，每日頭條，https://kknews.cc/zh-tw/health/4b9bxz2.html。（檢索日期：2020/05/13）

[55] 「川普：美國流感每年致死數萬人，致死率比新冠肺炎高，我都驚了」，每日頭條，https://kknews.cc/zh-tw/health/4b9bxz2.html。（檢索日期：2020/05/13）

[56] 「復盤美國新冠疫情：川普為何忽視警告、一錯再錯」，紐約時報中文網，https://cn.nytimes.com/usa/20200413/coronavirus-trump-response/zh-hant/。（檢索日期：2020/05/13）

[57] "The Lost Month: How a Failure to Test Blinded the U.S. to Covid-19", New York Times, https://www.nytimes.com/2020/03/28/us/testing-coronavirus-pandemic.html (2020/05/14)

幾週來在美國長驅直入。[58] 另外，美國情報部門在1、2月就接連提出全球疫情危機的機密預警，但川普與國會議員不重視，無法採取有助遏止疫情蔓延的行動。[59]

此外，川普直接稱呼「新型冠狀肺炎」為「中國病毒」（Chinese Virus），從3月23日，首度改口不稱「中國病毒」，而是以「病毒」稱呼，他堅稱「中國病毒」只是為了清楚傳達病毒來自中國，也是對中國散播病毒是美軍帶入武漢的不實謠言的反擊。[60] 中國極度不滿起源於武漢的疫情被稱為「武漢肺炎」或是「中國病毒」，藉由「世界衛生組織」命名規則，痛斥取名不應污名化、地域化。2月11日，世衛正式將武漢肺炎命名為「2019年冠狀病毒疾病」（COVID-19）。其實，早在2月8日，中國國家衛生健康委員會就搶先公告，將武漢肺炎暫時命名為「新型冠狀病毒肺炎」。[61]

另外，中美之間也爆發「病毒起源說」的爭議，連帶要求中國負起「賠償責任說」。主要是4年12日，中國外交部發言人趙立堅推特上發文，引述美國「疾病管制暨預防中心」（CDC）主任芮斐德（Robert Redfield）於4月11日，在美國眾議院監督委員會承認，「一些似乎死於流感的美國人，在死後的診斷中被檢測出新型冠狀病毒呈陽性」。[62] 4月15日，川普在白宮被問到此事

---

[58] 「成為確診第一大國　美國是怎麼錯失疫情防控的黃金30天？」，康健雜誌，https://www.commonhealth.com.tw/article/article.action?nid=81284。（檢索日期：2020/05/13）

[59] 根據報載「紐時指出，美國衛生及公共服務部去年1到8月，執行代號「紅色疫災」（Crimson Contagion）的系列演習，模擬因應流感大流行。演習劇本是中國爆發呼吸道病毒疫情，因發燒旅客搭機而快速在全球擴散，美國首先在芝加哥發現病毒，世界衛生組織（WHO）在47天後宣告大流行，但為時已晚，美國估計會有1.1億人染病、770萬人住院、58.6萬人病死。」，請參見：「新冠肺炎疫情失控川普怪中國　美媒揭政府把預警當耳邊風」，聯合新聞網，https://udn.com/news/story/120944/4432929。（檢索日期：2020/05/13）

[60] 川普其實出於亞裔選票的考量，因為，在記者會上重申他的推文表示，「確實保護在美國與全世界的亞裔美國人社群十分重要，他們是一群很優秀的人，不論病毒是以何種方式、形式或管道傳播，都『不是』他們的錯。」，請參見：「改口不稱中國病毒　川普：病毒傳播非亞裔的錯」，中央通訊社，https://www.cna.com.tw/news/aopl/202003240020.aspx（檢索日期：2020/05/13）

[61] 中國將此病毒簡稱為「新冠肺炎」，英文則是 Novel Coronavirus Pneumonia，縮寫 NCP」，請參見：「武漢肺炎一詞被消失，中國官媒竄改過往新聞報導」，科技情報，https://technews.tw/2020/03/26/wuhan-pneumonia-china-media/。（檢索日期：2020/03/27）

[62] 趙立堅的推文表達：「美國疾控中心主任被抓了個現行。零號病人是什麼時候在美國出現的？有多少人被感染？醫院的名字是什麼？可能是美軍把疫情帶到了武漢。美國要透明！要公開數據！美國欠我們一個解釋」。趙立堅並在另一條英文推文向美國發出質疑，「美國當季有3400萬流感患者，2萬人死亡，請告訴我們，這其中有多少人與新冠肺炎有關？」「中國外交部連發推文 指美軍帶病

時表示，他確實讀過一篇文章，但不認為這篇文章具有代表性，也明確表示他在跟中國國家主席習近平會談時不曾談過這個議題，「他們知道它（病毒）從哪來，我們都知道它來自哪裡。」[63] 另外，自3月19日起，美國國務院發言人歐塔加斯（Morgan Ortagus）與中國外交部發言人華春瑩，也針對中國隱匿疫情的爭議，在社交媒體「推特」（Twitter）上「唇槍舌劍」。[64] 3月24日，華春瑩表示2019年12月，武漢衛生健康委員會就已經發出疫情通知，因此何來隱匿之說？歐塔加斯則反擊，說12月31日，台灣早就首次告知who疫情可能會人傳人。[65] 根據中國新華社一篇報導分析：「在美國疫情依舊嚴峻之際，美國政府卻在疫情政治化的道路上越走越遠，種種不負責任行為的背後是推卸自身責任、轉移輿論壓力和撈取選舉籌碼的精心算計。」[66] 易言之，中美兩國一定程度在操作病毒「甩鍋說」，除了國際聲譽考量之外，亦有各自內部民意支持需求。

此外，3月23日，根據歐洲對外事務部網站公布一篇「病毒大流行及其正在創造的新世界」文章，歐盟各國外長視訊連線討論目前情勢後，歐洲聯盟外交和安全政策高級代表波瑞爾(Josep Borrell)表示，2019年冠狀病毒將重塑世界，雖然不知道危機何時結束，但可以肯定的是，屆時世界將變得非常不同。[67]

最新一期德國「明鏡周刊」（Der Spiegel）報導，根據德國聯邦情報局（BND）情資，習近平於1月21日與「世界衛生組織」（WHO）秘書長譚德塞

毒進武漢」，中央通訊社，https://www.cna.com.tw/news/firstnews/202003130168.aspx。（檢索日期：2020/05/13）

[63] 「中國官員指美軍帶進病毒 川普：我們都知道來源」，中央通訊社，https://www.cna.com.tw/news/aopl/202003140024.aspx。（檢索日期：2020/05/13）

[64] 請參見：「100天疫情推特戰：從中國外交部推文解讀大外宣布局」，報導者，https://www.twreporter.org/a/covid-19-ccp-grand-external-propaganda-twitter。（檢索日期：2020/05/13）

[65] 「隱匿疫情」有沒有？ 美中雙方各持己見」，壹電視，http://www.nexttv.com.tw/NextTV/News/Focus/WorldNews/2020-03-24/143856.html。（檢索日期：2020/05/13）

[66] 「美國政府政治化疫情有何企圖」，中國評論新聞網，http://hk.crntt.com/doc/1057/6/0/1/105760191.html?coluid=7&kindid=0&docid=105760191&mdate=0507115824。（檢索日期：2020/05/07）

[67] 「繼美國後 歐盟外長直指中國隱瞞疫情訊息」，中央廣播電台，https://www.rti.org.tw/news/view/id/2056892。（檢索日期：2020/05/13）

（Tedros Adhanom Ghebreyesus）通話時，要求世衛不要發布病毒人傳人的訊息和延後全球大流行的警告，直到1月底才宣布2019冠狀病毒疾病（COVID-19）是「國際關注的公共衛生緊急事件」（PHEIC）。[68] 因此，西方國家普遍認為，如果中國不隱瞞訊息，北京不封鎖消息，全球可以增取4到6個星期時間對抗病毒。[69] 4月19日，川普再度指責中國隱匿疫情，若證實目前推論，中國必須面臨後果，而美國參眾兩院分別推出法案，允許美國民眾對政府求償。澳洲也跟進美國展開獨立調查，其外交部表示，將調查包括「世界衛生組織」（WHO）的處理方式是否失當，以及中國早期對應疫情的手段有沒有問題。[70]

所以，因應以美國為首的「究責說」，北京也展開反擊行動。2020年5月5日，中國駐美大使崔天凱在《華盛頓郵報》發表題為《指責遊戲該結束了》的署名文章，[71] 批評現在美國流行的「逢中必反」，呼籲結束「指責遊戲」，修復中美之間的互信，把力氣用到合作抗擊疫情上來。[72] 5月7日，川普更上一層批判：「新冠肺炎的爆發使得美國遭逢第二次世界大戰的珍珠港空襲、或是九一一國際恐怖主義攻擊。」因此，美國政府考慮基於中國早期對待全球緊急事件的處理態度，進行對中國政府的懲罰性行動。[73] 5月12日，美國共和黨參議員葛理漢（Lindsey Graham）提出「新冠病毒究責法案」（Covid-19 Accountability Act）將要求總統在60天內向國會說明，「中國對美國、美國

---

[68] 「武漢肺炎：世衛宣佈全球衛生緊急狀態　美國作出應急反應」，BBC中文網，https://www.bbc.com/zhongwen/trad/world-51312008。（檢索日期：2020/05/13）

[69] "Bundesregierung zweifelt an US-These zur Entstehung des Coronavirus", Der Spiegel, accessed at: https://www.spiegel.de/politik/deutschland/corona-krise-bundesregierung-zweifelt-an-us-these-zur-entstehung-des-coronavirus-a-51add7cf-96b6-4d04-a2d0-71ce27cff69c (2020/05/09)

[70] 「新冠肺炎／美國會助民眾向中國求償　澳洲獨立調查世衛責任」，中國廣播公司，http://www.bcc.com.tw/newsView.4149859。（檢索日期：2020/05/13）

[71] "Chinese ambassador: Ignoring the facts to blame China will only make things worse", The Washington Post, accessed at: https://www.washingtonpost.com/opinions/chinese-ambassador-cui-tiankai-blaming-china-will-not-end-this-pandemic/2020/05/05/4e1d61dc-8f03-11ea-a9c0-73b93422d691_story.html .(2020/05/07)

[72] 「崔天凱：結束指責遊戲修復中美互信」，中國評論新聞網，http://hk.crntt.com/doc/1057/5/9/6/105759663.html?coluid=1&kindid=0&docid=105759663&mdate=0507000917（檢索日期：2020/05/07）

[73] 其原文為："Mr Trump said the outbreak had hit the US harder than the Japanese bombing of Pearl Harbor in World War Two, or the 9/11 attacks two decades ago⋯ His administration is weighing punitive actions against China over its early handling of the global emergency", "Trump says coronavirus worse 'attack' than Pearl Harbor", BBC News, accessed at: https://www.bbc.com/news/world-us-canada-52568405(2020/05/07)

的盟國或像是世衛組織等聯合國附屬機構領導的調查提供完整、全面的資訊」。[74] 是以，根據北京「中國現代國際關係研究院」的一份內部研究，由中國國家安全部在4月初向包括中國國家主席習近平在內的北京領導高層提出的報告總結稱，全球反中情緒創1989年天安門鎮壓以來最高。[75]

　　此外，由於台灣面對新冠肺炎疫情，不管是疫情管控、醫療物資、社會經生活，變成一個世界典範。但是，台灣無法參加世界衛生組織大會就成為一個人類健康的缺陷。加上，川普也抱持「世衛組織」需為疫情負部分責任，並指控世衛組織「以中國為中心」。[76] 因此，美國極力支持台灣參與「世界衛生組織」活動。2020年5月6日，美國國務卿龐佩奧（Mike Pompeo）公開呼籲所有國家支持台灣，以「觀察員」身分參與「世界衛生大會」（WHA），及其他聯合國周邊組織的活動。美國國務院發言人歐塔加斯（Morgan Ortagus）接受《美國之音》專訪時說，美國不怕為台灣的權益挺身而出，也希望歐洲國家跟進。[77]

伍、代結語：中國安全的雙層賽局與安全化檢證

　　首先，針對「第二節」提出的三個「問題意識」總結提出以下的回應。習近平主政以來，針對國家安全治理規劃除了提出「總體國家安全觀」之外，也相當完備的「機構」與「法規」：「國家安全委員會」、「國家安全戰略綱要」、「國家安全法」等，並持續改革就有的各項特定「小組會議」為「委員會」，擴大專業決策、分工合作的理念。不過，大部份由習近平直接擔任主

---

[74] 法案內容還包括：「要求中國證明已經關閉所有可能會讓人類健康陷入風險的「濕市場」，釋放在疫情爆發後逮捕的香港民主派人士。該法案也授權總統對中國實施廣泛的制裁，包括凍結資產、旅遊禁令、撤銷簽證、限縮美國機構向中國公司貸款，以及將中國企業從美國股票市場除名」，請參見：「美議員推『新冠病毒究責法案』！授權川普制裁中國」，經濟日報，https://money.udn.com/money/story/10511/4559818。（檢索日期：2020/05/13）

[75] 「中國國安智庫：全球反中掀新高 中美恐兵戎相見」，中央通訊社，https://www.cna.com.tw/news/acn/202005050320.aspx。（檢索日期：2020/05/07）

[76] 「美媒：美國對付WHO 擬中止資助或成立新機構」，中央通訊社，https://www.cna.com.tw/news/firstnews/202004110045.aspx。（檢索日期：2020/05/07）

[77] 「新冠肺炎」美國務院籲各國支持台灣參與WHA　盼歐洲國家跟進」，上報，https://www.upmedia.mg/news_info.php?SerialNo=86928。（檢索日期：2020/05/08）

任委員或是小組長的委員會，業務繁雜、分工精密，潛藏是否能夠及時因應的「指揮鏈」問題。

其次，新冠肺炎期間，習近平審慎評估疫情，指出這是一場「打贏疫情防控的人民戰爭、總體戰、阻擊戰」，又提出「生物安全」，強調要納入「總體國家安全觀」的內涵，未來要進行「生物安全法」訴求，基本上就是一種「安全化」：「語言行為」的具體展現。習近平並沒有藉助已經成立的「國家安全委員會」，而是在「戰略上」，由「中央政治局常務委員會」指導，設立臨時「中央應對新冠肺炎疫情工作領導小組」，來指揮由中國國務院已經成立的「國務院聯防聯控機制」下的「新型冠狀病毒感染的肺炎應對處置工作領導小組」，協調有關部會執行後續防疫工作。（參考下圖：圖一：中國新冠疫情管控指揮管制圖）

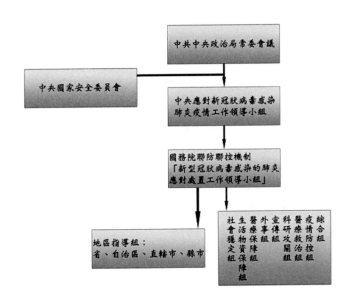

圖 1：中國新冠疫情管控指揮管制圖

資料來源：筆者自製。

第三、從「四、新冠肺炎期間的國際與國內因素」一節中的分析瞭解，中

國因對新冠肺炎防疫，不僅有「國內層面」考量，包括：維持疫情境內管控、感染境外阻絕，經濟產能正常，以維繫「底線思維」：兩個一百年戰略目標，以及小康社會攻關關鍵年，在「國際層面」方面，面臨美國繼「貿易爭議」之後的另外產生新的美中「議程設定」的對抗：「新冠病毒來源說」、「北京隱匿疫情說」、繼而「全球咎責賠償說」，以及由於台灣防疫的優異表現，牽動「美台關係」支持台灣參與世衛組織，所引發的兩岸關係的對峙。顯示出中、美兩國面對的「國際層面」與「國內層面」相互影響，從而牽動兩國後續的政策作為。

第四、根據「表一：「雙層賽局」的「底線集合」與「安全化」相互影響關係表」解析，從中國的「底線集合」角言，「國內層面」（LII）面對疫情，一切以中央「馬首是瞻」，融合其「支持度」與「一致性」，使得中國對外「國際層面」（LI）上，毫無制約力量，是以其「底線集合」超越美國的「底線集合」。華盛頓主要在於美國內部，包括各洲、國會、情報界對於新冠疫情的處理的上述「美中爭議說」現象，並沒影一致立場，一定程度削弱川普在面對中國的「國際層面」（LI）的「底線集合」受到限制。因此，中美兩方呈現國際與國內層面之間的相互影響，尤其是雙方的「底線集合」無法擴大雙方在「國際層面」或是「兩岸層面」的互動發展。主要在於北京存在牢不可破的「底線思維」，同樣的，華盛頓也必須堅持「美國優先」立場，以贏得民意支持率，任何失掉「民意」的政黨與團體，都無法在2020年11月選舉獲得勝利贏得政權，這也是川普的「底線思維」。

第五、從上述「雙層賽局」與「安全化」角度言，中國如何增加中、美兩方的「底線集合」，除了有賴雙方溝通互動，藉以避免誤判之外，還需思考「美台關係」發展對於「中美關係」的影響，或是進一步思考兩岸如何因應「美國因素」。因此，北京是否能夠充分發揮一般國家設置「國家安全委員會」總體國家安全戰略大政方針的決策、指揮、協調與整合的功能，與其它既存「專業委員會」：深化改革委員會、外事工作委員會、網路與資訊安全委員會等等的相互情資整合與決策參考。

　　第六、2017年美國川普上台以來，透過「2017年國家安戰略報告」定位中國為長期「戰略競爭對手」，目前美國主導兩方的「貿易戰」、華盛頓推動「公海自由航行權」與北京進行「內海化」下的「南海戰」，還有新冠肺炎疫情期間的「防疫戰」。加上，川普應用「台灣牌」：「台灣旅行法」、「亞洲安全在保證倡議法」與「台北法案」，從「低階協議」的累積，強化1979年以來基於「臺灣關係法」下的台美關係。基本上，已經牽動傳統美中台三角關係的結構，變成美國居於樞紐地位主導下的兩個平行的雙邊關係：「美中關係」與「美台關係」。基於地緣戰略思考，畢竟「遠親不如近鄰」，是以，台灣因應未來中美印太區域的「戰略競逐」，如何能夠扮演關鍵性的「戰略支點」：求取在美中之間的「戰略平衡」與「等距外交」。

Southeast and South Asian Countries' Responses to US-China Cybersecurity Confrontations

Yisuo Tzeng [*]

I. Introduction

As cybersecurity conflicts between the US and China are on the rise and getting intense against the backdrop of US-China trade wars and, to a greater extent, technology cold war in the making, many countries are under pressure to pick a side, either with the US or not. On the one hand, the US is a major source of security protection and technology transfer. On the other, China represents market opportunities and security concern to various degrees. Given that many countries are caught in this difficult position, it is imperative to raise the question regarding in what ways would Indo-Pacific regional countries, particularly those in southeast and south Asia, respond to the US-China cybersecurity confrontations. With the employment of documentary analysis, this essay addresses this complicated issue by focusing on such countries as India, Vietnam, Singapore, Malaysia and Thailand through the lens of geostrategic as well as policy-institutional analysis.

The rest of the essay breaks into the following sections. What comes first is an overview of the cybersecurity confrontations between US and China, with a specific focus on the ways US has been making the case to contain the adoption of Chinese Huawei's 5G apparatus in Europe, Five-Eyes allies, and Northeast Asia. Drawing on the back-and-forth confrontations in different regions, a couple of propositions come forward for subsequent examinations with case study on five aforementioned

[*] Assistant Research Fellow, Division of Cyber Warfare and Information Security, Institute for National Defense (Taiwan)

countries, covering each's standing position in the US's Indo-Pacific Strategy, domestic demand on 5G, views on Huawei 5G disputes, and actions related to the adoption of Huawei 5G systems. Final part presents preliminary findings of this study with recommendations made in an attempt to contribute to not just enrichment of literature but also real-world significance.

II. Overview of US-China Cybersecurity Confrontations

i. US-China cybersecurity confrontations

Cybersecurity confrontations between states or state-sponsored agencies present as the employment of malicious activities in cyberspace by either side– in such ways as malware implants, backdoor eavesdropping, cyber espionage, fabrication of disinformation, and cyberattacks in forms of distributed denial of service (DDoS), advanced persistent threats (APT), and so on – targeting the other side of the confrontation line, and the corresponding counter measures. Countermeasures could be many things, including counterattack with responses in kind, pre-emptive measures by defending forward, mitigation with quarantines, and blockades through boycotts, ban, and exclusion of software, middleware, and hardware that are likely to pose cybersecurity threats.

The confrontational activities in kind undoubtedly constitute "grey-zone conflicts" with the defining character of causing conflicts below the threshold of war. What arouses the alarm is the asymmetric warfare when one squares in the loss of trade secrets and intellectual property rights that fall prey to cyber espionage and thefts. The US-China "No Spy Agreement" came forth as a result of the compromise between US-China cybersecurity confrontations of the sort. Yet the no-spy agreement is far from effective in stemming China's thefts of intellectual property rights, which constitutes unfair trade in the eyes of the Trump administration.

Thefts of data through end user device – be it mobile as smartphones or fixed as critical information infrastructure – becomes a problem for not just loss of business intelligence and interests but, much more than that, a thorny issue for national security so far as thefts of trade secrets and covert surveillance are concerned. Given surveillance, with consumer privacy and possible state secrecy compromised, could be mediated by government's demand of access to data collected through private internet device/service providers/platforms, battlefields of cybersecurity confrontations go beyond state-to-state struggles. Therefore, cybersecurity concerns arising from public-private partnership (PPP) has become central to confrontations nowadays, rendering, something like US government versus Huawei 5G back-and-forth contentions, a central stage in state-business confrontations.

Since attribution, namely finger-pointing, remains problematic when it comes to cyberattacks, espionage, etc., the US-Huawei dispute over 5G provides an exemplary case in point to illuminate cybersecurity confrontations with ample official statements and discourse in place.

Huawei 5G system, in terms of either device, function or services, is admittedly by many users better at cost-performance efficiency when compared to other competitors – a result, the US government has argued, of Chinese governmental subsidy and twisted aid with intellectual property thefts. Unsurprisingly, Trump administration, along the line of US-China trade war, pushed ahead bans on the federal government's procurement and use of Huawei products and equipment. Yet the decision was made based on national security concern to uphold the Trump administration's tenet that economy is security, particularly so when the installed 5G apparatus accelerate not just the speed of data transmission but more importantly the connectivity of networks that certainly are concerned with national security.

ii. State of Five-Eyes Allies, European and like-minded countries

Further bans on the sale of US apps, operation systems, and related restrictions upon the Made-in-US elements were imposed to unfolding an image of US-China technology Cold War in the making, an endeavor meant to leave US allies and like-minded countries little to no room but to choose the US side. Notwithstanding the US pressured hard on Five Eyes allies, European countries and like-minded countries in northeastern Asia, responses thus far may have not met the minimum level of US expectations. Except for Japan, Australia, New Zealand, and Taiwan, no more country by the time of this writing has confirmed to follow the US suit and reject Huawei 5G system. It naturally begs the questions whether Southeast and South Asian countries would act likewise and keep options open for Huawei 5G bidding for their fifth-generation communication and information infrastructure.

Before proceeding to answer the above question, it is imperative to draw on the ways the US has been trying to persuade its friends and allies away from Huawei 5G, thereby making sense of whether the US would nail the success this time in different regions. Based on the sequence the US government has been exercising its strategic communications, During late 2018 and early 2019, a time of the US fired first shot at China and unveil the screen of trade war, Five Eyes allies, namely Canada, Britain, Australia and New Zealand, came first under the US influence and the communications seemed to work out when Australia and New Zealand opted out Huawei on not just 5G system but also undersea optical cable connection project. British and Canadian governments sat on the fence with each's intelligence agencies calling for bans in resonance with the concerns the US government raised regarding the trust that constitutes the cornerstone of intelligence sharing might swash away as long as the shared intelligence information goes through connections built and maintained by Huawei 5G networks.

British government put forward conditional adoption of certain Huawei 5G equipment with the division of core and peripheral networks. By drawing the core-noncore line, Huawei 5G antenna and base stations are categorized as peripheral networks that the British government argues are unlikely to transmit critical intelligence and thereby do no harm to national security. Around the same time, Huawei brought forth signing up "no-spy agreement" with state government as a guarantee not to share user and network information with Beijing. Chinese ambassador to Britain offered the official endorsement that Chinese government would not demand Huawei to deliver customer data. The kind of collaboration between the public and private sectors in the face of international pressure is indeed rare, if not unique.

The US responded to British core-periphery dichotomy with a clear-cut rebuttal that there is no distinction between core and non-core networks as long as one links to the network. Taking into account of the telecommunication sectors' sunk costs devoted into the integration of Huawei 5G, coupled with pretty much watered-down sense of threats posed by China, a rich giant located in faraway East Asia, the US government's counter-argument paled in the eyes of European countries, and thereby lends further support to shy away from the US-led containment of Huawei and Chinese-wise internet and communication technology (ICT) products and services. In addition to those countries already open to Huawei 5G – Spain, Italy, France and Portugal for instance, German followed the British suit with core-periphery distinction, thereby staying with the prior decision of inclusion of Huawei 5G with precautions laid out as passing the laboratory tests.

Canada's posture towards Huawei 5G steers in a similar direction to that of most European countries. Common to both Canada and most European countries in their openness to Huawei 5G is not domestic sunk cost incurred in telecom sectors but rather slim sense of cybersecurity threats posed by China. To them, identification of a common cybersecurity threat would definitely point to Russia, rather than China, owing largely to the former's notorious capabilities in exercising influence operations through cyberspace.

Hybrid warfare in east Europe, reflexive campaign in 2016 as well as 2018

US elections, and hack-n-leak operations in 2018 French election all illustrate a formidable, manipulative Russia. On the contrary, China's poor track records in either Hong Kong or Taiwan have paled so far as Russian trolls are concerned. This line of reasoning explains why Britain, one of the Five Eyes allies, would go on and embrace Huawei 5G at the potential cost of cut off intelligence sharing. If Britain is at odds with the US for this matter, let alone other European countries. It is no surprise that Huawei keeps sweeping across Europe. Up until February 2020, Huawei employs over 12,000 employees and runs 2 regional offices and 23 research & development centres in 14 European countries. The company in 2019 has secured more than 50 commercial 5G contracts worldwide, 28 of which are with European operators.[1]

Unlike Canada and European countries, Australia, New Zealand, Japan and Taiwan has shared common cybersecurity threats posed by China, thereby imposing restrictions/bans upon Huawei 5G regardless of large sum of foreign direct investments made possible by Huawei in the shaping of 5G ecosystem. As Australia and New Zealand, two Five Eyes allies located in South Pacific, have faced increasing Chinese presence within and surrounding their borderlines, it is only natural for them to clean the house and make sure no Huawei 5G in place in the near future. Needless to say, Japan as an ally and Taiwan as a like-minded no-NATO ally to the US face daunting challenge posed by Chinese cybersecurity threats, thereby following the track of the US to ban Huawei 5G in government procurement whatsoever.

III. An Exploration of the State of South and Southeast Asia

i. Propositions

Two propositions come forward in the aftermath of a brief scrutiny of US-

---

[1] "Huawei bringing 5G to Europe," *Thailand Today*, February 11, 2020, https://www.thailandtoday. co/11/02/2020/huawei-bringing-5g-to-europe/.

Huawei cybersecurity confrontations in North America, Europe, South Pacific and East Asia:

> *For one, countries facing high level of cybersecurity threats posed by China are likely to blacklist Huawei 5G regardless of domestic telecom sector's vested interests.*
>
> *For the other, countries facing low level of cybersecurity threats posed by China are likely to embrace Huawei 5G, particularly so given Huawei's vested interest in their telecom networks.*

Cases selected from South and Southeast Asian countries provide hard-case tests so long as they stay mixed in their strategic posture between China and the US – a long-lasting hedging strategy among southeast Asian countries with security positions leaning towards the US military power staying in the region to check against China's rising military while economically laying eyes on Chinese market and remaining cautious for China Belt-and-Road Initiatives-related investment in their domestic infrastructure out of fears of liability trap and loss of control over critical nodes. Moreover, these countries caught in the middle of US-Huawei cybersecurity confrontations realize the dilemma they face: the pick of one side may suffer backlash from the other in ways of diplomatic, security or economic sanctions.

What follows next is a brief study of five cases where 5G rollout is in fast development and deployment—India in South Asia, and Vietnam, Singapore, Malaysia and Thailand in Southeast Asia to explore the ways these countries respond to US-Huawei cybersecurity confrontations.

ii. Five Countries in South and Southeast Asia

1. India

India is of critical importance in the US's Indo Pacific strategy. Against the backdrop of China's Belt and Road Initiative, US State Department's *Indo-Pacific:*

A Shared Vision released in September 2019, stresses that India, infrastructure investment and 5G are, among others, three critical pillars. That said, following closely with its tenet of non-alliance, India braces own path to meet its national interests. India's decision making regarding the rollout of 5G networks is no exception.

Yet constrained by limited funding, investment, and technology capacity, the rollout of 5G is not something India's own private or state-owned telecom sector enterprises can accomplish on their own. The so-called Big Three, Bharti Airtel, Vodafone Idea, and Reliance Jio, which together hold a share of over 50% of India's total wireless market, look outward for international cooperation with multinational big shots – Huawei, Samsung, Nokia and Ericsson are all on the list. Reliance Jio has been working with Samsung first and then extending cooperation with Huawei, Nokia and Ericsson. Bharti Airtel works with Huawei, ZTE, Nokia and Ericsson at 5G trials, with the addition of Cisco in core network readiness. Vodafone Idea cooperates with Huawei, ZTE, Nokia and Ericsson, with Huawei 5G AI to upgrade it current 4G network.[2]

Given that indigenous 5G development is simply beyond domestic sectoral capacities, coupled with multiple combinations already in competitions with each other, India government is careful and shrewd in deciding which option for the building of 5G network best serving both national and sectoral interests. That sort of decision-making surely has to square in strategic and security considerations. Choosing Huawei 5G certainly satisfies cost-benefits equation, but domestic opposition out of either security concerns or business competition, or a mixture of both, remains robust. To be sure, fear of US pressure or being marginalized in the Indo-Pacific region is not a primary, mainstream opinion. Instead, it is mostly agreed

---

[2] Poojitha Jayadevan, "When will 5G be available in India?," *Computerworld*, April 29, 2020, https://www.computerworld.com/article/3540254/when-will-5g-be-available-in-india.html.

upon that India should not bow to the US pressure to exclude Huawei option. Under this circumstance, India granted permission to Huawei, along with other telecom companies, to launch 5G trials in late 2019.

Putting Huawei in the hopeful list has placed India in an advantageous bargaining position to carve the best benefits out of the 5G rollout deal in their demand for the service provider to meet the request of either technology transfer or security/privacy requirements. At the same time, India government also strategically creates a tentative space to buffer and balance domestic differences and US concerns at the moment. In so doing, the independent path of diplomacy this time featured by balancing between the US and China pressure stay firm while also serving the prerogatives of economic interest and technology development.

India-China border conflicts erupted in 2020 changed the above landscape, particularly so in the aftermath of cyber attacks suspiciously from China upon Indian power grids nearby the borderline during the conflicts. Incidents of the sort have pushed up the ladder of perceived cyber threat that Indians had never experienced before. Unsurprisingly, the aftermath has led to India's decision to kick out China's telecom giants of 5G infrastructure loop.[3]

2. Vietnam

Placed in the context of Southeast Asia, Vietnam stands out as a classic case in deciphering the hedging strategy towards US-China competitions in the region. With a history of Vietnam-China war-fighting and, more importantly, long-lasting disputes with China over South China Sea, Vietnam is undoubtedly the most salient among southeast Asian countries to voice strong security concerns over China's expansion, thereby siding with the US in the promotion of Indo Pacific strategy. On the other hand, Vietnam, a communist party-ruled country just like China, followed

---

[3] "Huawei and ZTE Left out of India's 5G Trials," BBC News, May 5, 2021, https://www.bbc.com/news/business-56990236.

the Chinese track of market economy model for decades, have been embracing Chinese market and investment. The latter is best showcased by China's Belt and Road Initiatives-related activities keeping strong and robust in Vietnam.

The combination of leaning towards the US out of security concerns with bracing for China driven by economic interests should have rendered problematic in the controversies over the issue of whether or not to blacklist Huawei 5G for cybersecurity concern. Not so problematic for Vietnam – Viettel, a Vietnamese military-run telecom company, also the lead in domestic subscribers market, expressed clear-cut stance to exclude Huawei 5G out of cybersecurity concern. To avoid technology reliance on a single telecom company, Viettel is reported to decide to develop its own 5G equipment,[4] with the initial use hardware supplied by Finland's Nokia and Sweden's Ericsson,[5] while continuing to pay patent royalties to Huawei, Ericsson, and others. Two other major Vietnamese telecoms, Vinaphone and Mobifone, also distance themselves from Huawei, with each relying on, respectively, Nokia and Samsung equipment.[6]

Relationships with Huawei in overseas 5G market is simply a different story. As Viettel eyes on the opportunities beyond merely Vietnamese domestic market and extends business in Southeast Asia, South America and Africa, adoption of Huawei equipment becomes a major impetus for profiting in overseas competitions. In three southeast Asian countries, namely Laos, Cambodia and Myanmar, Viettel maintains cooperation with Huawei in pursuing local market share. Huawei and another Chinese big telecom vendor, ZTE, are in competitive relations in Cambodian 5G base stations rollout. In this light, although Viettel is owned by Vietnam's Ministry

---

[4] Tomoya Onishi, "Vietnam carrier develops native 5G tech to lock out Huawei," *Nikkei Asian Review*, January 25, 2020, https://asia.nikkei.com/Business/Telecommunication/Vietnam-carrier-develops-native-5G-tech-to-lock-out-Huawei.

[5] Tomoya Onishi, "Vietnam's Viettel shuns Huawei 5G tech over cybersecurity," Nikkei Asian Review, September 6, 2019, https://asia.nikkei.com/Spotlight/Huawei-crackdown/Vietnam-s-Viettel-shuns-Huawei-5G-tech-over-cybersecurity.

[6] Ibid..

of National Defense, which sides with the US and oftentimes at military odds with China in South China Sea, Viettel's 5G expansion in southeast Asia and beyond is unlikely to follow the US path and blacklist Huawei, a stark contrast to their way in Vietnam. The domestic-overseas differentiation in the decision of exclusion of Huawei 5G is pretty pragmatic to reconcile security interests of connecting to the US Indo Pacific strategy with economic interests of involving into China's Belt and Road Initiatives.

3. Singapore

Singapore is of most importance in the US Indo Pacific cybersecurity strategy in Southeast Asia. In US State Department's *Indo Pacific: A Shared Vision*, Singapore is singled out as the cornerstone for the cybersecurity confidence building measures in southeast Asia. An endeavour to confidence building in cybersecurity realms signals the picture the US envision in the near future in which China's 5G may prevail in the region and thereby the imperative for the US to be ready to prepare for the secure communication channel with regional countries. Singapore best fits the function of the hub between the US and regional counterparts, given the nuanced, leading role Singapore has been playing in the region where most, if not all, regional states are caught in the US-China technology competitions representative in the form of US-Huawei 5G cybersecurity confrontation and the need for those countries to figure a way out while carving the most benefit out of it.

Among other southeast Asian countries, Singapore enjoys the leading edge in pursuit of the 5G rollout. In 2019 Singapore's Infocomm Media Development Authority (IMDA) announced the bids for 5G standalone rollout competitions. The rollout "standalone" networks is defined as networks that do not piggyback on existing mobile networks. Together with overall network resilience and security, vendor diversity is one of the three focal points of the rollout program. All three are perfect match to the US vision of 5G security.

Four telecom joined the bids with three proposal submissions in 2019. Singtel and TPG Telecom each submitted a proposal, while StarHub joined hands with M1 to put in a joint submission. TPG is the only one that clearly known to cooperate with Huawei 5G. IMDA announced in April 29, 2020 that two winning bidders, Singtel and a joint venture by StarHub and M1, would begin rolling out 5G standalone networks from January 2021 to cover for at least half of Singapore by the end of 2022. Nationwide coverage will have to be complete by the end of 2025.

TPG's fallout is widely interpreted as Singapore's lining up with the US to keep Huawei 5G off of the table. Yet IMDA responded with clarity that two winning teams also have Huawei 5G cooperation and Huawei equipment at their disposal. In addition, TPG can access these network services through a "wholesale arrangement," when by 2025 Singapore is expected to have two "full-fledged" nationwide 5G networks. Doing so meets the goals of resilience and vendor diversity that IMDA keeps emphasizing. In further response to the aforementioned US Indo Pacific strategy, IMDA intends to establish a 5G security testbed program for technology exploration to better safeguard the country against 5G network cyber threats and vulnerabilities. Altogether Singapore undoubtedly strikes a pretty good balance between the US security expectations and China's economic-technology expansion in the region.

4. Malaysia

Malaysia presents a typical case in Southeast Asian countries that not just try to hedge against US-China conflicts but also endevour to make the most of it from the US-China confrontations in either Indo Pacific Strategy versus Belt-and-Road Initiative game or US government versus Huawei 5G standstills. On one hand, Malaysia oftentimes entails in maritime conflicts with China over South China Sea disputes. On the other hand, either Japan or US Indo Pacific strategy sees Malaysia as critical nodal point in the maritime security for sea lines of transportation. When former Prime Minister Mahathir Mohamad came in power, Malaysia government, in avoidance of the debt trap as a result of China' s Belt-and-Road Initiative, backed out from the East Coast Rail Link (ECRL) project and, after renegotiation, cut the deal with China from previously 66 billion ringgit (US$15 billion) to 21.5 billion.[7]

For a small-to-medium sized power like Malaysia, preservation of national interest takes shrewed bargaining skills. The sort of skills needs some convincing interpretations to pave the ground. In that light, it is no surprise that in Malaysia there is abundance of rebuttal of US's legitimacy to blacklist Huawei on the grounds of espionage concerns. All that leads to former Prime Minister Mahathir Mohamad's counterargument of the US challenges on whether Malaysians hold concerns and suspicions towards Huawei 5G's cyber backdoor. The short and blunt answer is there is no such concerns in Malaysia in that if it is not China, the US would spy on Malaysia as well and Malaysia could do not much about it.[8] Admitting Malaysia's weak link in cybersecurity and counterespionage in the face of either China or the US, Malaysia opens great room for manoeuvring between the US and China by translating disadvantage and weakness into advantageous negotiation chips

Malaysia then unlocks tentative ban on Huawei and Maxis, one of Malaysia's major telecom companies, signed the agreement with Huawei for the rollout of 5G program. Critical to the deliberative message that Malaysia intends to convey to the US government is former Prime Minister Mahathir Mohamad's witness as a way of strong endorsement for Huawei's involvement in Malaysian nationwide 5G rollout afterwards. In return, Huawei is to offer technology transfer so that Maxis will obtain insights, standards, products and solutions that will run 5G deployment and operation in Malaysia.[9] Maxis stands alongside with former Prime Minister Mahathir Mohama

---

[7] Syed Saddiq Syed Abdul Rahman, "Mahathir is not bowing to China on Huawei. He's standing up to US bully: Malaysian minister Syed Saddiq," South China Morning Post, June 7, 2019, https://www.scmp.com/week-asia/opinion/article/3013585/mahathir-not-bowing-china-huawei-hes-standing-us-bully-malaysian.

[8] Ibid..

[9] P Prem Kumar, "Huawei officially lands role in Malaysia's 5G rollout," *Nikkei Asian Review*, October 3, 2019, https://asia.nikkei.com/Spotlight/5G-networks/Huawei-officially-lands-role-in-Malaysia-s-5G-rollou.

and express full acknowledgement and awareness of potential security risks of the sealed deal with Huawei, leaving little to no space for the US government to further press pressure on Malaysia.

5. Thailand

Under the architect of US Info Pacific strategy, Thailand is close to the US with a security alliance relationship in place. Yet Thailand is closer to China in economic interactions with the boost of Belt-and-Road Initiative. New China-Thailand cooperation in the East Economic Corridor (EEC) include high-tech industries, digital commerce and automobiles. China's Huawei is deeply investing and involved in Thailand's 5G. Five Thai companies – Advanced Info Services (AIS), True Corp, Total Access Communication (DTAC), TOT and CAT Telecom – have submitted bidding documents for participation in the country's 5G spectrum auction in February 16 2020. Compared to the transparent auction process, the building of the 5G infrastructure will be conducted by state agencies with the support of internationally recognised companies including Huawei.

In February 2019, Huawei was one of three international companies invited to join Thailand's first 5G trial. The trial invited Huawei, Nokia and Ericsson, to test self-driving cars and remotely operated robots at Kasetsart University in Chonburi. Much of the 5G equipment used by the private companies for end users' purposes will also be licensed from or directly produced by Huawei.

Yet while the world is deliberating the merits of Huawei's involvement with their respective 5G sectors, Thailand seems set to embrace the controversial firm without debate. Thai Deputy Prime Minister Somkid Jatusripitak explicated the economic benefits 5G brings about – Thailand's economy would expand at a faster pace while 'private investment' on the technology could exceed 110 billion baht in 2020. Nevertheless, there is not much of a debate regarding Huawei's active investment and involvement in Thailand's 5G infrastructure. According to the

President of Huawei Southeast Asia, the company had invested 160 million baht to test 5G systems and deploy their equipment in the country. Huawei Technologies Thailand indicated the company would be looking to support eco-friendly investment and supports 5G related laws that would 'benefit the public sector in Thailand.' But in parliament, 5G related laws have not been debated, not one press briefing from opposition parties mentions the capitulation of 5G networks to such foreign interests as Huawei. [10]

IV. Conclusion

i. Analytics

This short essay raised the question regarding how South and Southeast Asian countries cope with US-China cybersecurity confrontations represented by US blocking of Huawei 5G. After a brief review of the European countries and US allies, five countries – India in South Asia, and Vietnam, Singapore, Malaysia, Thailand in Southeast Asia – come under preliminary scrutiny to see if they fit the proposed tenets that upholds those who felt under China's cybersecurity threats are likely to follow the US calls and put bans on Huawei, while those with invested economic ties with China are likely to turn their back on the US for that matter.

A succinct analysis paints a mixed picture in which countries under scrutiny indeed behaved as expected. This tentative result came as no surprise when one takes into account the complex historical context and current state of political-economic dynamics in South and Southeast Asia. Yet the answer of "yes" indeed come with caveats – it is not those who are under China's cybersecurity threats go with the US side, but rather those who are under China's overall security threats are likely to extend their suspicions over the cybersecurity risks Huawei might pose, thereby

---

[10] Cod Satrusayang, "Thailand's political immaturity means its 5G network could be compromised without debate," *Thai Enquirer*, February 4, 2020, https://www.thaienquirer.com/7940/thailands-political-immaturity-means-its-5g-network-could-be-compromised-without-debate/.

siding with the US calls to put Huawei in the blacklist.

However, the size of the power might play a role in determining how they respond. India and Vietnam are the two under China's security threats, but India's grand market size renders luxury to sit on the fence and see how to carve the most benefits out of the game. Suspicious cyber attacks from China serve as a wake-up call for Indians to ban China's telecom companies from 5G buildup. Vietnam, on the other hand, takes precautions to exclude Huawei in domestic market while the military-run telecom company upholding flexibility and profit-lurking spirit switches to cooperative relations with Huawei in pursuit of neighboring 5G rollout projects. As a tiny city-state, Singapore is nuanced in the way it copes with Huawei 5G – no appearance of Huawei 5G in the surface with Huawei 5G device admittedly embedded in every bidding team, since Singapore is a country more concerned with retaining a balanced position than holding any specific grinding attitude towards either the US or China.

By embracing Huawei 5G with open arms, Malaysia and Thailand present quite another category that seems to match the expectation for those seeking economic benefits from China. What is noteworthy is that there remains a salient difference between these two Huawei-friendly countries. Malaysia has made great use of domestic debates to boost its negotiating position vis-à-vis China and thereby carved benefits in either Belt-and-Road Initiative-related Railroad project or 5G-related technology transfer deal with Huawei. That does not seem like to be the case for Thailand. Somehow one may still argue that the difference exists simply owing to the fact that Malaysia still faces security threats posed by China over South China Sea disputes, which is rarely a situation Thailand might have to encounter.

ii. Caveat and Suggestion

Compared to the current state that no European country sides with the US and places Huawei 5G in the blacklist, South and Southeast Asian countries present policy choice mixed with security and economic concerns while those in Europe by no means hold the similar sense of security threats arising from China, a country too far away to become a real concern in their mindset. As for Taiwan, Japan, Australia and New Zealand, China's security threats are real and hence cybersecurity threats posed by Huawei 5G naturally becomes risks to be avoided in ways of locking out in the very beginning. In this light, one should be careful in placing too simplistic a conclusive comment regarding the ways certain country likely to respond to US-China cybersecurity confrontations.

Given the mixture of differed levels of security concerns over Huawei 5G cybersecurity risks, what US Indo Pacific strategy still possess in hand as a pretty good selling points is to push ahead 5G connectivity under the architect of cybersecurity confidence building measures to prepare with pre-emptive thinking for a future of a dichotomous 5G world defining by Huawei vs non-Huawei camp. While Singapore currently is designated as the regional cybersecurity hub in Southeast Asia, Taiwan may base on its well-noted track record and serve as the counterpart in Northeast Asia to connect and build confidence regarding cyber activities among the US, Taiwan, Japan and South Korea cyberspace.

梅克爾政府的難民政策與危機管理之研究

龔隆生 *

壹、前言

德國總理梅克爾（Angela Merkel），這位號稱「歐陸的柴契爾夫人」已定於2021年9月將卸任，即便是如此，當德國面臨國際安全議題時，或是受到非傳統攻擊狀況時，要如何面對與應變，仍是德國政府當前重要的課題。事實上，國際安全的擴大化與複雜化客觀上要求世界各國加強安全合作，廿一世紀的德國，在歐洲區域安全中扮演何種的角色？為因應國際政經環境的劇變，國際關係相互依賴程度增高，以及非軍事性因素的影響，許多新型安全概念應運而生，因為就世界經濟充分結合的「運作核心」（functioning core）國家而言，仍有一些失能國家所構成的「非整合差距」（non-integrating gap）」[1]，對區域或者是國家安全產生了些許的威脅。有鑑於在國際秩序的約制之下，非但國際關係中傳統安全困境依然存在，而且在非傳統安全方面的合作也是困難重重。

2015年發生了大舉難民跨海偷渡到歐洲的情形，造成當時國際社會與歐洲國家皆束手無策的困境。德國政府本身在國際社會的堅定與努力，以及經濟實力與外交政治地位來看，梅克爾政府在國際危機管理中所承擔的責任是逐次增加，直言之，國際危機管理是德國未來幾年外交政策的核心。冷戰結束後，由於歐洲已經失去前蘇聯軍事威脅的來源，聯盟向心力已日漸式微，因此有遠見的國家已開始思考所謂的「新威脅」在那裡？本文區分兩個層次來探討德國

---

* 淡江大學國際事務與戰略研究所博士生

[1] 「非整合差距」（non-integrating gap）由美國軍事地緣戰略家托馬斯・巴內特（Thomas P. M. Barnett 生於1962年）所提出，他發展了地緣政治理論，將世界劃分為「職能核心」和「非整合差距」。美軍介入伊拉克戰爭時，他曾寫了一篇文章支持軍事行動，題為〈五角大樓的新地圖〉（後來成為詳細闡述他的地緣政治理論的書的標題）。他的地緣政治理論的中心論點是，全球化帶來的國家之間的聯繫（包括網絡連接，金融交易和媒體）與那些擁有穩定政府，這些地區與尚未滲透全球化的地區形成鮮明對比，全球化是政治壓制，貧窮，疾病，大規模殺人和衝突的代名詞。這些領域構成了非整合差距。

在肩負歐洲非傳統安全政策的角色定位，第一個是「區域層次」，以歐盟為中心，針對各成員國的政策差異與效果來論述，第二個是「國家層次」，德國身為歐盟的主導國，是如何將不同歷史與文化對歐盟成員國家說服，並如何利用決策機制與危機處理之能力，來處理討敘利亞「難民」（refugee）問題。本文並非僅是在闡述危機與威脅評估的適切性，而是強調其本質與特色，尤其是德國政府如何承當歐盟主導的角色定位，來面對「交織性的重疊認同與利益」（overlapping vertical identity）。德國人民所質疑的是，2016年梅克爾的一句話：「我們能夠應付得來，因為德國是一個強大的國家。」（Wir schaffen das, denn Deutschland ist ein starkes Land.）[2]，但事實結果，德國是很難應付這個國際危機。

從西方國際關係學者對於「機制」這一概念的定義，並非侷限於組織而言，德國的克制文化（Kultur der Zurückhaltung）[3]是否也陷入了政治、外交與軍事的困境？是否也產生了迷失？以往德國主要以國際組織成員的身份參與國際危機管理，其自身鮮少作危機管理行為體，但是隨著逐年德國國際地位的提升和歐洲區域的脈動，德國作為國際危機管理主體的作用逐漸引起關注。[4]迄今為止德國官方與學界以德國國際危機管理機制為研究物件的成果顯為稀少，本文借助德國國際危機管理相關文獻和官方資料，來整理德國危機管理的法律基礎，在德國現有國內危機管理機制的基礎上，來探索德國聯邦危機管理機制。

本文主要以「歷史研究」途徑，探究德國梅克爾政府在接續施若德

---

[2] NEUJAHRSANSPRACHE VON MERKEL: *"Wir schaffen das, denn Deutschland ist ein starkes Land"* VON GÜNTER BANNAS, BERLIN-AKTUALISIERT AM 31.12.2015 <https://www.faz.net/aktuell/politik/inland/neujahrsansprache-von-merkel-wir-schaffen-das-denn-deutschland-ist-ein-starkes-land-13991331.html> (2019/1/2)

[3] 連玉如著，《新世界政治與德國外交政策—新德國問題探索》（北京：北京大學出版社，2003年），頁17-18。「克制文化」（kultur der Zurückhaltung），是德國二戰之後特殊社會歷史條件下形成的一種特殊文化，另一種說法也是反軍國主義的文化，並優先以非軍事手段解決衝突。並請參閱約翰·米爾斯海默（John .Mearsheimer）著，李澤澤譯，《大幻想：自由主義之夢與國際現實》（上海：上海人民出版社，2019年），頁333。現實主義者並不是唯一克制倡導者，也有從非現實主義視角看待外交政策的克制倡導者，事實上，甚至還有一些贊成克制的自由國際主義者。

[4] Michale Brecher, Crises in Word Politics: Theory and Reality (Oxford/NewYork/Seoul/Tokey: Pergamon Press, 1993), p.43.

（Gerhard Fritz Kurt Schröder）總理的政權之後，來論述德國危機管理決策所產生的影響與發展進程，尤以2015年接收難民為例，輔以「文獻分析」的方法，從相關文獻中探得德梅克爾政府的政治行動，以及外交政策是如何因應與執行這危機處理手段等幾個面向，來探討德國在肩負歐洲非傳統安全政策的核心角色（zentraler Akteur）定位，以驗證梅克爾曾說的一句話「牆阻擋了我的機會，卻擋不住我的夢想」。[5] 梅克爾的難民政策與危機管理之處理，是引起了歐盟一些國家的抵制與不滿，無論是從區域利益或是大國之間的覬覦，都對歐盟一體化產生莫大之衝擊。其次，就是極右翼政治力量的崛起，而導致自由、民主、人權的價值觀受到嚴峻挑戰，甚至恐怖主義的延伸繼續對世界產生嚴重的威脅。大抵言之，本文有下列六個部分。第一部分、前言；第二部分、歐盟難民問題之發展；第三部分、梅克爾主政時期的難民政策；第四部分、梅克爾政府的決策機制與危機管理；第五部分、梅克爾難民政策的影響；第六部分、最後的結論，則是對此次危機的內涵提出心得與看法。

貳、歐盟對難民處理之發展

從實際觀察，歐盟（EU）在2015年之前，對難民危機管理上的作為是力不從心的，一言以蔽之，它既無法限定邊界，也無法在歐盟內部裏達成一致的共識，難以有效控制難民潮的衝擊。「多層次治理」（multi-level governance）早已成為歐盟治理的代名詞，歐盟各成員國為了維護自身利益，在決策過程上的確是阻力重重，因而在危機管理溝通上所付出的代價卻是曠日費時。多年來歐洲勞動市場原本就不是引入大量移民作為依存，在所謂「多元文化」（Multikulturalismus）之下，難民的問題所帶來之影響，無法融入歐洲整個社會，職是之故，就會影響社會治安，勢必也成為國家安全重大的隱憂。[6]

實質而言，難民危機伴隨著許多不確定的安全隱憂，2015年11月巴黎恐怖攻擊事件對整個歐洲民眾心理造成極大的恐慌與衝擊。因此，歐盟在推動難

---

[5] 原文是Die Mauer blockierte meine Gelegenheit, aber sie konnte meinen Traum nicht blockieren.「牆擋得了我的路，卻擋不住我的夢想」，德國總理梅克爾2019年對哈佛大學畢業生的演說。<https://www.storm.mg/article/1343589> 2019-05-31，檢索日期2020年12月4日。

[6] 鄒露著，《德國國際危機管理與實踐研究》（北京：社會科學文獻出版社。2020年），頁229。

民分攤政策時更加艱難，其次則是難民安置需要龐大的資源，其中包括資金、物資、土地等等，對這些經濟體質原本就已疲軟，且又經歷了金融風暴的歐盟國家而言，更是雪上加霜。難民危機之勢，就是因為在申根協定（Schengen Agreemen）[7] 中不設邊界檢查，人民擁有自由遊動、遷徙的權利，對這一群難民的不確定性，或許就是削弱歐洲一體化的因素之一。任何國際社會的成員都會有一些理念與行動相衝突（conflict）的利益因素，這在國際危機中都不足為奇，因為它必然不會在初次的協議中就能達成共識。再者，難民問題已挑戰了歐盟統籌管理的能力，其中包含了對價值觀、權力、地位和資源的訴求，這些難民風險因素一旦強加分攤給各成員國身上，這些沉重的負擔，勢必引起某些國家的反彈、壓制、甚至對抗。[8] 這些都會有損歐盟組織的合作與互動，甚至更有可能選擇抵制現有政治架構，來對抗失衡不平等的國際秩序。

一、歐盟一體化政策受到威脅

在歐盟多層政治複合體中，從傳統的國家利益、決策機構的部門利益、乃至於公眾參與要求，都凸顯出歐盟各成員國抗拒之下，對所謂一致決行為形成了嚴重的制約，在不同議題情況下出現政治意識上不同的意見。實際上，歐盟成員國絕大多數都已加入了聯合國的《關於難民地位公約》（Convention relating to the Status of Refugees）和歐盟的《都柏林公約》（Dublin Regulation）[9]。事實上，這些歐盟成員國對難民申請者的情況有所不同，相對地接收難民的認知與對待也有不同，例如：身份（identity）、社會（society）、文化（culture）、群體（group）等，其中又以「身份」和「文

---

[7] 1985年法德荷比盧所簽的申根協定是第一代Schengen Agreement，1990年的申根執行協議，是第二代Schengen Agreement。

[8] 請參閱翁明賢，常漢青著，《兵棋推演—意涵、模式與操作》（台北：五南出版公司，2019），頁151。

[9] Auswäri Zehn-Punkte-Plan für europäische Flüchtlingspolitik (Offizieller Artikel, 2015)，<http://www.auswaertiges-amt.de/DE/Aussenpolitik/Globale-Fragen/Fluechtlinge/Aktuelles/150823_BM_BM一Gabriel_FAS_node.html.>（2020/12/3）區分為1997年都柏林協定、2003年都柏林第二代規則、2013年都柏林三代規則—進行補充與調整，包括庇護程序、權責、和簽證規範。所謂「歐洲共同庇護系統」（CEAS）中的「都柏林規定」（Dublin Regulation）。不少國際人士都認同，「都柏林規定」是在目前難民潮衝擊下，也已經近乎沒有執行，甚至數年的歐洲成員國之政府也會大舉修訂、甚至以新法取代「都柏林規定」。

化」兩個概念又難以界定。[10] 以義大利與希臘兩國為例，即便是已給予難民了身份和資格，但難民仍會根據申根簽定的條件，自由遷徙走向生活條件更好的德、英、法等國家。

　　歐盟過去的案例而言，在涉及「共同外交與安全」（Common Foreign and Security Policy，簡稱CFSP）[11] 時，都會使用協商溝通方式，採取一致決的決策模式，但在難民配額的方案，歐盟各成員國都持反對意見，一直無法達成共識，這項決策的確是艱困難行。職是之故，歐盟毅然放棄民主協商與一致決的表決機制，改由多數決來做最後的裁決，雖然最後如期通過難民之配額，但似乎也得看出端倪，歐盟過去所追求一體化內部的決策機制與模式似有所失靈，無法順利一致執行的窘境。

二、反移民與反歐洲的極右翼政黨再興起

　　過去雖然反移民和反歐洲的聲浪一直都存在，但當時勢力微弱皆不足以撼動歐盟一體化的發展，直到2008年的全球金融危機之後，所伴隨著越來越嚴重的移民進入歐洲，致使歐洲經濟發展遲緩以及歐洲各國面臨債務危機。相形之下，更使得歐洲失業人口增加，中下階級民眾反對歐洲一體化的聲浪逐漸增加，此現象直到2014年難民問題開始醞釀，此一契機才讓反歐洲的極右派政黨再次興起。[12]

　　過去十年間，極端右翼政黨在歐洲各國異軍突起，如德國另類選擇黨（Alternative für Deutschland, AfD）[13]、法國國民陣線（Front National）、英國

---

[10] 約翰・米爾斯海默（John J.Mearsheimer）著，李澤澤譯，《大幻想：自由主義之夢與國際現實》，頁23。

[11] 歐盟三大支柱，根據1992年的《歐洲聯盟條約》，分別為：第一支柱為「歐洲共同體」，涉及經濟、社會、環境等政策。第二支柱為「共同外交與安全政策」，涉及外交、軍事等政策。第三支柱為「刑事領域警務與司法合作」（支柱前身是「司法與內政合作」），涉及共同合作打擊刑事犯罪。

[12] 約翰・米爾斯海默（John J.Mearsheimer）著，李澤澤譯，《大幻想：自由主義之夢與國際現實》，頁47。

[13] 德國人民另類選擇黨（Alternative für Deutschland, AfD），是一個在2013年歐元危機時建立的反歐元團體。2015年難民到達高峰時，該團體展現其風潮。三年之後，這一黨派已然成為了聯邦議會中的第三大黨，且在地方議會中也都坐擁席位。

獨立黨（UK Independence Party）、荷蘭自由黨（Party for Freedom）等政黨，都在各國選舉中的表現格外突出，且在大選中成為最大贏家，這些政黨都獲得了相當的民意與核心基礎，而這些政黨的核心政策主張，就是反對外來移民和歐洲一體化為主軸。針對這波是二戰以來第二大的難民潮，歐盟成員國家對難民的處理態度始終無法達成共識，也讓歐盟成立的初衷「一個融合的大歐洲」受到嚴峻的考驗。

三、歐盟國家對難民之處置

歐盟於1992年成立了「歐洲共同體人道援助機構」（European Commission European Civil Protection and Humanitarian Aid Operations，簡稱ECHO），它主要功能就是針對歐盟以外，若遭到天災或是衝突危機所苦的難民能提供緊急援助，但由於該機構本身欠缺一些軟（硬）體資源，因此與國際關懷組織（CARE International，簡稱CI）、無國界航空組織（Aviation Sans Frontières）、國際紅十字會（International Committee of The Red Cross，簡稱ICRC）、世衛組織(WHO)等多個非政府組織及專業機構合作，資助災害預防與訓練計畫。歐盟更於2001年在歐執會（European Commission）內部建立了一個「民防機制」（Civil Protection Mechanism, CPM），就是希望能在緊急應變時可自行採取行動；但如果是非歐盟境內的災害，則必須與「歐盟輪值主席國」取得密切徵詢，才得以行動。[14]（如圖1）

2015年9月9日，時任歐盟執行委員會主席容克（Jean-Claude Juncker）在史特拉斯堡（Straßburg）的歐洲議會，呼籲各成員國能共同分擔收容難民，並依照GDP、人口數量、失業率來決定難民人數。對於這項配置方案，德、法、西班牙等國初期都願意承擔責任，但也有些國家築起高牆禁止難民進入。[15]事實上，歐盟國家對於難民處置主要分為三類，一是德國的「歡迎文化」

---

[14] Charles M. Perry,Marina Travayiakis, Bobby Andersen,Yaron Eisenberg編，國防部印譯，《臨危不亂—救災外交、國家安全與國際合作》（Finding the Rright Mix Disaster Diplomacy, National Security,and International Cooperation）（台北：國防部，2009年），頁234-237。

[15] 蔡孟翰，「無國界的漂流者：國際法的難民規範」，《法律白話文》，2015.9.22，<http://plainlaw.me/2015/09/22/refugee/>，檢索日期2020年12月11日。

(Willkommenskultur)廣收難民;二是適度接收難民的英、法兩國;三是消極且不接收的克羅埃西亞、匈牙利、斯洛伐克、捷克、波蘭等東歐國家。因此,歐盟國家在難民問題上始終無法達成共識,甚至因彼此政策不同而造成東、西歐分歧日漸明顯,尤其是匈牙利、捷克、波蘭等東歐國家,更批評西歐國家主張的多元文化主義,不但無法減緩難民危機,更會招致內政治安的問題。[16]

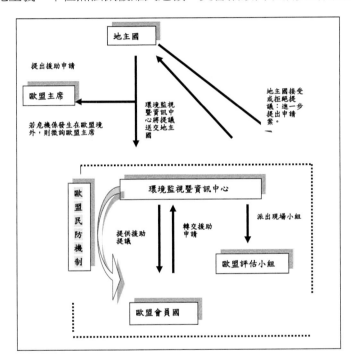

圖1　歐盟民防機制的援助申請流程

資料來源:Charles M. Perry,Marina Travayiakis, Bobby Andersen,Yaron Eisenberg編,國防部印譯,《臨危不亂—救災外交、國家安全與國際合作》（*Finding the Rright Mix Disaster Diplomacy, National Security,and International Cooperation*）（台北:國防部,2009年）,頁237。

---

[16] 約翰・米爾斯海默（John J.Mearsheimer）著,李澤澤譯,《大幻想:自由主義之夢與國際現實》（上海:上海人民出版社,2019）,頁118。請參閱文本,「組成一個民族國家所有個人共享相同做法和信仰是不切實際的。相反存在著重大的共同性,這種共性因具體情況而異。區分為厚文化（thick cultures）與薄文化（thin cultures）,是有意義的。這反應了一個國家的文化多樣性。厚文化具有顯著的文化同質性。而薄文化則是具有多樣性。

參、梅克爾主政時期的難民政策

梅克爾在2020年年終談話時說今年是她執政15年來最艱困的一年，不僅是難民議題，而且COVID-19疫情的問題，都讓她在決策時艱困難熬。從歷程來看梅克爾政治生涯中，是經歷了許多國際安全的抉擇，如加入「反恐陣營」、「反核家園」等措施，她都堅持一貫的政策，以展現德國是以「自由、多元、包容」的精神來面對國際社會。[17] 因此，國際社會當時都讚譽她是「歐陸的柴契爾夫人」、「俾斯麥第二」、「歐洲的嚴師」等美名，但相形之下，她也承擔了許多責難的惡名。對於一位曾經為德國與歐洲穩定付出心力的總理來說，國際社會總是抱持著愛莫能助，且持續緘默的態度來看待；即便是她坦然面對那些惡意批評的聲浪，且勇於面對而不迴避，但是終究抵擋不住國內強勢反對的聲討。

2020年梅克爾所主導的德國基督教民主聯盟黨（Christian Democratic Union of Germany—CDU簡稱基民盟）已票選出北萊茵—西發利亞邦的拉謝特（Armin Laschet）出線當選執政黨黨魁，他也或許會在2021年9月份大選後，接棒成為下一任的德國總理，這也意味著梅克爾在任15年之後，象徵著國家行為的價值觀或許將要更迭。可以很明確地說，難民危機不僅是國與國之間所存在的危機評估有所不同，在不同時代的政權也會有所差異，這證明了一點，就是在國際關係理論中，所謂國家利益是瞬息萬變很難去用絕對值來定論與定義的。

一、歐盟層面

儘管梅克爾政府一直強調人權及人道的維護，但在歐盟的層面上，一旦當難民危機挑戰到國家安全定、宗教認同以及社會安定時，歐盟成員國之間就會產生了相異的主張，一種是支持廣納；二是限制接收。實質上德國政府是支持前者，而大部分歐盟成員國則是支持後者，這二分法的結果，勢必注定會分

---

[17] Merkel sieht keinen Kurswechsel in der Flüchtlingspolitik", Süddeutsche Zeitung, 2016.10.1, <http://www.sueddeutsche.de/politik/1.3187733.> (2020/12/3)

裂歐盟一體化的主張。2015年德國政府的「歡迎文化」，腦怒了歐盟某些成員國，面對來勢洶洶的難民潮，更對德國的開放政策產生極度的不滿，尤以義大利、匈牙利、希臘等國家造成了極大的困擾，各國紛紛採取抵制措施，相繼頒訂不同的邊境管理政策。

　　理論上，歐盟國家的一些邊境管制、移民政策、庇護等政策，都是受制於歐盟運作條約（Treaty on the Functioning of the European Union）第77At-80At，以及里斯本條約（Treat of Lisbon）23, 24號議定書的規範。因此，匈牙利等東歐國家，為防止本國出現伊斯蘭化傾向，則在邊界修築圍欄阻止難民入境，捷克更是對難民強制羈留42天。[18] 自從保加利亞和匈牙利實施邊境管制之後，在其他面向上歐盟成員國尚能達成一致，奧地利和其他國家則相繼也在申根地區重新啟用臨時邊境管制，實施更嚴格的共同對外邊境管制，對歐洲國際邊界管理署（Frontex）進行擴編，增加一萬名邊境管制人員強化其作用。[19]

　　面對源源不斷的難民分配要求和不斷加深邊境保護合作，歐盟一些成員國欲想重新建立起「都柏林體系」（Dublin Regulation）或建立一個可以替代它的新體系，作為歐盟執行的法源依據，但這構想幾乎是微乎其微，畢竟難民（Flüchtling）是無法等同於移民（Einwanderung）的作法。難民危機讓歐洲人民意識到認同需求與實際狀況之間落差太大，在面對大舉難民湧入，歐洲成員國不得不在接收與安置方面要做好因應，若僅憑歐盟主導國德國一己之力是無法應對的；更重要的是，能否說服其他歐洲盟國達成共同的解決方案，才是德國在歐洲危機管理中協調能力的要務。

　　面對難民危機來勢洶洶，歐盟—這個歐洲最大的經濟體卻難以招架，因而更需要德國出面充當協調者，來作推展管控的進程。在歐盟的架構之下，梅克爾當下所認知的是，難民危機管控上涉及了兩方面的重要協調，一是分攤配額

---

[18] Peter-Christian Müller-Graff and Friedemann Kainer, "Asyl-, Einwanderungs- und Visapolitik (Zugangspolitiken)", in Werner Weidenfeld and Wolfgang Wessels (eds.), Europa von A bis Z -Taschenbuch der europäischen Integration. 14 ed. (Baden-Baden: Nomos, 2016), pp. 81-83.

[19] Volker Perthes, Ausblick 2016: Begriffe und Realitäten internatwaalen (Berlin: Stiftung Wissenschaft und Politik&Deutsches Institut für Internationale Politik und Sicherheit, 2016) ,pp.23-25.

問題；二是「申根體系」邊境管控的問題。在難民危機發生之後，由於人蛇集團非法偷渡太過汜濫，無論是在國際法還是歐盟自身的法律中，都很難辨別確認其真實身份。因此，歐盟成員國有權依照現有法令，在一定之期限針對一般移民進行身份核 ，規定移民的身份核查期限為18個月。[20]

2015年梅克爾在柏林舉行「難民高峰會」，提出歐盟一個共同原則，要求歐盟成員國能夠合理分配難民人數，儘管波蘭、匈牙利、捷克和斯洛伐克等這四個東歐成員國反對，德國政府還是採取說服和施壓雙重管道的做法，尋求歐執會以及法國、西班牙等國支持，才在歐盟內政部長理 會上第一次促成難民配額方案。[21]（如表1）

梅克爾除支持國際行動，還加強歐盟外圍邊境的執法，減少入境的人數，這一政策都得到基民盟（CDU）內部的支持。因為難民問題涉及到歐洲整體價值觀的維護，事關歐盟的未來，歐盟是無法拒收難民，否則等同否認歐洲所奉行的「平等、公平、人權」等價值觀，這對歐洲一體化的發展是嚴重的打擊。[22] 職是之故，梅克爾才再三呼籲，歐洲不能在難民議題上退縮，否則捍衛普世公民權益及人道主義將遭到瓦解。從國際現勢論析，歐洲難民問題並非肇因於歐洲之動亂，而是境外所造成，歐盟各成員國本應調整其外交與內政之政策，主動協助面對敘利亞難民的第一線國家，如土耳其、約旦以及黎巴嫩等國，能提供更多資源在難民身上，以保障其安全及改善其生活。[23]

如前文所述，歐盟各國對國際危機的理解不同，進行危機管理的動機和目標也有差異，因此在歐盟層面，持開放態度的德國與持反對意見的國家對立非常強烈，認為難民危機是德國的問題也是歐洲的問題，但是對德國面對難民危機的認知與態度卻是完全不認同，使得歐盟共同體的價值，因拒絕難民問題而

---

[20] Demian von Osten, "Die Balkan-Flüchtlinge und die Vorurteile", *ARD-aktuell*, 2015.9.29, <http://www.tagesschau.de/inland/balkan-fluechtlinge-105.html.> (2020/12/11)

[21] 鄒露著，《德國國際危機管理與實踐研究》(北京：社會科學文獻出版社，2020)，頁233。

[22] Auswärtiges Amt, Mehr Sicherheit für alle in Europa—Für einen Neustart der Rustungskontrolle (Offizieller Artikel，2016)，〈https://www. Auswäertigesamt. de/de/newsroom/160826-bm-faz/282910〉(2020/12/11)

[23] BMEL, Neue kooperationsprojekte durch Staatssekretär Bleser in der Ukraine eröffnet(Offizieller Atikel,2016), <http://www.bmel.de/DE/Ministerium/IntZusammenarbeit/BilateraleZusammenarbeit/_Texte/Dossier-Europa.Html? notFirst=true&docld=790681 2.> (2020/12/11)

失去了可信度，也直接對歐盟一體化產生莫大的衝擊，也觸及了國家主權的核心領域。因此，從另一個角度看當前的歐盟一體化仍是艱辛的，在申根地區所實施的難民分配政策也是困境重重。

表1　歐盟難民政策的主要體系結構

體系政策 運作層面　＼　政策內容	總體政策架構	具體政策項目	實施政策的 歐盟機構
1.超國家和地區間層面（歐盟外部層面）	經過修改的「歐洲睦鄰政策」架構	「歐洲睦鄰政策工具」（ENPI）框架內的各種援助、貿易、交流與雙邊對話的項目	歐洲投資銀行(EIB) 歐洲復興開發銀行(EBRD) 歐盟對外行動署(EEAS)
	得到修改和加強的「地中海聯盟」架構	在經濟與社會事務領域開展的援助、交流與雙邊對話的項目體系。	歐盟與北非國家聯合運作的「地中海聯盟秘書處」。
	應對移民與人口流動的全球方案	「斯德哥爾摩進程」地區發展與保護專案(RDPPs)人口流動夥伴關係項目。	歐洲理事會的「避難與移民問題高級工作組」(HLWG) 歐盟對外行動署
2.帶有超國家色彩的國家間層面（歐盟內部與外部之間的層面）	歐洲共同邊界控制體系	「難民登陸熱點追蹤」機制。歐盟聯合邊境與海岸巡邏行動（波塞冬行動「特雷頓行動」、「索菲亞行動」）	歐盟邊境與海岸警衛局、「歐洲警察署」(Europol)
3.國家和次國家層面（歐盟內部層面）	「歐洲共同避難體系」設想（含有一些需要成員國間協作的內容）	「都柏林體系改革計畫」新要點「歐洲安置專案」(European Resettlement Scheme)「歐盟安置架構」(Union Resettlement Framework)	「歐盟避難支持辦公室」(EASO)

政策內容 體系政策 運作層面	總體政策架構	具體政策項目	實施政策的 歐盟機構
4.跨越多個層面的交叉綜合體系	「歐洲移民議程」架構	內部— 1.「歐盟民事保護機制」； 2.難民遷居與安置分配協調機制 外部— 1.「歐盟對非洲緊急信託基金」項目； 2.「歐盟地區信託基金」項目。	歐盟委員會 歐盟理事會 歐盟對外行動署 歐洲投資銀行

資料來源：忻華，〈歐盟應對難民危機的決策機理分析〉（上海：德國研究，第3期33卷，2018年），10-11頁。

二、德國政府層面

　　2015年梅克爾政府為了處理難民的問題，特於聯邦外交部成立了危機預防司（Kriesenabteilung）的機構，職司難民危機處理，並對其進行改組，以執行棘手又複雜的難民問題。德國總理梅克爾曾經向德國人民宣告且堅定地表示：「我們可以做到」（Wir schaffen das），她的理念是，在處理移民問題上一直是德國政府的首要工作，如果能在政府預算寬裕的情況下伸出援手，以應付這個國際危機問題，更能展現出國家能力與價值觀。[24] 梅克爾初始認為難民的問題，所需要的只是各國之間經濟利益的協調，而不須勞師動眾用到歐盟層面的決策，只要從內政方面著手，對難民的安置有更多接收和安置就足以應付，也能為其他歐盟成員國發揮領導模範之作用。固然在難民危機管理的決策，德國政府在道義上雖獲得了許多國家的認同，但是反對派認為她的保守作風讓德國無法在國際多邊體系中發揮作用，在政策上更是搖擺不定，國家內部隨即出現

---

[24] Hubertus Volmer, "'Wir schaffen das'. Merkel mag ihr Zitat nicht mehr so", n-tv.de, 2016.9.17, <http://www.n-tv.de/politik/-article18663961.html.> 德國執政聯盟在2016年時增加了60億歐元的預算，以因應德國接收難民之後所用之開銷。檢索日期：2021年2月19日。

了分歧裂痕。

梅克爾一直身兼德國總理和歐盟領導人兩種身份的角色定位,她領導德國和歐陸度過了接二連三的危機,例如歐債危機(Europäische Schuldenkrise)與烏克蘭危機(Ukraine-Krise),成為歐洲的主導力量,被西方視為「自由秩序的守護者」,也使新德國重歸往日的榮耀,在她執掌歐洲最大經濟體的黃金十年,經濟成長了五倍之多,失業率則下降至80年代以來的最低的指標,相較於美國以及脫歐的英國,甚至改革失敗的法國,在梅克爾領導下的德國,一直是被歐洲人民看作是一個穩定的國家。[25]

歷史檢証,在梅克爾之前所卸任的德國領導人,都留下非常明耀的政治遺產,而這些人都算是梅克爾的精神導師,成為德國統一後富強的象徵。然而卻在她做出接收百萬名難民的決策之後,使得歐洲現狀變得不穩定,再加德國財政部長蕭伯樂(Wolfgang Schäuble)對義大利、西班牙、葡萄牙以及希臘等債務國家施加緊縮政策,而招至成員國的怨懟與不滿,尤其是希臘財政部長瓦魯法克斯(Yanis Varoufakis)認為,梅克爾政府的緊縮政策,所作出懲罰性的經濟緊縮措施,羞辱了希臘也煽起了民粹主義的氣氛。[26] 張亞中教授曾在〈開放和平論:追求永久和平的另一個選擇〉中論述:「人類發展迄今,多元社會,國際的經濟市場都已經開放,唯獨政治市場是封閉的。沒有政治市場的開放,國際和平永遠無法脫離「自己政府」這個讓人民又愛又憂的魔咒。」[27]

肆、梅克爾政府的決策機制與危機管理

國際難民問題,對於一個國際社會中的大國而言,或許它可以不顧國際輿論的撻伐,但在地緣政治因素下權衡 害得失之後,它又得顧及自己大國形象,因此,即會產出一個決策來行使,這就是所謂的政治性偏差(political

---

[25] Auswärtiges Amt, Mehr Sicherheit für alle in Europa—Für einen Neustart der Rustungskontrolle (Offizieller Artikel,2016),〈https:/www. Auswaertigesamt.de/de/newsroom/160826-bm-faz/282910〉(2021/4/19)

[26] 璿玥,〈移民和緊縮政策下梅克爾的政治遺產〉,<https://plainlaw.me/2019/02/27/immigration/>,2019-02-27,檢索日期:2021年4月19日。

[27] 張亞中,〈開放和平論:追求永久和平的另一個選擇〉,《問題與研究》,第46卷2期,2017年,頁13。

bias)。而這種政治性偏差，它會展現在組織內的決策過程，或者是在防止衝突爆發而設計的機制上，[28] 如各種危機小組、安全幕僚群，或者是協商單位等，在強化所提 的現象。在理論上「機制」的定義可以視為國際關係領域的原則、規範、規則以及決策程序。若是逐一加以分析論述，「原則」是關於事實和公正的信念；「規範」則是權利和義務的行為標準；「規則」是對行動的特別指示；「決策程序」則是做出集體選擇的實踐和應用。[29] 從梅克爾政府在處理國際危機管理「機制」的實質上，它係經過規則和制度、設立組織機構、明確的責任分工，並採取正式與非正式的方式來進行的政治運作，這其中必然涵蓋了政府管理的組織機構與系統、法律基礎、決策手段與機制，以及國際組織的約制。

實質而言，德國是一個遵崇法制的國家，在危機管理中必然會遵循國際法的原則，特別是《聯合國憲章》的規定。因此，梅克爾政府在危機管理機制上有兩大論述支柱，一是政策基礎，二是組織機構。職是之故，德國政府在國際危機管理機制是建立在健全國家部門與協調機制的基礎上[30]，國家干預是解決政治難題，促進經濟與社會發展的手段。在特殊情況下，總理可以直接協調和領導各部門；主導部門危機處理小組則負責聯邦和各邦層面的協調工作，危機管理小組必須派遣對話之聯絡人員，以確保其在工作時隨時保持待命狀態，並將其聯絡方式與結果告知聯邦總理和聯邦內政部危機管理處（das Refeat Krisenmanagemen）中的共同會報與情勢中心（das Gemeinsame Melde-und Lagezentrum，簡稱GMLZ）。[31] 其執行理念就是改善與國家組織，或超國家組織以及德國和其他國家之間的合作，促進聯邦和各邦之間的情報資訊交換，還

---

[28] 請參閱郭大文，〈美國對台海 岸關係的偏差動員分析〉，《台灣國際季刊》，第10卷第4期，2014年，頁123。Elmer Eric Schattschneider 於1960 發表〈美國民主的現實觀點：準主權的人民〉（The Semisovereign people: A Realist's View of Democracy in America）的專文指出：「組織本身就是一種為動而準備的偏差動員（mobilization of bias in preparation for action）；這種偏差直接涉及特定的公共益訴求、相關 為者間的權 關係、特定的 益分配方式，及相關成本的分擔辦法。

[29] 楊洁勉，《後冷戰時期的中美關係—危機管理和實踐》，（上海：上海人民出版社，2004），頁12。

[30] 鄒露著，《德國國際危機管理與實踐研究》（北京：社會科學文獻出版社，2020），頁134。國家機構和組織工具涵蓋了國家安全、警消、醫療、衛生、交通、社會保障等龐大的體系部門。所謂危機管理就是根據危險和損失的具體情況，由相應領域的主要部門來負責統一協調。

[31] 鄒露著，《德國國際危機管理與實踐研究》，頁150-151。

必須要協調和促進與鄰國的合作、發展和救援的合作。

從沙特施耐德（Elmer Eric Schattschneider）的政治性偏差理論中，他提出非常精典的論述，他說「任何形式的政治組織都隱含著特定的偏差。這種偏差僅 於某種衝突的呈現而壓抑其他衝突的表面化，因為「組織」本身即是某種偏差動員（mobilization of bias），故某些議題會被安排進入政治 域，而其他的則排除出去」[32]。因此，在梅克爾政府的決策機制與組織中，從國家權 和國家 益的角 來看，是希望從國家內部的組織與機制決策能滲入、控制外部的各個成員國，如難民配額與難民對待；相反的，各個成員國也同時會由外來的組織與機制滲入控制德國內部組織與個人，如難民員額無上限的聲浪與邊境管制，這兩者都是相互對應關係，合則利，不合則害。可以解析的是，梅克爾的決策機制與組織，加上組織內部原已存在的各種次級組織或個人，沙特施耐德認為組織經常只是複雜的權 與 益的集合群（constellation）中的一部分。組織內外的各種組織與個人，亦競相 用或透過該組織而追求極大化與極佳化的政經目的。[33]

一、決策機制與危機處理

在國際危機決策環節當中，有三種概念模式：一、「理性行為體模式」（rational actor model）認為政府是單一行為體，強調國家互動和國家意圖；二、「組織行為模式」（organizational behavior model）強調從國家內部機制解釋國家行為，包括構成政府大型組織的特殊邏輯、能力、文化和程序；三、「政府政治模式」（governmental polities model）側重政府內的政治活動，認為外交事務並非是某種單一行為體的選擇，也非組織的輸出，而是政府中各種的博奕者（players）之間相互討價還價的合成物。[34]（如圖1）

國際關係學者麥克爾‧布雷徹（Michael Brecher）所提出的「危機中的國

---

[32] 以上請參閱郭大文，〈美國對台海 岸關係的偏差動員分析〉，《台灣國際季刊》，頁123。

[33] 郭大文，〈美國對台海 岸關係的偏差動員分析〉，《台灣國際季刊》，頁123。

[34] 葛雷夢姆‧艾利森&菲利普‧澤利科著，王傳光，王云萍譯，《決策的本質：還原古巴導彈危機的真相》，（上海：商務印書館，2015年），頁22-23。

家行為模型」[35]（見圖2）與德國危機預防機制特點較為一致。審慎觀察其模型，任何一種國際和國內環境變化會對決策者都會造成無形的壓力，因而要在有限的時間內做出決策來解決危機，避免發生戰爭或衝突。因此有關難民議題，梅克爾需要更多的資訊情報為其提供全面決策的依據，附帶借鑒智庫及專業幕僚諮詢與建議，並透過決策中樞系統的協商，在安全穩定前提下進行優質的決策，才能進一步優化危機決策機制。就現實來看，當前德國聯邦政府在國際危機管理的行動中，似乎很難做到預防危機的引發，因此德國政府的危機決策的重點，都顯現在危機發生後的管控階段和危機結束後的善後階段。如此論述，不得不逆向論述梅克爾政府在國際難民危機管理決策過程中，考慮了哪些層面因素？

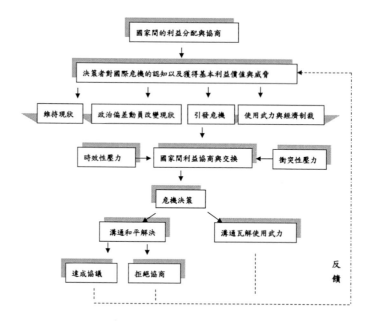

圖2　決策者參與國際危機決策

資料來源：參考韓召穎，趙倩，〈國際危機中的領導人決策行為分析—基於多元啟發理論視角〉，《國際政治科學期刊》，頁1-2。其餘筆者自行整理

---

[35] 見Michael Brecher, "*State Behavior in International Crisis*, "Journal of Conflict Resolution", Vol.23, No.3, 1979, pp.446-447. <https://www.press.umich.edu/14982/study_of_crisis> (2020/11/3)

　　麥克爾‧布雷徹曾對決策者提出三個論點：一、其基本價值受到嚴重威脅；二、需要在有限時間作出回應；三、存在捲入嚴重軍事衝突的可能性。具體而言，基本價值是國家與決策者的根本利益所在，因此，國際體系中的利益分配調整可能會對行為體的基本價值產生不同程度的威脅，從特徵上，國際危機是時效性壓力與衝突性壓力的國家互動情境，前者是決策時間急迫限制因素所在，後者是意為著國家間互動強度爆發衝突的機率增大。[36] 從前文論述，就辯證的角度看，三種決策模式之間是相互補充、相輔相成的關係。梅克爾政府在國際危機管理的決策機制需要考量這三種模式中的相關因素，有鑑於決策機制內部的複雜性，不應該簡單化將國家看作單一行為體，簡單的說，大舉難民問題，也絕非只是單一國家內政問題，它牽涉到地緣關係、經濟互利關係、國際安全關係等因素，因此梅克爾政府在制定決策過程中，國家之間的協調和國家內部都要達成共識，才能在一定之目標下，取決於一致之行動。政府的組織能力和程序，是解釋危機決策機制的重要因素，而堅實有力的組織、協商和有效的程序，才可以解決危機，並提 決策的進程與效果。例如德國在面對烏克蘭與難民危機，這些重大決策必然會出現在政府內部與歐盟體制上意見的分歧，德國國內的鷹派和鴿派之間，必然也是分歧與抗爭不斷，因而梅克爾在危機決策分析時，勢必考慮到政府內部不同立場之間的博弈。

　　羅伯特‧派特南（Robet D Putnam）曾提出雙層博弈理論（Two level games theory），所謂一國領導人要兼顧國際與國內兩個層次的博弈局面，其決策既要爭取國際體系層面的國家利益，又要滿足國內單元的利益要求，據此獲取雙贏整合，參與到其他國家互動中內。充分強調討價還價的國家中的互動，以及國內層次的雙向互動。即便是雙層博弈並非危機決策模型，但對建立行為體的危機決策模型具有借鑒意義 [37]。梅克爾政府基於國際危機特點，會將該模型危機結果與恐怖行動連結在一起，認為危機管理的最終目標是避免衝突

---

[36] 韓召穎，趙倩，〈國際危機中的領導人決策行為分析—基於多元啟發理論視角〉，《國際政治科學期刊》，頁1-2。<http://qjip.tsinghuajournals.com/article/2017/2096-1545/101393D-2017-4-102.shtml>，檢索日期：2021年4月1日。

[37] 韓召穎，趙倩，〈國際危機中的領導人決策行為分析—基於多元啟發理論視角〉，頁1-2。

與不確定可變因子的發生，這是站在國際安全層面的考量，一個領導者不得不有前置佈署的能力與遠見。

　　平實而言，德國政府在難民危機決策過程中，梅克爾的性格與其執政風格也是考評的因素，畢竟在危機預防階段中，一個決策者冷靜、理性、堅定與協調能力的性格，是有助於他做出適當的決策，進而採取有效惜施，才能避免讓矛盾升級為危機，防患於未然，將危機消弭於無形，古巴飛彈危機、1996台海危機就是個典範實例。反之，如果在該階段決策者不夠靈活，缺乏合理性及有效協商、對話與溝通，無法獲得國際的聲援，且無法從異中求同達成共識，因而錯失危機預防最佳時機，危害因子就會隨之接踵而來。即便是在危機發生之後，決策者若還能保持沉穩及時做出合理判斷，制訂合理的危機應對計畫，靈活採取危機管理措施，必能減少危機所帶來的損失。

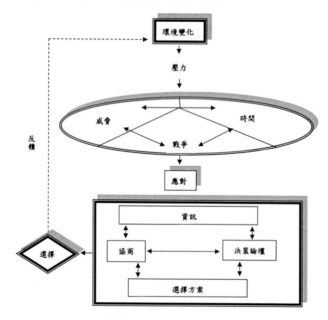

戰爭=軍事敵對行動的預估概率(非戰爭危機)
　　=軍事平衡逆轉的預估(內部戰爭危機)

圖3：危機中的國家行為模型

資料來源：Michale Brecher, "*Crises in Word Politics:Theory and Reality*" (Oxford/New York/
　　Seoul/Tokyo::pergamon Press, 1993), P.43.

有鑒於此，梅克爾政府的國際難民危機決策行為的分析，可以臚列以下幾個因素：一、國家的危機管理目標，二、危機決策組織模式，三、危機決策組織流程，四、決策者的執政風格和性格特徵。危機決策機制就是以擔負危機管理的國家政治機構為核心，在社會公共系統其他重要因素影響下，依照相應組織結構運作，從而對危機事態進行預警、應對和恢復的組織體系。而這個組織體系主要包括中樞指揮系統、支援與安全系統（危機的直接處置機構）和資訊管理系統。[38]（如圖4）

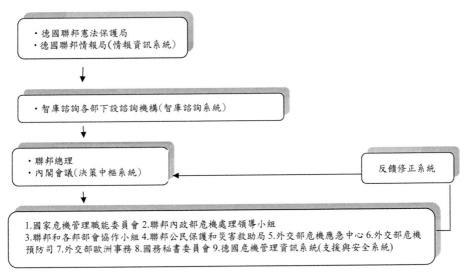

圖4：德國聯邦政府危機管理決策機制

資料來源：鄒露，《德國國際危機管理與實踐研究》（北京：社會文獻出版社。2020），頁170。

二、情報資訊系統之運作

戰略情報資訊是提供給決策者最重要的一個參考依據，從全方位的專業情報資訊來看，有為的決策者在面對危機時，其決策體系的神經系統，就是要依

---

[38] 楊洁勉，《后冷戰時期的中美關係—危機管理的理論和實踐》（上海：上海人民出版社，2004），頁45。

靠有價值且正確的情報資訊，來提供他最後的決心下達。這些都要依靠專業組織機構收集、彙整、分發和傳達的程序，即時提供給中樞決策機構作出完整的決策依據。梅克爾政府在國際難民危機管理中，所使用的情報資訊來源機構，主要包括對內的聯邦憲法保護局（Bundesamt für Verfassungsschutz, BfV），和對外的聯邦情報局（Bundesnachrichtendienst, BND）。

A.聯邦憲法保衛局（Bundesamt für Verfassungsschutz,BfV）一成立於1950年，其總部位於科隆（Köln）近郊的艾倫費爾德（Ehrenfeld），是對內情報資訊機構，主要負責國內安全情報工作，其隸屬聯邦內政部，各邦的分部都歸內政部門所管轄。現今有鑒於地緣政治和安全形勢變化，其部門結構及其主要的功能包括 ：[39]（見表2）

1. 職司搜集、分析、轉發、運用及國家安全的情報，並監控國外情報機關和激進團體在德國內部的行動，及極端和恐怖組織的行為；

2. 主掌政治和經濟機密的保密工作之外，並參與制定各項保密措施及涉密領域人員審查的工作；

3. 積極從事反間工作，主要任務包括偵防、特務、破壞、暗殺等活動；

表2　聯邦憲法保衛局組織結構

局長、副局長領導 首席技術官	
Z處	管理處
IT處	情報技術和特殊技術處
S處	內部安全、預防機密洩漏與破壞行動、專家審查、內部審查處
O處	監測處
C處	網路防衛處
1處	負責支持聯邦憲法保衛局正常工作
2處	負責反極端主義（左、右翼）
3處	負責反間諜，和保護國家安全

---

[39] 鄒露著，《德國國際危機管理與實踐研究》，頁170-171。

4處	負責反外國極端分子對安全所構成的威脅
5處	負責反伊斯蘭極端主義和恐怖主義
AfV處	法制處

資料來源：請參閱鄒露著，《德國國際危機管理與實踐研究》（北京：社會科學文獻出版社，2020），頁171。

B.德國聯邦情報局（Bundesnachrichtendienst, BND）一成立於1956年4月，職司搜集和分析外國的政治、軍事、科技和經濟情報，為聯邦政府和國防軍海外派兵提供有價值之情報，透過國外相關情報，也為政府在安全和外交領域決策上提供情報支援。[40]

三、危機管理原則及手段

德國在國際危機管理各階段中，主要是凸顯聯合國、歐盟、歐洲安全與合作組織、北約等國際組織為危機管理主體性之地位，避免觸及國家政治意識和利益的權衡，這一體系或許是可以理解為，德國聯邦政府對以往歷史的克制文化與慣例。很明顯的是在德國國際危機管理各階段及其危機管理手段、行為主體及原則下，可看出其整體階段及所使用之手段與引發的結果（見表3）。

表3 德國國際危機管理的各階段

階段	手段	主體	原則
★和平或無武裝衝突	・危機預防 ・裁軍和軍備控制 ・建設和平 ・小型武器控制 ・政治代表 ・制裁 ・特別代表 ・安全部門改革 ・選舉監督	・聯合國 ・歐盟 ・歐洲安全與合作組織	・衝突敏感性 ・本地所有權 ・人類安全 ・第1325號決議 ・保護平民 ・保護責任
★危機升級	・友好團隊 ・強制和平	・聯合國 ・歐盟	

---

[40] 鄒露著，《德國國際危機管理與實踐研究》，頁171。

階段	手段	主體	原則
	・維護和平 ・共同安全與防衛政策 ・衝突調停 ・制裁 ・危機快速反應部隊 ・特別代表	・北約	
★武裝衝突	・衝突管理 ・友好團隊 ・強制和平 ・維護和平 ・共同安全與防衛政策 ・人道主義援助 ・危機快速反應部隊 ・民事—軍事合作	・聯合國 ・歐盟 ・北約	
★危機善後	・鞏固和平 ・解除武裝、復原和重返社會 ・推進民主 ・友好小組 ・強制和平 ・維護和平 ・共同安全與防衛政策	・聯合國 ・歐盟 ・北約	
★危機善後	・國際法庭 ・小型武器控制 ・調解和衝突調停 ・政治代表 ・警察代表 ・安全部門改革 ・特別代表 ・和解和過度司法 ・選舉監督 ・經濟重建 ・民事—軍事合作	・聯合國 ・歐盟 ・歐洲安全與合作組織	

資料來源：Claudia Major, Tobias Pietz, Elisabeth Schondorf, Wanda Hummel,Toolbox *"Krisenmanagement-Von der zivilen Krisenpravention bis zum Peacebuilding: prinzipien, Akteure, Instrumente"* (Berlin: Stiftung Wissenschaft und Politik, Zentrum fur Internationale Friedensansatze,2013), p.6.

　　德國身為歐盟主導國，在國際危機中要發揮協調者的角色。因此，從難民危機管理手段中可以看出，梅克爾政府是想以民事危機管理手段為主，以遵循和平解決危機的原則，她不倡議強制使用軍警高壓抵制手段來面對難民潮，可以看出其在國際危機管理手段中，主要是在國際組織架構下，對於衝突敏感性之原則採用溫和手段，旨意在避免衝突且不違背人道主義為原則，進而更著重人道援助與危機後的重建。就聯邦政府實際狀況而言，則是透過法律與文化對話等方式，協助國家鞏固政治體制和社會結構，來預防衝突之發生。德國危機管理區分為四個階段—危機預防、危機升級、武裝衝突、危機善後，將危機發生後管理階段劃分為，危機升級與武裝衝突兩個階段。[41]（見圖5）

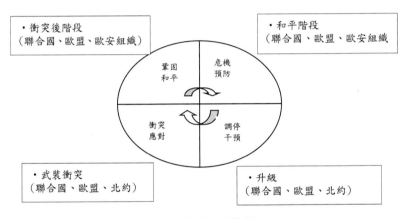

圖5　危機管理階段

資料來源：Claudia Major, Tobias Pietz, Elisabeth Schondorf,Wanda Hummel, *Toolbox Krisenmanagement-Von der zivilen Krisenpravention bis zum Peacebuilding: prinzipien, Akteure, Instrumente* (Berlin:Stiftung Wissenschaft und Politik, Zentrum fur Internationale Friedensansatze, 2013),p7.

　　當危機發生時，一般理性行為模式則會立即採取和平方式來應對危機，盡可能降低危機與傷害；最後當危機結束後，經過整體救助和救援手段，再進行協助當地進行危機恢復與重建。如今，一些衝突敏感性原則是德國緊急救援，發展合作和危機管理最高指導原則，該原則適用於德國聯邦外交部（Auswärtiges Amt）、德國聯邦經濟合作發展部（Bundesministerium für

---

[41] 鄒露著，《德國國際危機管理與實踐研究》(北京：社會科學文獻出版社，2020)，頁157。

wirtschaftliche Zusammenarbeit und Entwicklung）、德國世界饑餓救助組織（Deutsche Welthungerhilfe e. V.）等機制。（如表4）

2015年有上萬的難民湧入了歐洲，匈牙利和奧地利等國，陸續向梅克爾總理請求協助，梅克爾當下偏差動員理念（the mobilization of bias）的決策中，其是在所不辭讓新德國有一次救贖的機會，伸展二次戰後新德國的形象。42 因此，德國人民用掌聲歡迎接難民進入，這也顯見當時德國人民是支持梅克爾的「歡迎決策」，只是因為她顧此失彼的財政緊縮政策，為自己的國際形象與執政帶來不少的困境。43 梅克爾的功勞帶領了德國在歐洲重新崛起為主導的力量，但是她的疏失，卻是沒能讓德國人民充分認識到德國在國際政治意涵中產生了什麼困境？德國人民在認知上一直認為梅克爾的軟實力（Soft Power）建構了當前國際政治的實力，加上其行事作風謹慎沉穩，使得德國能更自如地重返歷史在歐洲統治的地位。但是難民危機，使得她的執政風格和現今強勢極右派反對者相互抗衡的景象，更使民族主義聲浪再次萌發，亦顯得她在決策意志上格外孤獨。

表4 德國國際危機管理實施機制

實施機制	任務	監控、情報機構	決策機構	執行機構	危機管理手段	戰略方針
國家危機預警機制	危機預測↓發出預警↓提供應急方案	1.諮詢智囊系統 2.聯邦憲法保衛局和聯邦情報局 3.聯邦和各州共同報告與形勢中心 4.德國緊急計劃資訊系統	決策中樞系統（聯邦內閣）	1.國家危機管理職能委員會 2.聯邦內政部危機處理領導小組 3.聯邦和各邦部協作小組 4.外交部危機預防司	1.裁軍和軍備控制 2.建設和平 3.小型武器控制 4.政治代表團 5.制裁 6.特別代表 7.安全部門改革 8.選舉監督	1.建立可信的國家結構（法治國家、民主、人權和安全） 2.推動和

---

42 BMBF, Die Ukraine als Partner in Bildung und Forschung (Offizieller Artikt 2016), <https://www.bmbf.de/de/die-ukraine-als-partner-in-bildung-und-forschung-3107.html> (2021/1/19)

43 Wenke Börnsen, "Merkel bei 'Anne Will'. 'Ich habe einen Plan'", *tagesschau.de*, 2015.10.8, <https://www.tagesschau.de/inland/merkel-anne-will-103.html> (2021/1/11).

實施機制	任務	監控、情報機構	決策機構	執行機構	危機管理手段	戰略方針
		5.衛星緊急系統 6.模組化預警系統 7.外交聯動體系		5.危機管理演習系統 6.危機管理培訓系統		平潛力(市民社會、媒體、文化與教育) 3.保障生存機會(經濟與社會、環境與資源)
國際危機管控機制	控制和止損			1.國家危機管理職能委員會 2.聯邦內政部危機處理領導小組 3.聯邦和各州部會協作小組 4.外交部危機應急中心 5.外交部危機預防司 6.危機管理演習系統 7.歐盟危機情況指導小組	1.友好團隊 2.強制和平 3.維護和平 4.共同安全與防衛政策行動 5.衝突調停 6.制裁 7.危機快速反應部隊 8.特別代表 9.民—軍事合作 10.人道主義援助	
國際危機反饋機制	危機後救援 ↓ 危機後恢復 ↓ 危機後重建	1.諮詢智囊系統 2.聯邦憲法保衛局和聯邦情報局 3.聯邦和各州共同報告與形勢中心 4.德國緊急計劃資訊系統 5.衛星緊急系統 6.外交聯動體系		1.外交部危機警中心 2.外交部危機預防司 3.危機管理演習系統 4.危機管理心理輔導 5.危機管理培訓系統	1.解除武裝、復原和重返社會 2.促進民主 3.友好小組 4.強制和平 5.維護和平 6.共同安全與防衛政策行動 7.國際法庭 8.小型武器控制 9.調解和衝突調停	1.建立可信的國家結構（法治國家、民主、人權和安全） 2.推動和平潛力（市民社會、媒體、

實施機制	任務	監控、情報機構	決策機構	執行機構	危機管理手段	戰略方針
	↓ 資訊反饋和改善機制				10.政治代表團 11.警察代表團 12.安全部門改革 13.特別代表 14.和解與過渡司法 15.選舉監督 16.經濟重建 17.民 ─軍事合作 18.人道主義援助 19.派遣專家、顧問	文化與教育） 3.保障生存機會（經濟與社會、環境與資源）

資料來源：鄒露著，《德國國際危機管理與實踐研究》（北京：社會科學文獻出版社，2020年），頁180-181。

伍、梅克爾難民政策的影響

　　難民危機與歐盟整合之間關係，其最主要的癥結在於難民議題與歐洲整合一體化產生了因果關係。但是如果要深究這些所謂「歐盟在瓦解的危險邊緣」，那麼烏克蘭危機，又何嘗不是所謂歐盟第二支柱「共同外交與安全」窒礙難行的衝突點，要能從一定之方針，取決於一致之行動，那更是困境重重。[44] 歐洲所面臨的難民問題，其實就如全球許多國家在面對移民與移工問題的縮影一樣，不是只有歐洲有此問題，在亞洲、美洲也都是如此，只是難民身份這其中錯綜複雜的因素，在有限時間內是非常難以處理。

---

[44] Hanns W. Maull, "Deutsche Außenpolitik—Verantwortung und Macht", Zeitschrift für Sicherheits- und Außenpolitik 8/1 (2015), pp. 213-237.

一、德國內部政治衝擊

回顧1953年以來，二戰後的德國經濟是處於恢復期，從文獻中顯現，從1953～1989年難民申請數量（90萬份）僅占1953~2014的23%。隨著兩德統一後，德國社會福利制度相對使經濟難民的吸引力劇增，一直到2014年受中東局勢不穩定之影響，難民申請數量更高（202834份），比2013年增長了59.7%，而2015年（425035）比2014年更增加了約110%。[45]

依照歐盟的政策規定，對第一波難民進入歐盟任何一個國家，該國就須負責接收、處理。儘管德國政府的相關政策法規訂立的很嚴謹，但因難民數量湧入龐大，帶來許多社會安全問題，對德國社會秩序造成影響，甚至也帶來宗教與文化的衝突，更影響到政治的穩定和社會安全層面的問題，尤以2016年法國國慶日發生恐怖攻擊事件，最後蔓延到德國境內，給德國人民與政府帶來新危機感。

德國的「開門政策」和大部分歐洲國家持反對態度形成對立，因此難民湧入歐洲，分裂成為兩條路徑，一是「德國路徑」，二是「維西格勒集團」（Visegrad Group）[46] 路徑，使得危機逐步內化為「東西矛盾」，造成東、西歐國家之間形成重大分歧。亦使得梅克爾政府在全國民調當中聲望直直下滑，最主要是她一直堅守拒絕設置難民人數上限，堅持推動國內和國際行動協助才能共渡難關，這是梅克爾決策之下難民危機管控的理念與手段，也是對梅克爾政府在歐盟危機管理能力上提出嚴峻的挑戰。[47] 職是之故，最後迫使梅克爾對自己的開放邊界政策的予以改變策略，這或許正是呼應了德國聯邦內政部長澤霍費爾（Horst Seehofer）的政策，從而結束了德國聯合政府避免了破裂的命運，

---

[45] 鄒露，《德國國際危機管理與實踐研究》（北京：社會文獻出版社，2020年），頁223。

[46] 維西格勒集團（Visegrad Group），或稱維西格勒集團四國（Visegrad Four，簡寫為V4），是由中歐的捷克、匈牙利、波蘭、斯洛伐克等4國所組成的政治及文化合作組織，以匈牙利城市維謝格拉德命名。其成員國都在2004年5月1日加入歐洲聯盟，亦為北約成員國。

[47] 〈跌跤的鐵娘子：梅克爾的領導危機〉，2016/09/29，張福昌。梅克爾其難民哲學定調為以下四點：我們辦得到（Wir schaffen das!）；人權無上限（Menschenrecht kennt keine Grenze）；堅守歐洲價值（Wir sollen Europäisches Wert halten）；尋找共同的歐洲辦法（Wir müssen eine gemeinsame Europäische Lösung schließen），<http://global.udn.com/global_vision/story/8663/1985411>，檢索日期：2020年12月3日。

也穩定全局作用，讓梅克爾得以迅速召集歐盟高峰會，並促使歐洲就新的移民和難民政策的架構達成一致。使歐洲政治一體化的進程，又再一次經歷嚴峻的考驗。[48] 從國際現實上看，各成員國是不可能一開始就就認同德國的決策，因此難民危機，反應了歐洲一體化還不夠完善，似有失靈之虞。從整體國家利益及決策來看，不可否認的是梅克爾太過於理想化，不僅是無法感動別人，也無法成就自己。

二、歐盟難民政策的檢討

梅克爾的難民政策是其在任期間最為關鍵的議題，這一決策弱化了她堅強的形象。也承受了「強行支配歐洲其他成員國，不顧友邦的利益」的指責。尤其是歐洲大陸經濟與難民雙重危機的南歐國家，致使2012年希臘因經濟危機，進而退出貨幣聯盟這事件一起併發。對於一個緊縮政策的策劃者而言，債務危機讓德國設下了關鍵的考驗，這些歐盟國家都對梅克爾政府的決策發出嚴厲地批評，在她領導下的德國，歐盟許多成員國都質疑，她究竟是一個要將歐洲利益視為自身利益的「歐洲的德國」（Europäsich-Deutschland），還是一個想要擁有「德國的歐洲」（Deutsch-Europa）。[49] 很多經濟學家都呼籲梅克爾要採取寬鬆的政策，來緩和歐盟區域經濟，但終究是事與願違，梅克爾卻反道其行，堅持嚴格把關，強調有必要透過財政緊縮來保護歐洲貨幣。這些決策思維上就花費更長的時間去建立歐洲整體共識。

當時梅克爾政府政策走向是如何，茲檢討分析有下列幾點：[50]

第一、梅克爾初始的難民危機管理理念出發點是值得肯定，但卻忽略歐盟某些國家難民接收的能力與負擔，對各種利益分配也不夠均衡，再加上各國

[48] 「難民政策遇瓶頸？僅3成德國人贊同梅克爾」，《自由電子報》，2015.10.15，<http://news.ltn.com.tw/news/world/breakingnews/1476051>，檢索日期：2020年12月10日。

[49] 邱海穎，《德國國家身份構建：安格拉‧梅克爾演講辭的批評話語分析2005-2013》，（北京：光明日報出版社，2017），頁69。梅克爾在其演講中，曾提出「統一的德國」、「歐洲的德國」、「世界的德國」的國家形象。「統一的德國」強調了德國國內的穩定和諧，有利於安撫民心、鼓舞士氣，增強國家凝聚力。「歐洲的德國」強調德國的屬性，代表著歐洲的利益突出德國在歐盟的核心地位。「世界的德國」顯示了德國開放、包容的態度，昭示德國有能力為世界和平與安全助力。

[50] 鄒露，《德國國際危機管理與實踐研究》，（北京：社會文獻出版社，2020年），頁247-248。

歷史文化、宗教背景、政治型態、經濟發展懸殊等情況，讓一些歐盟成員國家帶來沉重的壓力。第二、梅克爾肩負著各方國際的壓力，在危機管理強調多方合作，作為歐盟創始國的立場而言，梅克爾一直尋求與各成員國能達成共識來應對危機。第三、梅克爾政府想為勞動力增長注入新的力道，但對大部分歐洲國家而言，難民危機讓歐洲社會帶來沉重損失和後續融入的負擔，才是難以接受的因子。第四、由於社會層次加劇，民粹式的民族主義的激化，致使2017年歐洲各國政治發生了極大的變動，右翼民粹主義政黨趁勢崛起，在社會中支持率不斷的上升。第五、在難民政策方面，梅克爾在價值觀方面過分的執著與理想化，對人權、自由等價值觀很堅持，即便是在國際人道主義上得到了讚譽，但是在現實當中，卻被恐怖攻擊事件面臨重重危機。第六、歐盟的難民政策法令尚不完善，在世界局勢不斷變化之下，現存的國際公約早已不符現實環境所用，致使歐洲各國的難民政策分歧之大，甚難達成一致共識。

　　二戰後，德國有意識地融入西方民主和自由的價值體系，對參與建構全球化、文明化、法制化，經濟社會重建乃至發展，都非常積極支持國際組織的改革與發展，也惟有在價值觀、原則、文明力量（Zivilmacht）或意識型態等層面上融入歐洲與德意志民族之間實現新的平衡（balance），才能獲得歐洲人民的認同，成為一個值得信賴的國家。因此，如何保持和維護德國的價值觀，才是成為德國國家利益不可分割的一部分。德國著名的政治學家毛爾（Hanns W. Maull）曾述說，在全球化的國際體系中，各種文明之間的瞭解與互動不斷地加深，而歐洲區域以及德國都以平等、人權、民主、自由、博愛為主的價值觀，德國作為一個穩定的民主國家，在經濟和福利上，都具備了文明力量的物質基礎，自由的力量，思想的領地，世界的責任，而未來的力量，是應共同取決於這一點。[51] 梅克爾政府在敞開胸懷迎接難民之時，其福利制度卻是難以招架，也無法停止削減，所謂平等、公正、自由等價值觀，在現實的危機管理中

---

[51] Hanns W. Maull, "Deutschland als Zivilmacht", in Siegmar Schmidt, Gunther Hellmann,Reinhaid Wolf, Handbuch zur deutschen Außenpolitik (Wiesbaden:VS Verlag für Sozialwissenschaft/GWV Fachverlage GmbH, 2007), pp.73- 84。原文是Die kraft der Freiheit,das Land der Ideen,Verantwortung in der Welt-das machte Deutschlands Stärke vor 16 Jahren aus und das macht sie heute aus.Davon hängt unsere gemeinsame Zukunft ab.

帶來了嚴峻地挑戰。因此，可以肯定的說，難民危機對德國政府而言是一場價值觀的危機。[52]

三、未來難民政策的發展

德國外交政策考慮到三個層面：一、對於德國社會；二、對於歐盟；三、對於全球體系。德國政府在難民危機管理上表現的非常地包容，這是出自於梅克爾政府對價值觀的一種堅持，然而這必須建立在國家利益和價值觀問題上，經過國際合作解決國際問題是符合德國國家利益，但前提是要有一定的經濟基礎之上。隨著國際難民數量的增加，不論是在政治、經濟、社會各層面的壓力不斷增大，梅克爾才警覺到要收緊這一政策，這也就顯示了在價值觀與國家利益發生衝突時，決策者最終仍會選擇國家利益。可見梅克爾長期以來奉行的價值觀外交，也是一種外交手段，更是一種實現戰略目標的手段之一而已。[53] 難民危機已經威脅到歐盟整體之團結，德國及其他國家所提供的支援已經達到極限，已無法再接收難民的能力，必須採取必要措施調控難民湧入速度。因此2015年德國經濟部長與外交部長曾為了要強化歐洲難民政策，聯合提出十點計劃：[54]

第一、歐盟制定統一標準，保障難民在接收國得到人道的安置。

第二、有統一的庇護標準，以保障難民護地位。未來也能會促成歐盟難民政策的結合。

第三、有一個公平客觀的接收難民分配方案。

第四、有一致的邊境管理。

---

[52] Auswärtiges Amt, Botschafterkonferenz 2016: "Verantwortung, Interessen, Instrumente" (Offizieller Artikel, 2016), < https://www.auswaertiges-amt.de/de/aamt/160829-boko/282954>，(2019/12/2)

[53] STERN PLUS,"Was geschah heute vor vier Jahren? ZDF-Doku rekonstruiert Merkels Flüchtlingsentscheidung", 2015/9/4, <https://www.stern.de/kultur/tv/angela-merkel-und-die-fluechtlinge--zdf-doku-ueber-den-4--september-2015-88 84392.html> (2020/12/1)

[54] Auswärtiges Amt. Zehn-Punkte-Plan für eurpäische Flüchtlingspolitik (Offizieller Artikel, 2015), <http//www.auswaertiges-amt.de/DE/aussenpolitik/Globale-Fragen/Fluechlinge/Aktuelles/150823_BM_Gabriel_FAS_node.html> (2020/12/2)

第五、有效向負擔過重的歐盟成員國提供協助。

第六、強調人道主義的價值，不能對危及生命的難民置之不顧。

第七、將難民遣送回國是處理難民與來源國的源頭，德國會向這些國家提供技術和財政支持。

第八、在歐盟組織內達成安全共識。

第九、德國需要完善的移民法來減輕庇護負擔。

第十、積極鞏固國家體制、遏制暴力和內戰，改善經濟和福利條件，消除中東和非洲國家難民的根源。

　　德國的價值觀—民主、平等、自由，為其贏得了國際社會的認同，但也成為難民危機中的最大障礙。梅克爾在宣揚了多年的自由、平等、人道主義等價值觀之後，德國政府卻無法拒絕這些難民的湧入以及控制難民的數量。[55] 除此之外，許多難民信奉的是伊斯蘭教，這與西方社會的文化認同與價值觀是有很大的差異。因此，要識別難民身份，且要促進難民融入德國社會的確是一個難題，而湧入的難民更可能會由於宗教、信仰、價值觀等差異而易被極端組織或勢力所利用，成為社會不穩定的因素，他們能否接受政教分離、法制社會、宗教自由等思想，對具有伊斯蘭教背景的難民而言，想要融入德國社會更是一項艱困的挑戰。[56]

　　從以上幾個面向來觀察，梅克爾政府的難民政策或許是對歐盟一體化最期待，但國際關係瞬息萬變，任何不可抗拒之因素以及不確定因素也都存在，更是對現階段歐洲經濟一體化和政治一體化發展不平衡的另一種詮釋。2017年2月，梅克爾政府才開始推行「Starthilfe Plus」計劃[57]，以金錢資助鼓勵尋求庇

---

[55] International Organization for Migration, "Missing Migrants Project. Mediterranean Update",2015.10.27, <http://missingmigrants.iom.int;Anna Reimann>, "Asyl und Einwanderung: Fakten zur Flüchtlingskrise–endlich verständlich", Spiegel Online, 2015.10.8, <http://www.spiegel.de/politik/deutschland/a-1030320.html.> (2020/12/12)

[56] 約翰‧米爾斯海默（John J.Mearsheimer）著，李澤澤譯，《大幻想：自由主義之夢與國際現實》，頁31。

[57] <https://www.dw.com/de/mit-starthilfe-plus-zur%C3%BCck-in-die-heimat/a-37361288>（2021/4/12）原

護者自願返回母國，每名自願撤回庇護申請的人最多可獲發1200歐元，符合若干指定條件者有額外資助。就目前而言，德國聯邦政府與各邦難民接收計畫已造成一定的財政負擔，如何促進難民融入德國社會以及有效進入勞動市場，這恐怕是繼任梅克爾政府之後持續要面臨的重大課題。[58]

陸、結語

很務實地說，梅克爾在內政上並非十全十美，2015年的難民危機發生初期，梅克爾政府對於難民危機管理上的態度是較為被動的，此階段並沒有將難民危機管控制定一個明確的應對方案，這當然可歸結到梅克爾總理的執政風格與態度，她一向是在冷靜觀察之後再因勢利導的採取行動。然而，在此難民危機質疑聲中，她依然是堅持地開放難民這項棘手的政策，最後隨著難民人數激增後，卻給德國政治、經濟、社會安全、文化帶來了整體的影響。從政治角度來看，難民問題的確是影響到一些歐洲國家的內政和政策走向；從經濟角度來看，也干擾了歐洲難民接收國家的經濟；從勞動力市場和國家社會安全角度來看，更加劇了歐洲接收國的失業現象，對社會安全構成了一定程度的威脅；從社會文化角度來看，難民的文化往往與難民接收國的主流文化產生衝突和撞擊，給社會融合與社會一體化帶來了很大的困難；最後從國際關係角度來看，這不僅僅影響著歐洲難民接收國家之間的關係，也影響著歐洲接收國家與難民轉向國家、難民輸出國家之間的國際關係。

梅克爾在沒有完善危機管理的情況下，就讓將120萬名尋求庇護者進入德國，因而造成德國社會與其他歐盟國家的反彈，也促成德國另類選擇黨（AfD）的崛起；也讓德國選民選擇離開兩大傳統主流政黨，更讓德國人民對兩大政黨政治局勢的界定，感到極度的不滿。一個有智慧的決策者，必須在短

文：＂Die Bundesregierung will Flüchtlinge, die sich freiwillig und frühzeitig für eine Rückkehr in ihre Heimat entscheiden, finanziell extra belohnen.＂ Starthilfe Plus＂ heißt das Programm, das an diesem Mittwoch anläuft.聯邦內政部長Thomas deMaizière (CDU)認為，在新的返回支持計劃下，尋求庇護者自願離境的人數大大增加。對於沒有希望留在德國的難民來說，自願離境比驅逐出境是更好的方法。

[58] 〈Flüchtlinge und Schutzsuchende in Deutschland〉,<https://www.infomigrants.net/en/tag/starthilfe%20 plus/ 2019/11/21> (2021/01/22).

期內化解國內憂懼情勢及具有迫切情況之下，更須理解國際政經與歷史脈絡，若以宏觀角度來剖析這些看似不相關聯，實則息息相關的難民問題，梅克爾所領導之決策機制與危機管理之能力尚待加強，如何在國際社會以及歐盟體系中，對成員國妥善尋求共識，取決一致之行動。實質而言，難民和歐盟區之危機都非單一國家的問題，唯有透過「雙邊」、「區域」、「全球層次」的合作才能有效處理。我們可以做得到（Wir schaffen das），相信歐洲人民普遍是可以接受，但是在面對國際與國內即刻政治壓力時，她後面所說的這句話，德國是一個強大的國家（Deutschland ist ein starkes Land），卻讓其它成員國為之卻步。梅克爾政府若僅以自己國家利益為優先，卻還便宜行事地走回民族主義，枉顧歐盟共同發展之利益，絕非適切的政策選項。換言之，當今難民危機的問題固然令人憂心，但基於歐盟以往傳統都是以協商、談判、溝通等方式來處理爭端，難民問題必須不斷折衝、談判，才不至於中斷歐洲整合的前景，以及撕裂歐洲人民對自由與人道價值的信仰，即使英國已於2021年脫離歐盟，以目前歐盟整合前景而言，這或許是危機，也是轉機。

目前，歐盟幾個經濟狀況較好的國家，都已在敘利亞內戰日趨平靜之時，以相當之遣返資金護送難民安然返回家鄉。於此同時，歐盟主導的德、法兩國，也聯合多數成員國，透過國際合作與協商，儘早促使戰事平息。展望未來，歐洲難民的問題勢必無法在短時間內得到曙光與解決，現又因為全球Covid19的疫情肆虐，更讓歐洲非傳統安全又面臨嚴峻的考驗。梅克爾在這政治生涯落幕前，真正是渡過了最艱困的一年，但是她為德國與歐洲社會所貢獻15年的心血與成果，國際社會是無法對她一筆勾銷的。

Debates and Choices for EU's Taiwan Policy

Pei-Shan Kao [*]

I. Introduction

Since Joe Biden took office on January 20, 2021, the relationship between the United States and China still has not progressed towards a positive direction as some scholars predicted. Conversely, US-China relations have fallen into and maintained a "new cold-war" since the Trump administration. If one can still remember, during the Trump era, on its first *National Security Strategy* (NSS) that was released on December 18, 2017, the Administration claims that "the U.S. must operate in a global strategic context, wherein adversaries often compromise American interests using nonmilitary tools···some actors, particularly Russia and China, have exploited international institutions in a manner that has compromised American economic security···the U.S. military needs significant investment to maintain superiority against adversaries such as China and Russia".[1]

The second important document, National Defense Strategy, was issued on January 19, 2018, on the first page of the NDS, it argues that "the United States must bolster its competitive military advantage" to face the threats posed by China and Russia, and that "inter-state strategic competition, not terrorism, is now the primary concern in U.S. national security".[2] That is, to maintain the U.S. strategic competitive edge relative to China and Russia is the US top priority. The two document both

[*] PhD in Government, University of Essex, UK; Associate Professor, Central Police University
[1] Kathleen J. McInnis, "The 2017 National Security Strategy: Issues for Congress," *CRS Insight*, December 19, 2017, pp. 2-3. 〈http:// https://fas.org/sgp/crs/natsec/IN10842.pdf〉.
[2] Kathleen J. McInnis, "The 2018 National Defense Strategy," *CRS Insight,* February 5, 2018, p. 1. 〈https://fas.org/sgp/crs/natsec/IN10855.pdf〉.

deliver a notion of "peace through strength", and ask for the "joint force". The third important paper released by the Trump administration was the *Nuclear Posture Review* that was issued on February 2, 2018, it describes that "China is developing capabilities to counter U.S. power projection operations in the region and to deny the United States the capability and freedom of action to protect U.S., allied, and partner interests...Direct military conflict between China and the United States would have the potential for nuclear escalation."[3] Therefore, the United States is ready to firmly respond to Chinese non-nuclear or nuclear aggression although it still will continue to seek a meaningful dialogue with China on their respective nuclear policies.

On March 3, 2021, the Biden Administration released an *Interim National Security Strategic Guidance*, the document argues that revitalizing the US core strengths is necessary but not sufficient, and that the United States must be prepared to "answer Beijing's challenge."[4] The INSSG explains that the United States is facing international strategic challenges from a "pacing threat", China, and that this will "require a return to coordinated, if not collective, international action,"[5] even though allies and partners have different interests with those of the United States. On US foreign deployment and design, Europe always has played a very important role just like Asia. If the United States is looking for a coordinated international action, it must pay much more attention on its allies in the world to enhance their relations. The United States has attempted to influence the European foreign policy; however, the 27 member countries of the EU not only have divergent opinions on their policies to the United States and China, but also have their own thoughts and considerations to develop relations with Taiwan. To reach its aims, that is to say, to make a grand "anti-China" alliance, the United States has sent many high-level officials and

---

[3] "Nuclear Posture Review," *US Department of Defense*, February 2, 2018, p. 31. 〈https://media.defense.gov/2018/Feb/02/2001872886/-1/-1/1/2018-NUCLEAR-POSTURE-REVIEW-FINAL-REPORT.PDF〉.

[4] Kathleen J. McInnis, "The Interim National Security Strategic Guidance," *CRS Insight,* March 29, 2021, p. 2. 〈https://crsreports.congress.gov/product/pdf/IF/IF11798〉.

[5] Ibid.

delegates to Europe and Asia since March 2021. On the other hand, China's "Wolf-Warrior Diplomacy" is viewed by the world as aggressive and threatening, "the China threat" and "containing China" have quickly spreading not only inside Asia but also in Europe. Whether from the perspective of US-China new cold war or US grand alliance to counter China, the Taiwan issue is always the hot spot under great powers' competition. This paper therefore wants to explore the EU-Taiwan relations and also the influence of the United States to EU's foreign policy. It will then examine the debates on EU's Taiwan policy, including the official standpoint so to make the conclusion.

II. The Development of EU-Taiwan Relations

According to the European Commission, Taiwan was the EU's 15th trading partner in 2018 while the EU is Taiwan's fourth trade partner. The major products exported from the EU to Taiwan are semi-finished products, machinery and transport equipment, etc. The most traded commodities between the two are office telecommunications equipment, machinery, transport equipment and chemicals. The bilateral trade reached to 50.5 billion Euros in 2019. The detailed developments of EU-Taiwan relations can be observed from annual brochures published by the European Economic and Trade Office (EETO). For instance, according to the "2020 EU-Taiwan Relations Brochure," in 2019, the EU still is Taiwan's fourth trading partner after China, the USA, and Japan; the bilateral trade reached a total of 50.5 billion Euros as Table 1 showed. If one compares it with the previous year, the annual growth rate is 9.1%. In Asia, Taiwan is the EU's 5th largest trading partner, after China, Japan, South Korea and India. Taiwan ranked the EU's 12th largest trading partner. The EU has trade deficits with Taiwan but they have gradually decreased. Taiwan mainly exported manufacture products such as Office and telecommunication equipment to the EU, this occupied 94.6% of the bilateral trade in 2019. [6] The EU exports to Taiwan focused on machinery and transport equipment included office and telecommunication equipment, other machinery, non-electrical machinery and transport equipment, etc.[7] On European member states, Germany, Netherland, France, Italy, Belgium and Spain are Taiwan's major trading partners in good.

Table 1: EU-Taiwan Trade in Goods (2010-2019)

Unit: billion (Euro)

Year	EU's exports	EU's imports	Total trade	Annual Growth Rate(%)	Trade Balance
2010	13.6	20.8	34.3	39.9	-7.2
2011	14.7	20.4	35.2	2.5	-5.7
2012	14.5	18.7	33.2	-5.6	-4.2
2013	15.1	18.1	33.2	0.1	-3.0
2014	15.7	19.3	34.9	5.1	-3.6
2015	16.8	21.2	38.0	8.9	-4.4
2016	17.6	23.0	40.6	6.7	-5.3
2017	19.4	25.3	44.7	10.1	-6.0
2018	20.1	26.2	46.3	3.7	-6.1
2019	**23.6**	**27.0**	**50.5**	**9.1**	**-3.4**

Source: "The 2020 EU-Taiwan Relations Brochure", *European Economic and Trade Office*, September 10, 2020, p.12. 〈https://eeas.europa.eu/sites/default/files/2020_eu-taiwan_relations.pdf〉.

---

[6] "The 2020 EU-Taiwan Relations Brochure", *European Economic and Trade Office*, September 10, 2020, p.16. 〈https://eeas.europa.eu/sites/default/files/2020_eu-taiwan_relations.pdf〉.
[7] Ibid.

Table 2: EU-Taiwan Trade in Services (2011-2018)

Unit: billion (Euro)

Year	EU's exports	EU's imports	Total	Balance
2011	3.4	2.1	5.5	1.3
2012	3.7	2.4	6.1	1.3
2013	3.5	2.8	6.3	0.6
2014	3.8	2.8	6.6	1.0
2015	4.0	3.0	7.0	1.0
2016	4.1	3.0	7.1	1.0
2017	4.3	3.2	7.5	1.0
2018	**4.6**	**3.5**	**8.1**	**1.0**

Source: "The 2020 EU-Taiwan Relations Brochure", *European Economic and Trade Office*, September 10, 2020, p.21. 〈https://eeas.europa.eu/sites/default/files/2020_eu-taiwan_relations.pdf〉.

On trade in services between the EU and Taiwan, in 2018, the total amount was 8.1 billion Euros as Table 2 showed, Taiwan ranked the EU's 27th trading partner around the world. However, this just accounted for 0.5% of the EU's total trade in services, there is still space for growth. The major services Taiwan exported to the EU are sea transportation and business services while European exports to Taiwan mainly focus on transportation, business, travel and financial services. On foreign direct investment, the EU remains Taiwan's largest Foreign Direct investor who has already accumulated up to 48.6 billion Euros in Taiwan in 2019; occupied 31% of Taiwan's inward FDI.[8] Over half of the European investment (57.3%) to Taiwan focused on manufacturing sectors such as electronic parts and components manufacturing. On EU's FDI to Taiwan, Netherland (64.6%), Denmark (19.4%), Germany (13.4%), Luxembourg (1.2%) and France (0.9%) are the major investors.[9]

---

[8] Ming-Yen Tsai, "Why Deepening EU-Taiwan Economic Ties Matter," *The Diplomat,* November 12, 2020. 〈https://thediplomat.com/2020/11/why-deepening-eu-taiwan-economic-ties-matter/〉.

[9] "The 2020 EU-Taiwan Relations Brochure", *European Economic and Trade Office,* pp. 29-30.

However, Taiwanese investments in the EU still has abundant room for growth as it only accounted for 6.1% of Taiwan's outbound investment; in 2019, the approved Taiwan investment to the EU was $0.7 billion. Most of Taiwanese investment in Europe flew to Luxembourg (88.9%), Germany (4.5%), Denmark (1.9%), Poland (1.8%), and Austria (1.5%); and mainly covered financial and insurance industry, amounting to 89.1%; followed by manufacturing (9.4%), wholesale and retail sector (1.2%). [10]

The two also have close cultural, educational and social contacts and exchanges. For instance, in 2019, there are 6,430 European students studying in Taiwan, most of them are from France (33.8%), Germany (22.2%), Italy (6.1%), Spain (5.6%), and Holland (5.5%).[11] Seven Taiwanese universities established in 2009 "the European Union Centre in Taiwan" to promote European studies and hold academic and social activities in Taiwan. Similarly, there are over 7,300 Taiwanese students staying in Europe, and most of them study in Germany, France, Poland, Spain and Netherland.[12] The EU also designs and create much research and innovation founding and programs for international cooperation with Taiwan such as Horizon 2020 programme, Marie Sklodowska-Curie Actions, European Research Council, etc. On cultural exchanges, the EU and Taiwan hold many activities and for such as the Taiwan-European Film Festival and Speaking Dating. Taiwan is also a popular travel destination for Europeans, for example, in 2020, 59,512 people from Europe visited Taiwan, including overseas Chinese.[13] Most of European visitors come from the United Kingdom, Germany and France. [14] Taiwanese visitors like to travel to Europe as well; in 2020, 59,773 Taiwanese visited Europe; France, Germany, Netherland, and Austria

---

[10] Ibid.

[11] bid, p.66.

[12] Ibid, p.65.

[13] "Annual Statistical Report on Tourism 2020", *Taiwanese Tourism Bureau*, March 24, 2021, p.5. 〈https://admin.taiwan.net.tw/Handlers/FileHandler.ashx?fid=00caeef4-ddf9-4d0a-8455-db6d2d279b20&type=4&no=1〉.

[14] Ibid.

are the most popular destinations.[15] Although the EU follows the "One-China policy" without any diplomatic relations with Taiwan, it still supports to develop relations with Taiwan and Taiwan's meaningful participation in international fora. The EU and Taiwan not just have many structured dialogues but also have multidimensional consultation regimes and mechanism on issues such as intellectual property rights, trade, and climate changes, etc.

III. Debates on EU's Taiwan Policy

i. EU-Taiwan Relations under "One China Policy"

The development of EU-Taiwan relations just like US-China relations always faces the pressure from China, that is to say, the "One China Policy". Therefore, inside the Europe, some countries support to engage with China so to keep distance with Taiwan while some countries do promote their relations with Taiwan. For example, in 2003, Chris Patten, then External Affairs Commissioner was interviewed by a media and expressed his views on EU-Taiwan relations. He argued that "One China Policy" could accommodate EU relations with Taiwan although he also explained that this did not mean the EU did not want to develop relations with Taiwan, who was an important economic partner for the EU. [16] He claimed that Taiwan not just had great economic performance but also was a "like-minded" partner of the EU; therefore, the EU still wanted to work closely with Taiwan. On the tension between the Taiwan Strait, he suggested that the two sides should resolve the disputes on their own by resuming dialogues and putting aside political positions.

On the report "Mapping Europe-China Relations: A Bottom-Up Approach"

---

[15] Ibid, p.27.

[16] About his interview, please check *Politico*, "'One China' policy can still accommodate EU relation with Taiwan, says Patten," January 22, 2003. 〈https://www.politico.eu/article/one-china-policy-can-still-accommodate-eu-relation-with-taiwan-says-patten/〉.

edited by Mikko Huotari, Miguel Otero-Iglesias, John Seaman and Alice Ekman,[17] published in 2015, the authors point out that since the rise of China, the course of international affairs has been shaped and shook, Europe-China relations therefore are more relevant and much more complex. The authors hence anaylsed and examined these complexities from the various state-level bilateral relationships, that is to say, a bottom-up approach, hoping to provide policy recommendations to better coordinate the EU's foreign policy towards China. In this report, the authors used the cases of Czech, Finland, France, Germany, Greece, Hungary, Italy, Netherlands, Poland, Portugal, Romania, Spain, Sweden and the UK to explore their relations with China. The authors summarise that as the EU member states' economic interdependence with China has sharply increased, the bilateral political relations therefore have gained in maturity and depth. There are much more high-level exchanges between the two, the context of EU-China relations has dramatically changed over the past five years. China's interests in Europe not just have expanded geographically to central, eastern and southern Europe but also with substantial investment. The EU "is encountering a much more proactive China on the diplomatic front and the contours of the relationship are increasingly designed in Beijing".[18] For the authors, since the EU member states did not formulate or coordinate a common China policy, their relations with China are asymmetrical so their relative influence over Beijing are waning. They are divided and even compete with themselves. The authors consider that this kind of disunity and competition stem exactly from the EU itself, not from China; and this has provided China with 28 'gateways to Europe'. That is why the relationships between China and individual European partners are of difference. This gives China a good opportunity to follow its own different and flexible foreign policy approach to deal with Europe.

---

[17] Mikko Huotari, Miguel Otero-Iglesias, John Seaman and Alice Ekman (eds), "Mapping Europe-China Relations: A Bottom-Up Approach," European Think-Tank Network on China, 2015, pp.1-88. 〈https://www.ifri.org/sites/default/files/atoms/files/etnc_web_final_1-1.pdf〉.

[18] Ibid, p. 5.

For Mikko Huotari (et al.), most of European national strategies towards China are controlled and dominated in terms of economics. Under this circumstance, many European states hence face dilemma and sway between political concerns such as China's democracy and human rights performance and potential economic and commercial interests. Each country has its own consideration and strategy to deal with China. For instance, in the UK, the debate about China is much more moderate and less nervous compared with the United States. On Sweden-China relations, environmental technology is the major cooperation section between the two, advanced manufacturing, life science, information technology and urbanisation are also included. For Spain, it is very calm to deal with its relations with China due to its business interests in China. On Sino-Romanian relations, the report explains that the special relations can be traced back to the 1950s when the two cooperated to oppose the Soviet Union. However, during the 1990s, due to the process of Romania's democratisation and cooperation with the West, the relationship between Romania and China became marginal. Since China opened to the West in the mid-2000s, the bilateral relations were closer. The report argues that it is unlikely to have significant development and progress in the short term time if there is no major change in Romania's economic situation. On the relations between Portugal and China, the authors explain that due to its important geographical situation, China can enhance its cooperation and exchanges with the EU, Africa and Latin America by means of Portugal. Therefore, Portugal is increasingly attractive for Chinese investment and trade. Poland also has special and close relations with China although their relations were insignificant when Poland struggled to join the NATO and the EU like Romania. But Poland reinvigorated its relations with China since the two signed a strategic partnership in 2011; the bilateral relationship therefore was comprehensive and multidimensional. Netherland is the first country Xi Jinping visited when he became China's President; Sino-Dutch relations therefore have been intensified over the past few years although some obstacles have ever happened between the two such as Dutch's submarines sales to Taiwan in the early 1980s and the different political-economic systems they are.

In 2004, Italy and China established a comprehensive strategic partnership in 2004 and the two celebrated in 2020 the 50th anniversary of the establishment of formal diplomatic relations. However, similarly, the difference of values still remains a problematic issue between the two. Just like Romania and Poland, Hungary also has developed good relations with China in the past few decades and has received many Chinese investments. For China, it is more interested in infrastructure building but many EU states prefer positive economic and trade cooperation to create and bring more job opportunities. Being a logistics hub for China, and its own fiscal and economic predicament, Greece considers China as a major partner although its understanding to China's strategy and intentions is still very low. Inside Germany, there are still many debates on the priority of German's foreign policy, that is to say, whether it should focus on its relations with China or a coordinated EU's China policy. For France, China still is the core of France's Asia policy, particularly on economic cooperation while human rights remain a major concern of France's foreign policy. Similar situation happened on Finland-China, Czech-China relations when the Finnish and Czech business and political sectors support to deepen their economic cooperation with China, people still have doubts and concerns on China's human rights issues. In sum, since the early 2000s, China has enhanced and engaged with Central and Eastern Europe (CEE) at the bilateral and multilateral levels. They therefore established the China-Central and Eastern Europe Forum (the so-called 16+1 format). On the other hand, China also promoted its relations with the northern, southern and western European countries.

On James Lee's article, "The Taiwan Question in US-China Relations and its Implications for the European Union," published in 2019, he attempts to provide a point of reference for the EU's Taiwan Policy. By means of examining the US position under the "One China Policy" from a historical perspective, he tries to clarify the strategic and political problems of that policy. He argues that whether the policy should change depends on whether the policy can adapt to the new strategic and political conditions after the rise of China. Lee explained that the Taiwan question is a long-standing dispute not just in Chinese politics but also in East Asia politics.[19] Although Taiwan enjoys a prominent position in world economy and boasts a standard of living, the military threat from China has never been resolved; conversely, it has been intensified greatly since China has expanded its economic and military capabilities. The predicament Taiwan faced has attracted the attention of the world, the European Parliament therefore issued many statements calling for dialogue between Taiwan and China to avoid escalating tensions. For Europe, it is cautious of the rise of China white China also represents a potential market to the European countries; that is why the EU must weigh the risks and advantages on its relations with China. Lee explained the Taiwan-China relationship from the historical background and discussed current issues between the two, including the US'S Taiwan policy and Taiwan's foreign relations, he therefore concluded some implications for the EU's Taiwan Policy. He argued that due to geographic distance, Taiwan at times has been considered as a distant concern for Europe. But Europe is very important for Taiwan as European opinion represents world opinion, for Lee, the United States and the EU should maintain a natural position on Taiwan's sovereignty and pay attention to specific words such as 'the Taiwanese government' or 'the Taiwanese people' to avoid misunderstanding and distrust.

Maaike Okano-Heijmans on her paper, "A Trade Diplomacy Triangle? Cross-Strait Relations and EU-China/EU-Taiwan Relations in the Past Year," published in 2016,[20] examined the developments in EU-China-Taiwan relations from the

---

[19] See Jame Lee, "The Taiwan Question in US-China Relations and its Implications for the European Union," European Union Insitute Working Papers, January 2019, pp. 1-12.

[20] Maaike Okano-Heijmans, "A Trade Diplomacy Triangle? Cross-Strait Relations and EU-China/EU-Taiwan

perspective of trade diplomacy. That is to say, she wants to explore how cross-Strait relations affect EU-China and EU-Taiwan relations and what the EU's main aims and considerations in its trade talks with the two, respectively. She claims that the EU is making efforts to develop a coordinated, long-term perspective on trade and economic diplomacy to "ensure a level-playing field for European companies; to ensure reciprocity in economic relations, and to uphold standards and values that based on its own lessons learned,"[21] Asian countries therefore play a very important role in the EU' s trade and economic diplomacy. Being an important trade partner of China and Taiwan, the EU certainly is a relevant player. For Okano-Heijmans, although the EU confirms its "One China Policy", it still could contribute to the stability of China-Taiwan relations by being a partner in China's economic reform and by supporting Taiwan's participation in international organisations. To conclude, these papers and articles all indicated the EU' s position on its relations with China and Taiwan. For the EU member states, although China signifies a huge economic market with many trade and business benefits and interests for European enterprises, they are still very concerned about China's human rights performance and arbitrary political system. Even under its confirmation of the "One China Policy", the EU still is most eager to advance its relations with Taiwan, a mind-liked partner.

ii. Time for a Positive Policy on Taiwan

Although the EU does not have formal diplomatic ties with Taiwan, the two still have many close economic and trade contacts and exchanges; in addition, both of them strongly support freedom, the rule of law, democracy, and human rights. Due to the rise of China, and the practice of its aggressive diplomacy, more and more

---

Relations in the Past Year," paper presented at the 13th Annual Conference on "The Taiwan Issue in China-Europe Relations," October 9-11, 2016. 〈https://www.swp-berlin.org/fileadmin/contents/products/projekt_papiere/Taiwan2ndTrack_Maaike_Okano-Heijmans_2016_01.pdf〉.

[21] Maaike Okano-Heijmans, "A Trade Diplomacy Triangle? Cross-Strait Relations and EU-China/EU-Taiwan Relations in the Past Year," p7.

European countries, politicians and scholars call on the EU to deepen economic ties with Taiwan and pay more attention to Taiwan. For example, on her paper, "Time for a Positive EU Policy on Taiwan," Dr. Zsuzsa Anna Ferenczy argues that the EU must have its own Taiwan policy that should not be decided by the United States and China; similarly, Taiwan should avoid to be the bargaining chip or leverage on US-China strategic competition.[22] Seeing Taiwan's successful pandemic response, medical professionalism and science, public trust and transparent governance, Ferenczy indicates that the "Taiwan Model" could be used as a "soft power" instrument for Taiwan's international integration. She considers that the EU should show support for Taiwan's soft power and that embracing Taiwan would be a double win situation for the EU and the United States. She therefore claims that it's the moment for the EU to play an active role now and to rethink its China policy.

Brigitte Dekker, the Junior Researcher at the Netherlands Institute of International Relations "Clingendael",[23] also calls for the EU to support Taiwan. In her paper, "Why the EU Should Pay More Attention to Taiwan," she suggests that the EU can focus on a strategic economic partnership with Taiwan in the long term by existing economic ties and by the norms and values shared by the EU and Taiwan.[24] She emphasises the importance of Taiwan's position for the world economy; for example, Taiwan controls 74% of world chip manufacturing. She also explains close and intensified economic cooperation between the two; for instance, the bilateral trade reached to a historic high of 51.9 billion Euros in 2018. However, she argues that this kind of business-focused EU-Taiwan relations is not suffice, practical initiatives are needed such as support for Taiwan's participation in international

---

[22] See Zsuzsa Anna Ferenczy, "Time for a Positive EU Policy on Taiwan," *The Diplomat,* October 21, 2020, p. 6. 〈https://thediplomat.com/2020/10/time-for-a-positive-eu-policy-on-taiwan/〉.

[23] Clingendael is the Netherlands Institute of International Relations that is a leading think tank and academy on international affairs.

[24] See Brigitte Dekker, "Why the EU Should Pay More Attention to Taiwan," *Clingendael Alert,* January 9, 2020. 〈https://www.clingendael.org/publication/why-eu-should-pay-more-attention-taiwan〉.

organisations.

Ryan Hass, on his article, "For Taiwan, diplomatic green shoots are emerging in Europe," points out that Europe is one of Taiwan's great opportunity to strengthen its global role and standing.[25] Hass explains that since Germany, the key player on EU's China Policy, submitted its first-ever Indo-Pacific strategy in September 2020 to pursue Asia strategy to compete with China, this provides Taiwan with a good chance to enhance its global standing. He suggests some tips for Taiwan to win the EU's support, first of all, Taiwan needs to be clear on what counts as progress; that is to say, Taiwan needs to make contributions to issue-based groupings. Secondly, Taiwan should be a provider of solutions to the challenges some European countries faced such as providing personal protective equipment. Thirdly, Taiwan can show its power to help to tackle transnational threats originated from the ISIS and Afghanistan or challenges from the outbreak of Ebola in 2014. Finally, Taiwan should demonstrate patience and predictability to deepen its relations with other countries so to increase Beijing's price and cost to attack Taiwan.

To echo these appeals, Taiwan's Representative to the EU, Dr. Ming-Yen Tsai, published an article in the *Diplomat* to explain why deepening EU-Taiwan economic relations matter.[26] He points out that it is the time for the EU and Taiwan to further strengthen bilateral trade relations that can be started with bargaining on a bilateral investment agreement. For Tsai, the Europe-Asia Connectivity strategy signifies the EU's will to deepen and strengthen EU-Asia links; Taiwan's being the EU's major trading partner in Asia should not be ignored. He argues that the EU and Taiwan are like-minded partners who both committed to freedom, an open and market-driven economy, human rights and the rule of law. Under the systemic rival of China to the

---

[25] See Ryan Hass, "For Taiwan, diplomatic green shoots are emerging in Europe," *The Brookings Institute*, October 5, 2020. 〈https://www.brookings.edu/blog/order-from-chaos/2020/10/05/for-taiwan-diplomatic-green-shoots-are-emerging-in-europe/〉.

[26] See Ming-Yen Tsai, "Why Deepening EU-Taiwan Economic Ties Matter," *The Diplomat*.

EU, Taiwan can be a natural partner if the two can start negotiations on a Bilateral Investment Agreement (BIA). He indicates what happened in Hong Kong could damage the EU's investment there, Taiwan can be an ideal alternative for the EU; and in addition, Taiwan's efforts on green energy business also presents a great opportunity for European enterprises and companies. This meanwhile provides the EU with an "Open Strategic Autonomy".

Another scholar, the Cambridge educated Dr. John Y. P. Chang in his paper, "The EU and Taiwan: the Un-official Relations incredibly connected," that is re-written from his Ph.D. Thesis, examines the development of informal diplomatic EU-Taiwan relations and explains why the relationship has been so rocky. According to Chang, the unstable and controversial EU-Taiwan relations originated from the dramatic change of Taiwan's international political position and international status in the 1970s; and the war of diplomatic attrition across the Taiwan Strait for recognition as the sole Chinese government.[27] Under the pressure of Beijing, the EU therefore could not launch any constructive engagement with Taiwan until Taiwan's successful economic development and achievements in the 1980s. The two therefore established many financial, commercial and trade links and official talks and visits and the European Community realised Taiwan's economic power since the mid-1980s. By 1991, 12 EU member states established offices in Taiwan so that trade relations between the two have been progressed smoothly without any problems and difficulty. He concluded that Taiwan's popularity as an important EU trading partner goes beyond what China's political influence can blockade.

IV. The EU's Official Position to Taiwan

The European Chamber of Commerce Taiwan (ECCT) was establish in 1988 and it's the largest foreign chamber in Taiwan. It represents the interests and rights of

---

[27] See John Y. P. Chang, "The EU and Taiwan: the Un-official Relations incredibly connected," *Taiwan Research Institute.* 〈https://www.tri.org.tw/research/impdf/266.pdf〉.

over 860 individuals and 400 companies and orgnisations who have already invested over 40 billion Euros in Taiwan. On its official website,[28] it explains its relations with Taiwan can be traced back to the mid-eighties by some European business people to discuss founding a chamber in Taiwan. Few months later, the inaugural general meeting was held in January 1988 and the first board was organised. The ECCT was the second European chamber of commerce to be established outside Europe and it signified a huge step between Europe and Taiwan. Since then, the two have close contacts and exchanges so that the European Commission established the "European Economic and Trade Office in Taipei" in 2003 to further intensify the bilateral relations. In the past few decades, the EU-Taiwan relations have progress very smoothly. In April 2020, the European Commission President Ursula von der Leyen expressed her appreciation in her Twitter for Taiwan's donation of 5.6 million face masks to the EU to help fight the coronavirus. In addition, Representatives from the Czech Republic, France, Germany, the Netherlands, Poland and Slovakia all also expressed their appreciation to Taiwan. This is the first time an European Commission President and so many EU countries addressed and recognised publicly Taiwan. For Filip Grzegorzewski, the EU Representative in Taiwan, Taiwan's COVID-19 fighting experience is an inspiration for the Europe. In the interview with Taiwan's CommonWealth Magazine, Grzegorzewski, the head of the European Economic and Trade Office (EFTO) points out that there are now already many projects, consultations, working groups, seminars, dialogues, events going on between the EU and Taiwan such as the European Innovation Week, European Investment Forum, Energy Business Roundtable and Connectivity Seminars.[29] In particular, the European Business and Regulatory Cooperation (EBRC) is a special program for Taiwan funded by the EU to improve business and regulatory

---

[28] See the website of the *European Chamber of Commerce Taiwan,* 〈https://www.ecct.com.tw/ecct-history/〉.

[29] Liu Kwangyin, "EU representative: Taiwan's COVID-19 fighting experience is an inspiration for Europeans," *CommonWealth Magazine,* April 20, 2020. 〈https://english.cw.com.tw/article/article.action?id=2699〉.

cooperation between the two. Grzegorzewski explains that there are many areas of cooperation between the EU and Taiwan such as the digital agenda, green energy and investment, and that the two will work more on the supply chains. He believes that multilateral cooperation is very important and Taiwan has already proved itself by practical actions.

On September 22, 2020, the first-ever EU Investment Forum joined by 15 EU members was held in Taiwan to raise the investment level to bilateral trade levels. The EU representative in Taiwan Filip Grzegorzewski claimed that although the two have very close trade contacts and exchanges, Taiwan has very low investment in the EU, only 1.7% compared with 56.1% Taiwan's FDI in China.[30] However, 25% of European investment went to Taiwan. To response, President Tsai Ing-wen expressed that Taiwan is ready to become one of the top partners of the EU in the ICT, biotech, health and mobility sectors, and she hoped that the two can negotiate a bilateral investment agreement (BIA) soon. [31] She also pointed out that Taiwan has introduced "Three Major Programs for Investing in Taiwan" to boost investor confidence; until September 2020, the investment programs has received 650 applications, attracted more than 1 trillion NT dollars and created more than 92,000 jobs.[32] The Taiwanese government has made many efforts to provide a fair and stable business environment, hoping international companies and enterprises can see Taiwan as a safe and reliable place for investment. Taiwan now is developing the Taiwan-EU Dialogue on Digital Economy and the Industrial Policy Dialogue. By means of bilateral and multilateral cooperation, Taiwan can make donation to the world and also play a key role in global supply chains. Due to Taiwan's successful achievement to fight the pandemic, in November 2020, 644 congressmen from the European Parliaments and

---

[30] Matthew Strong, "Taiwan to host EU investment forum," *Taiwan News,* September 17, 2020. 〈https://www.taiwannews.com.tw/en/news/4011010〉.

[31] "President Tsai attends EU Investment Forum 2020," *Office of the President, Republic of China,* September 22, 2020. 〈https://english.president.gov.tw/NEWS/6045〉.

[32] Ibid.

25 European countries' national parliaments signed a joint letter to support Taiwan's bid to join the World Health Assembly. [33] This shows the EU's positive views and support for Taiwan and also demonstrates what John Y. P. Chang indicated that 'the EP was then the only indirect channel accessible to Taiwanese officials in its contacts with the commission." [34]

The ECCT released its "Position Papers 2021" on November 11, 2020 to urge for the expedited signing of EU-Taiwan BIA to promote their trade and investment cooperation. [35] According to the papers, the ECCT proposed 170 recommendations for the EU from five perspectives to promote its relations with Taiwan, including to push for Taiwan's internationalisation, its image as a talent haven, also its competitive edge as a service industry hub, and Taiwan's healthcare innovations and its development of a green economy.[36] The ECCT Chairman Giuseppe Izzo also expected Taiwanese government to pursue standards in line with the EU in motor vehicle emission standards and product labeling rules for better trade practices. n November 26, 2020, the 6th Taiwan-EU Industrial Policy Dialogue was held by a virtual conference. [37] The Taiwanese representative Minister Mei-Hua Wang of the Ministry of Economic Affairs, and her counterpart Director General Kerstin JORNA of the DG GROW of the European Commission exchanged views on the development of industrial policy and searched for much more opportunities for cooperation on innovation. The two sides discussed on industrial policy changes, possible cooperation on offshore green energy, and the restructuring of global supply chains after the pandemic. They also exchanged views on technological innovation,

---

[33] Ming-Yen Tsai, "Why Deepening EU-Taiwan Economic Ties Matter," *The Diplomat*.

[34] John Y. P. Chang, "The EU and Taiwan: the Un-official Relations incredibly connected," *Taiwan Research Institute*, p. 17.

[35] See Huang Tzu-ti, "ECCT calls for inking of Taiwan-EU investment agreement," *Taiwan News*, November 12, 2020. 〈https://www.taiwannews.com.tw/en/news/4051775〉.

[36] Ibid.

[37] "The Taiwan-EU Industrial Policy Dialogue Focuses on Responses to Pandemic and Green Transformation," *Bureau of Foreign Trade, Ministry of Economic Affairs*, November 26, 2020. 〈https://www.trade.gov.tw/English/Pages/Detail.aspx?nodeID=86&pid=709485〉.

including artificial intelligence, smart machinery and robotics, SMEs and the internationalisation of industrial clusters to enhance their partnership. These all show the close contacts and cooperation between the two sides.

V. Conclusions

Being the world's largest trading bloc and the major investor in global market, the EU actually has more leverage than it assumes. The EU and Taiwan could make much more efforts and works in the areas of public health, high-technology, human rights and democracy. Due to the global shortage of chips, the EU representative Filip Grzegorzewski claimed in April 2021 that they will organise another investment forum later to attract and encourage much more Taiwanese investment on digital and semiconductors to Europe. He explained that the needs to be "more of Taiwan in Europe" will also be a good chance for Taiwan to strengthen its role in the global supply chains in the post-pandemic world.[38] Whether from the economic or political perspectives, the EU and Taiwan have same systems and share common values such as freedom, democracy and the rule of law. This certainly can enhance much more mutual cooperation and collaboration.

---

[38] "EU seeks greater Taiwanese investment as chip shortage bites," *Reuters*, April 20, 2021. 〈https://www.reuters.com/world/asia-pacific/eu-seeks-greater-taiwanese-investment-chip-shortage-bites-2021-04-20/〉.

專業叢書PS033　　　　　　ISBN 978-626-7032-20-6

後疫情時代印太戰略情勢下的臺灣安全戰略選擇

主　　　　編：翁明賢

發　行　人：葛煥昭
出　版　者：淡江大學出版中心
主　　　任：林雯瑤
行　政　編　輯：黃佩如
地　　　址：新北市淡水區英專路151號
封　面　設　計：淵明印刷有限公司
排　版　印　刷：維中科技有限公司
總　經　銷：紅螞蟻圖書有限公司
展　售　處：淡江大學出版中心
　　　　　　地址：新北市淡水區英專路151號海博館1樓
　　　　　　電話：02-86318661
　　　　　　淡江大學麗文書城
　　　　　　地址：新北市淡水區英專路151號商管大樓3樓
　　　　　　電話：02-26220431

出版日期 2022年6月 一版一刷

定　　　價　600元整

國家圖書館出版品預行編目(CIP)資料

後疫情時代印太戰略情勢下的臺灣安全戰略選擇 / 翁明賢主
編. -- 一版. -- 新北市 : 淡江大學出版中心, 2022.06
　面；　公分
部分內容為英文
ISBN 978-626-7032-20-6(平裝)
1.CST: 全球戰略 2.CST: 國際關係 3.CST: 區域研究 4.CST:
文集
592.407　　　　　　　　　　　　　111007415